Tourism in the Age of Globalisation

The revolutionary progress achieved in information and communication technology is gradually transforming the world into a global village. This volume, edited by internationally eminent specialists, evaluates the nature and resilience of the emerging global economy and its implications for tourism. The subject is examined in four parts:

- The age of globalisation – including contextual overviews of territorial economic integration.
- The globalisation of tourism demand and marketing – including the new tourist and consumer, and the effects of globalisation on tourism promotion.
- Globalisation and competitiveness in tourism – including inter-firm alliances, safety and national security, quality management and productivity.
- The globalisation of tourism's approach to sustainability – including environmental policies, cultural identity and sustainable development.

Cooper and Wahab's concluding chapter draws together the work to suggest that economic globalisation may ultimately fuel a new active global competition between tourist destinations.

This work will prove a valuable resource not only for students and researchers in the field, but anyone involved in tourism who is interested in the effects of globalisation upon their industry.

Salah Wahab is Professor of Tourism Management at the University of Alexandria Faculty of Tourism and Hotels. He is also president of Tourismplan, an affiliate member of the World Tourism Organisation, and a member of the International Academy for the study of Tourism. He is the author and editor of a number of internationally published books on tourism, as well as a member of the editorial boards of several tourism journals.

Chris Cooper is Foundation Professor of Tourism Management in the Department of Tourism and Leisure Management, UQ Business School, University of Queensland. He has an honours degree and PhD from University College London and before beginning an academic career worked in market planning for the tourism and retail sectors in the UK. Cooper has authored a number of leading textbooks on tourism and worked closely with the World Tourism Organisation in developing the status of tourism education on the international stage. He is editor of the *International Journal of Tourism Research*, author of many academic papers on tourism and has worked as a consultant and researcher in every region of the world.

Routledge Advances in Tourism
Series Editors: Brian Goodall and Gregory Ashworth

Tourism in the Age of Globalisation

Edited by
Salah Wahab
and Chris Cooper

London and New York

First published 2001
by Routledge
2 Park Square, Milton Park, Abingdon, Oxfordshire OX14 4RN

Simultaneously published in the USA and Canada
by Routledge
711 Third Avenue, New York, NY 10017

Routledge is an imprint of the Taylor and Francis Group, an informa company

Transferred to Digital Printing 2005

© 2001 editorial material and selection, Salah Wahab and Chris Cooper;
individual chapters, the contributors

Typeset in Baskerville by Florence Production Ltd,
Stoodleigh, Devon EX16 9PN

British Library Cataloguing in Publication Data
A catalogue record for this book is available from the British Library

Library of Congress Cataloging in Publication Data
Tourism in the age of globalisation / edited by
Salah Wahab and Chris Cooper.
    p. cm.
    Includes bibliographical references (p.).
    1. Tourism. 2. Globalization. I. Wahab, Salah.
    II. Cooper, Christopher P.
    G155.A1 T59215 2001
    338.4′91—dc21

ISBN 978-0-415-21316-5 (hbk)

ISBN 978-0-415-75818-5 (pbk)

Contents

Figures

Tables

Contributor

Philip Alford, Business School, University of Luton, Putteridge Bury, Hitchin Road, Luton, UK. philip.alford@luton.ac.uk

J. Appelman, Erasmus University Rotterdam School of Management, Centre for Tourism Management, PO Box 1738, 3000DR Rotterdam, The Netherlands. f.go@fac.fbk.eur.nl

Tom Baum, Scottish Hotel School, University of Strathclyde, Cathedral Street, Glasgow G4 0LG, Scotland. t.g.baum@strath.ac.uk

Aurora Pedro Bueno, Department of Economics, University of Valencia, Valencia, Spain. turismo@esp.upv.es

Dimitrios Buhalis, University of Westminster, 35 Marylebone Road, London NW1 5LS, UK. buhalid@wmin.ac.uk

Peter Burns, Business School, University of Luton, Putteridge Bury, Hitchin Road, Luton, UK. peter.burns@luton.ac.uk

Chris Cooper, Department of Tourism and Leisure Management, University of Queensland, 11 Salisbury Road, Ipswich, Queensland, Australia. c.cooper@mailbox.uq.edu.au

Eduardo Fayos-Solà, World Tourism Organisation, Capitan Haya 42, Madrid, Spain. educa@world-tourism.org

William C. Gartner, Tourism Center, University of Minnesota, 116D Classroom Building 1994 Burdorf Avenue, St Paul, MN 55108–6038, USA. wcg@umn.edu

F. M. Go, Erasmus University Rotterdam School of Management, Centre for Tourism Management, PO Box 1738, 3000DR Rotterdam, The Netherlands. f.go@fac.fbk.eur.nl

C. Michael Hall, Centre for Tourism, University of Otago, Dunedin, New Zealand. cmhall@commerce.otago.ac.nz

Donald E. Hawkins, School of Business and Public Management, George Washington University, 22061 Street NW, Washington DC 20052, USA. dhawk@gwis2.gwu.edu

Christopher Holtz, School of Business and Public Management, George Washington University, 22061 Street NW, Washington DC 20052, USA. dhawk@gwis2.gwu.edu

Nick Johns, 15 Collingwood Close, Steepletower, Hethersett, Norwich N99 3QE, UK. nick_john@bigfoot.com

Darren Lee-Ross, School of Business, James Cook University, Townsville, Queensland, Australia. darren.leeross@jcu.edu.au

Gui Santana, Centro de Ciencias Tecnogicas da Terra do Mar, Universidadae do Vale do Iajai – Univali, Av. Uruguai, 458 – Cx. Postal 360, CEP 88.302–202, Itajai-SC, Brazil. gui@cttmar.univali.br

A. V. Seaton, Business School, University of Luton, Putteridge Bury, Hitchin Road, Luton, UK. tony.seaton@luton.ac.uk

John Swarbrooke, Sheffield Hallam University, Sheffield, UK. j.mitchell@shu.ac.uk

Norbert Vanhove, WES, Baron Ruzettelaan 33, B-8310 Assebroek, Brugge, Belgium. wes@innet.be

Salah Wahab, Senior Professor of Tourism, Faculty of Tourism and Hospitality Management, Alexandria University, President of Tourismplan, Affiliate member of WTO, Mammal Al Sukkar Street 16, Garden City, Cairo, Egypt 11451. smmmwahab@link.com.eg

Preface

Tourism is one of the most international of industries, with both the public and private sectors increasingly concerned with issues of international competitiveness and benchmarking. This is no less so for domestic tourism where international standards are derived for service delivery and facilities – indeed in destinations such as the former Eastern European states, international benchmarks have been used to drive-up the standards and the quality of the domestic industry. Yet despite the obvious international focus of tourism from a geographical point of view, rather less attention has been given to the process of internationalisation in the tourism sector – both from a government and an industry viewpoint. It was for this reason that Salah Wahab perceived the need for a book drawing together leading authors to provide a state of the art review of the process of globalisation and its consequences for the various sectors of tourism. Since 1997, when the book was first mooted, the need for such a volume has become more acute as *globalisation* has begun to rival *sustainability* as an organising concept for the way we approach tourism.

We have organised this book into four parts. The first part sets the scene in terms of an introduction to globalisation and tourism by ourselves, followed by Hall and Fayos-Solà and Bueno's overview of global trends of territoriality, economic integration and tourism policy. In the second part, the authors focus on the impact of globalisation upon tourism demand, in particular the *new tourist* and the consequences of globalisation for both tourism promotion and the distribution system. It is clear from these chapters by Buhalis, Seaton and Alford and Vanhove that it is in the area of tourism demand that globalisation has, to date, had the most impact. In the third part, the chapters focus on elements of competitiveness and their relationship with globalisation. These elements comprise the destination itself (Swarbrooke), the role of alliances and mergers (Go and Appelman), education (Baum), safety and security (Santana), and quality management (Lee-Ross and Johns). The authors in this section disagree as to the impact of globalisation on tourism, with a consensus that perhaps the consequences for tourism are not yet felt to the same degree as in other sectors of the economy. In the fourth part the chapters review issues of sustainability in terms of the environment (Hawkins and Holtz), culture (Burns) and developing countries (Gartner). Here, the consequences of globalisation are clear: the impacts upon culture underline the paradox of local distinctive tourism products set within an increasingly uniform tourism landscape; the need for broader environmental policies and management points up the difficulty of gaining international agreements; and developing countries are seen to be particularly vulnerable in an increasingly competitive, global tourism market. In the Conclusion, we attempt to draw together the threads of the 14 chapters and conclude that whilst the impact of globalisation is variable across all sectors of tourism, there is no doubt that tourism is both embracing global concepts of business and policy, whilst being itself a driving force of globalisation by its very nature of exposing individuals, businesses and governments to practices elsewhere.

We are very grateful to all the contributors in this book for generously sharing their expertise and thoughts with us in their chapters. Peter Burns merits a special thanks for standing in at the eleventh hour to provide us with Chapter 13 on culture; but we also owe our thanks to everyone who has helped us with this book including Gerardine Cooper, Mona Wahab, the office staff at both Bournemouth University and the University of Queensland, and of course Routledge who commissioned the book and have waited patiently for the manuscript.

Salah Wahab and Chris Cooper
April 2000
Cairo and Brisbane

Part I
The age of globalisation

1 Tourism, globalisation and the competitive advantage of nations

Salah Wahab and Chris Cooper

Conceptual framework

We now live in a world of incredibly rapid and revolutionary change where many remarkable achievements have been reached. The reunity of Germany, the turn to democracy in the former USSR and other Eastern European countries, the peaceful end of racial discrimination in South Africa, the still cherished peace steps in the Middle East, the successful endeavours towards regional economic integrations and the economic progress achieved by many developing countries confirm this truth.

However, common sense dictates that there are still some important impediments on the way to a better future for humankind. Examples of these are:

- applying suitable and appropriate technology to improve productivity in the third world;
- organisational and managerial know-how that could contribute to create effective administrative, political and economic systems in inter-mediate and emerging nations;
- improvement of the performance of international institutions through reshaping their structure and operations; and
- coordinating efforts to render the world safer and more secure.

Above and beyond such goals, humankind needs a new vision, innovative and new attitudes about what can and must be done in various walks of life in response to the changing conditions economically, politically, technologically, socially, culturally and environmentally.

The close of the second millennium marks the emergence of new global trends to reshape the world economy and unveil the new, expected and unexpected happenings that would alter many of our established norms. Among these trends are:

- earnest trials to reach free trade among nations;
- territorial integrations in Europe, North America, South-East Asia; and

- alliances, mergers and acquisitions that create giant conglomerations practically in every business facet, achieve sustainable and balanced growth in third world countries, accomplish equitable distribution of resources and attain quality in most of the human endeavours.

This *marked direction towards globalisation* is a by-product of the revolutionary progress achieved in information and communication technology (CRS and GDS) that is gradually transforming the world into a global village. Today, we witness a global free-market economy, a global growth or recession, global companies, a global democratisation process, a global awareness of the importance of protecting the physical environment and a healthy habitat, and a global revitalisation of local communities' expectations and participation in socio-economic change – particularly in developing countries (Axford 1995). Virtually, an international global society is in the making and is expected to accelerate. Change is proceeding at such a rapid rate that policy-makers are being forced to change their attitudes, their previous fixed ideas and review long-held policies in order to be able to keep their grip on basic socio-economic, cultural and political values and norms (Gartner 1996).

Some authors, such as Kenichi Ohmae, asserted that globalisation is the product of many converging forces – technology, transnational corporations, new methods of communication and distribution systems as well as new vistas of competition – all of which are instrumental in creating a global market, an economy conceived of as mainly the same in all its parts (Ohmae 1991 as referred to in Berger and Dore 1996).

Therefore, globalisation is an all-embracing term that denotes a world which, due to many politico-economic, technological and informational advancements and developments is on its way to becoming borderless and an interdependent whole. Any occurrence anywhere in the world would, in one way or the other, exert an impact somewhere else. National differences are gradually fading and being submerged in a homogeneous mass or a single socio-economic order (Duming and Hamdani 1997; Castells 1993).

The prevalence of globalisation raises the prospect of a world which will soon be governed by international integrated economic forces that started to shape the destiny of various countries in Europe (European Union) and North America (NAFTA) and South-East Asia (ASEAN). It is likewise spearheaded by a few hundred giant corporate entities, some of which are bigger than many sovereign states (Barnet and Cavanagh 1995).

Thus, globalisation is based on the emerging interdependence or interconnections between states of different levels, socio-economic and political systems at various developmental stages. *It describes the process by which events, decisions and activities in one part of the world come to have significant outcomes for communities and individuals in quite distant parts of the globe.* Accordingly, globalisation has two distinct phenomena: scope (or stretching) and intensity (or deepening). On the one hand, it defines a set of processes that operate worldwide, which means a spatial dimension; on the other hand, it implies an intensification in the levels of interaction and interdependence between the states which constitute the world community. This does not mean that globalisation entails the political unification of the world, the necessary homogenisation of culture worldwide or the total economic integration of all or most countries of the world (McGrew 1992).

Another important aspect of globalisation that is worth researching is how democratic pluralism, the basis for the balance between the state and the market, could function globally. In economic theory and practice, markets allocate resources efficiently only when they are competitive and when firms, large and small, pay for the social and environmental impact of their activities as part and parcel of the costs of their production. This is usually done because strong governments do enforce the rules of the game that aim at balancing market and community interests. This is because government power should be at least equal to market power. When extending the boundaries of the market beyond national boundaries through economic globalisation, market power would, at least theoretically, go beyond the reach of national governments and fall into the hands of global firms (Daremus *et al.* 1998).

These global firms would be inclined to serve the interests only of their dominant shareholders (Corten 1996). Moreover, there is a marked change of the role of scientific and professional knowledge which is to be utilised as a powerful source of a new wealth-creation system. This could be called the 'new knowledge economy' which serves as the new force fuelling an active global competition between advanced countries. Developing nations, some of which some are still struggling with their traditional economic strategies, may find themselves still dislocated in this new global encounter.

The impact of globalisation as a megatrend on tourism

Tourism is not a clear-cut sector but an all-embracing and pervasive domain of service and industrial activities. It touches upon almost all spheres of national life within the country and that is particularly the reason why a sound state policy of tourism should be essentially formulated before any significant tourism investment projects are launched. As rightly put by MacCannell, tourism is an ideological framing of history, nature and tradition; a framing that has the power to reshape culture and nature to its own needs (MacCannell 1992).

Some writings already maintain that globalisation, as a megatrend, is changing the nature of international tourism (Keller 1996). It has become a landmark in human activities reaching 635 million international tourist visits in 1998 and registering receipts totalling US$444 billion without the cost of international transport which may reach US$180 billion. The continued expansion of tourism in the world due to world population growth, increasing affluence of many nations, the expansion and diversification of travel motivations and expectations, great technological achievements in information and communication, the fierce competition between an increasing number of tourist destinations, and deregulation movements, is an important playground for global forces. The new technologically advanced distribution channels permit anyone to receive the most up-to-date multimedia information on the best connections, and at the best prices, for most attractive destinations in the world (Keller 1996).

Populations of various countries respond to this globalisation of economies, markets, systems and cultures by looking at their own identities, as in contrast to globalisation lies *localisation* which is an opposing force. These two adverse forces cannot be averted by the state, the market or communities by acting alone (Cleverdon 1998).

Moreover, increased awareness of physical and cultural heritage safeguards have induced various tourist destinations to engage in the complex planning process for sustainable development in tourism. In addition, the Free Trade in Services Agreement is another difficult tunnel for globalisation to go through (Wahab and Pigram 1997).

Cultural differences between individual tourist destinations will continue to play an important role, among other factors, in the choice of a holiday destination. However, a transcending global cultural understanding might eventually emerge cutting across various cultural diversities with each having its local flavour.

Quality, production conditions, the role of public authorities, corporate structure and price strategies in tourism are likewise going to exert profound reciprocal influences on globalisation trends in tourism – but in varying proportions. Therefore, researching the future orientation of tourist-generating markets should be the point of departure in any analysis of the globalisation process. The first question to ask is what would be the behavioural patterns of potential tourists in the face of a universal supply whose various products and prices are becoming more and more transparent through the new multimedia (Keller 1996).

Competition, however, would take up a new course under the pressures of globalisation which would reshape the production conditions in various tourist destinations and change their marketing strategies.

It remains now to ask whether globalisation would finally succeed in tourism in the face of some strong anti-thesis to turn to localisation and what would be the fate of the global firm in a world that is economically based upon competitive forces and comparative advantages under an umbrella of anti-trust, anti-cartel and anti-monopoly laws prevailing in various countries.

The competitive advantage of nations within a global context

Nations of the world were classically categorised into four levels of development, namely:

- advanced or developed countries;
- countries in the intermediate level;
- developing; and
- underdeveloped or least developed countries.

This categorisation which was mainly based on monetary criteria represented by per capita income, is usually influenced by two types of factors; economic and non-economic. The economic development of a country is dependent upon its natural resources, human resources, capital accumulation, entrepreneurship, technology and scale of production and export capabilities. While these economic factors are indispensable for development, they cannot result in development as long as social institutions, political conditions and moral values in a nation do not encourage development. Some economists go even further, pointing out that the main determinants of economic development are 'notably aptitudes, abilities, qualities, capacities and faculties, attitudes, mores, values, objectives and motivations; and institutions and political arrangements' (Bauer 1973).

Rostow's five stages of economic development

Walt Rostow proposed one of the most important theories of economic development distinguishing between five stages of economic growth, viz:

- the traditional society;
- the pre-conditions for take-off;
- take-off;
- the drive to maturity; and
- the age of high mass-consumption.

Without going into details, it suffices to say that according to Rostow, the rapid growth of the leading sectors of any national economy depends upon the presence of four basic factors:

1   First, there must be an increase in the effective demand of their products generally brought about by dishoarding, reducing consumption, importing capital, or by a sharp increase in real incomes.
2.   Second, a new production function along with an expansion of capacity must be introduced into these sectors.
3   Third, sufficient initial capital and investment profits for take-off in these leading sectors.

4   Fourth, these leading sectors must cause expansion of output in other
     sectors through technical transformations.

While Rostow's theory, 'The Stages of Economic Growth' is the most
widely circulated and highly cited piece of economic literature in the last thirty
years, economists are almost agreed in doubting the authenticity of the division of economic history into five distinct 'stages of development' as presented
by Rostow. To maintain that every economy follows the same course of development with a common past and an identical future is to over schematise the
complex forces of development and to give the sequence of stages a general-
ity that is unwarranted. Many nations gave valid examples of the unnecessary
sequence of Rostow's stages of development. Japan, which became the third
advanced economy in the world, achieved this in less than half a century;
South Korea became one of the leading economic powers in South-East Asia
in almost thirty-five years. Singapore, Malaysia and Thailand likewise provide good examples of exemplary economic standing irrespective of the recent
economic difficulties that some of them have endured.

In general, we could safely say that Rostow's theory of stages of development may soon fall into obsolescence if not amended to accommodate the
intervening global events represented by the accelerated technological and
scientific knowledge of the information age. In the wake of the present technological upheaval, entire industries and lifestyles are being overturned, only
to give rise to entirely new ones. But these rapid, bewildering changes are not
just quantitative. They qualitatively mark the birth pangs of a new era.

Porter's stages of national competitive development

Michael Porter introduced his theory of competitive advantage to explain
economic development within nations and national differences in growth
and prosperity. He said that there are many conflicting explanations for
why nations are competitive and others are not.

Some think of national competitiveness 'as a macroeconomic phenomenon, driven by such variables as exchange rates, interest rates, and government deficits', while 'others argue that competitiveness is a function of cheap
and abundant labour'. Another viewpoint sees competitiveness as dependent
upon owning 'bounteous' natural resources. A more recent theory resides in
the argument that 'competitiveness is most strongly influenced by government policy'. All these explanations, and others, fall short of rationalising the
competitive position of a nation's industries. As put by Porter 'each has some
truth but will not stand up to close scrutiny' (Porter 1990).

The acceleration of science, technology and the expansion of territorial
economic integrations into the third millennium will necessarily exert vast
repercussions on the wealth of nations and our quality of life. As enunciated
before, in the past three centuries, wealth was usually accumulated by
those nations which were endowed with natural resources or by those

which colonised them. The rise of the great powers of Europe of the nineteenth century and the United States of America in the twentieth century follows this classical textbook principle.

Some leading economists consider that in the twenty-first century there will be a great movement in wealth away from nations with natural resources and capital. Information, brainpower and imagination, the organisation of new technologies and the management of international global communication systems will be the key strategic ingredients. In fact, many nations which are richly endowed with abundant natural resources will find their wealth greatly reduced because, in the marketplace of tomorrow, despite all disparities between nations, consumer goods will be cheaper, trade will be global, and markets will be linked electronically. Many nations which are devoid of natural resources will flourish in the next century because they placed a premium on those technologies which can give them a competitive edge in the global marketplace. Some writers assert that in the modern society of the twenty-first century, knowledge, skills and information management should stand alone as the only source of comparative advantage (Thurow 1996: 279).

Porter identifies a four-stage development process the characteristics of which are summarised in Table 1.1.

National prosperity, in Porter's view, is highly associated with the 'upgrading' of competitive advantage. In the outset, nationalities try to exploit their factor conditions to drive their development. At a later stage, they use different ways to attract foreign technology and invest in capital equipment, while encouraging more savings. Labour and resource-intensive industries are replaced by industries that are more capital and technology intensive. The most successful companies are able to produce higher value-added through product and service differentiation.

At a further stage, nations turn to innovation as a major driver of their national wealth. Upon fulfilling their aims, they may move still further to a stage marked by the effort to manage and preserve its existing wealth. Their investment and innovation activity might slow down and the nation's competitive advantage may begin to erode (Kotler *et al.* 1997).

Tourism, as the largest service industry in the world, greatly contributes to wealth creation as it virtually did to Spain in the 1960s and 1970s, to Greece since the early 1970s and to Turkey since the mid-1980s. With the advent of globalisation, tourism, as a dynamic force, has already become an effective tool of competitive advantage among global organisations through mergers and acquisitions. However, like all other sectors of the economy, tourism undergoes the product life cycle that starts with the take-off stage, then growth, then maturity and finally decline, unless rejuvenation occurs and retards the decline stage (Figure 1.1).

Travel groups based in Germany, the UK and France lead the way in the development of pan-European operations. Examples of globalisation trends are as follows (*Euromic*, May 1999).

Table 1.1 Porter's four-stage development process

Driver of development	Source of competitive advantage	Examples
Factor conditions	Basic factors of production (e.g. natural resources, geographical location, unskilled labour)	Australia, Canada, Singapore, South Korea before 1988
Investment	Investment in capital equipment, and transfer of technology from overseas; also requires presence of and national consensus in favour of investment over consumption	Japan during the 1960s, South Korea during 1980s
Innovation	All four determinants of national advantage interact to drive the creation of new technology	Japan since the late 1970s. Italy since the early 1970s. Sweden and Germany during most of the post-war period
Wealth	Emphasis on managing existing wealth causes the dynamics of the diamond to reverse, competitive advantage erodes as innovation is stifled, investment in advanced factors slows, rivalry ebbs and individual motivation wanes	UK during the post-war period. USA, Switzerland, Sweden and Germany since 1980

Source: Robert M. Grant, '"Porter's competitive advantage of nations", an assessment', *Strategic Management Journal,* 12, table 1: 540.

Germany – German market leaders prepare fresh take-overs

German travel groups will continue to seek opportunities to grow at home and expand abroad, under the combined pressure of consolidation and growing Europeanisation of the leisure travel market. This follows a period which changed the face of the German industry – with the Preussage take-overs of Hapag-Lloyd and TUI, the merger of Condor and NUR, the take-over of LTU by Swissair and the arrival of the first large UK operator, Airtours, through its stake in FTI. Given low growth, the fight for market share will become the big two, HTU (Hapag-Lloyd/TUI) and C&N are likely to look for further acquisitions to power their growth within Germany. Strong regional companies and specialists could be the top targets. Abroad, both will seek to secure their positions in the Benelux and German-speaking countries, but could also move into markets such as Italy and possibly France or Spain.

LTU, Germany's first integrated travel group, faces a tough situation after losing market share and seeing profit dwindle over the past few years. New owner, Swissair, will probably be too focused on securing LTU's position in

Germany before making any move abroad. There are question-marks against the future strategies of Deutsches Reisburo (DER), ITS and Alltours, the other large agency and operator companies. Hapag-Lloyd has acquired a 50.1 per cent stake in TUI, bought First Reiseburo (approved by the German anti-trust office) and taken a 24.9 per cent stake in UK's Thomas Cook.

C&N Touristik Group, the second European tourist group is equally owned by Lufthansa (50 per cent) and Karstadt, one of the major groups of department stores in Europe and in the world. C&N means Condor and Nur Touristik, an integrated tourist group with its own airline Condor (42 aircraft), 51 hotels and clubs and a travel organisation NUR Touristik, with a number of tour-operator brands in several countries in Europe. The group employs 8,500 people.

UK – Airtours leads UK's top four in overseas buying

UK tour operators are likely to expand abroad shortly following a frenzy of consolidation. The country's top four travel groups have spent millions of pounds buying up smaller operators and agency chains. The buying spree left the big four – Airtours, Thomson, Thomas Cook and First Choice – holding three-quarters of the market. The smallest of the four, First Choice, is now eight times the size of the UK's fifth largest operator, Cosmos. In order to expand further, the market leaders will have to look abroad.

Airtours is best-placed to do so, with its Scandinavian Leisure Group the number one operator in Denmark, Norway and Sweden. It also has a strong presence in Belgium through Sun International. But, most significantly, it has a foothold in Germany with a 36 per cent stake in operator FTI, which it is transforming into a full travel group with its own airline and hotels.

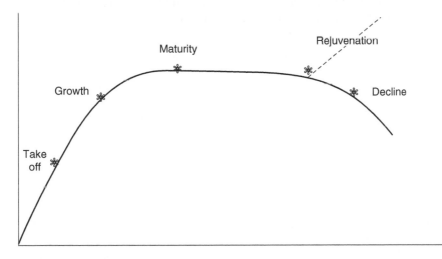

Figure 1.1 The tourism area life cycle (Source: Butler 1980).

Thomson has also begun expanding outside the UK with the launch of a Scandinavian branch and the take-over of Swedish-based Fritidsresor. The company has admitted Germany is a key target.

Thomas Cook is well placed to expand, since it is owned by German giant Preussag, which also owns Hapag Touristik Union, comprising TUI and Hapag-Lloyd. But First Choice probably still has work to do in the UK. It had no retail arm until recently and was losing out to its vertically integrated rivals. It is now starting to build a retail arm.

In terms of operators, Thomson Holidays is the UK market leader and its carrier Britannia is a leading charter airline. It also owns direct-sell operator Portland Direct and Holiday Cottages Group.

Airtours is the second British tour operator. The group is present in 16 countries, has 8.4 million clients, employs 20,000 people, is represented by 825 travel agencies, has 37 aircraft, 41 brands of tour operating, 21 hotels and 11 cruise ships (Carnival Cruises). Its tour operators are Airtours Holidays, Aspro Direct, Bridge Cresta, Tradewinds and Eurosites. Its charter airline is Airtours International. In 1998 it bought UK agency chain Travelworld in December and Irish operator Panorama in September.

Thomas Cook has various brands – operators Sunworld and Flying Colours are the main ones. Thomas Cook Holidays is the long-haul operator and Time Off the short-breaks specialist. Its airlines Flying Colours and Airworld are being merged. In terms of retail there are 385 Thomas Cook shops and it has merged with Carlson Leisure Group's 412 agencies and 650 affiliated shops. In terms of foreign operations, Thomas Cook is controlled by Preussag of Germany and therefore linked to Hapag Touristik Union. It has retail business in Australia, New Zealand, Canada, Hong Kong, Mexico and Egypt. Recent acquisitions include a controlling stake in Preussag. Its UK operations have merged with Carlson Leisure Group and bought Flying Colours, the UK's fifth largest operator and charter airline.

France – France trails its rivals in race to consolidate

The tour operating industry in France is set for a major wave of consolidation as operators aim to fight off take-overs from neighbouring countries. The French market has yet to undergo the consolidation that has occurred in Germany and the UK. It is still fragmented, with dozens of companies holding small shares of the market. Observers in France warn that, if there is no move to consolidate, the industry could be engulfed by its UK and German rivals. The leading French operators are two or three times smaller than market leaders in some neighbouring European countries and cannot achieve the same economies of scale.

Several major European companies have already expressed an interest in moving into France. NUR Touristic Benelux, a division of German giant C&N, bought into Aquatour, an operator based in north-east France. Preussag, the owner of TUI, is also looking at French operators. But it has denied that it plans a bid for Club Mediterranee. In their current position, many French companies may be easy targets for giants.

However, there are signs that some companies are preparing to consolidate within France. Hotel giant Accor has begun to build an operator arm. It bought tour operator Frantour from rail operator SNCF and is looking at other acquisitions and partnerships. Accor plans to focus more on tourism, and recently took over the Frantour hotel chain from French Railway SNCF. Jean-Marc Espalioux, Accor chief executive, recently declared: 'We are very interested in cooperating with Preussag, although no negotiations are taking place at present.'

Club Med also aims to develop its tour-operating arm. It is one of the largest travel companies in Europe, but its size is based mainly on its ownership of holiday villages rather than on tour operating.

Nouvelles Frontieres is also preparing for competition with the giant travel groups of Europe. It is due to float on the French stock exchange shortly, but any expansion is likely to be within France. Havas Voyages, Club Med and Accor are public companies. Fram, the third tour operator in France, comes far behind Nouvelles Frontieres and Club Mediterranee.

Globalisation impacts

Globalisation would undoubtedly lead to unwarranted decline for destinations and also for companies that are not sufficiently competitive on the international market, or have ceased to become competitive either because of deteriorating attractions, facilities or lack of service quality or high prices. Today's world market is a buyer's market. The expanding territorial integration of tourism destinations has contributed to the growth of intra-regional tourism in various regions (Keller 1996). Moreover, such integrations would cause a similar increase in inter-regional traffic because as already mentioned, globalisation has already contributed to the enhancement of air travel movement through alliances and mergers of large carriers, tour operators and lodging organisations.

Whereas horizontal and vertical integrations in the travel and tourism industry started in the 1960s and were intensified throughout the 1960s and 1970s, the globalisation era is mainly characterised by *diagonal integration*. This diagonal integration 'is meant to get closer to the consumer and reduce costs through economies of scope, system gains and synergies' (Sessa 1996).

Tourism and the competitive advantage of nations

Any tourist destination desiring to create competitive advantage in the tourism industry, which is a complex process of matching supply and demand within a rather highly competitive marketplace, should start by defining the importance of tourism and determine its place in the national economy. Such determination is in its turn a complex process that is based upon profound management and marketing decision-making. Following such determination, the tourist destination has to go through a seven stage process of applied research, namely:

1   Market research to ascertain all or most influential forces at work in the market and its conditions whether legal, economic, socio-cultural, environmental, political or psychological. Moreover, an analysis of international tourist movements in various regions is an important problematic issue for research. Its structure, components, motivations and future trends are key indicators that should be clarified.
2   International competition analysis to unveil the competition existing between various tourist destinations.
3   Diamond assessment including:
     • factor conditions;
     • related, emerging and supporting industries;
     • strategy, structure and rivalry exigencies; and
     • role of government.
4   Demand and matching supply analysis.
5   Cultural globalisation and diversity.
6   Strategy design.
7   Ascertaining competitive advantage.

The most influential globalisation factors in tourism

In the ongoing globalisation process, it is necessary for any tourist destination striving to achieve competitive advantage in tourism to define its own vocation with reference to the changing nature and characteristics of tourism. If the tourist destination is endowed with significant and diversified tourist attractions and facilities, it is only logical that such a destination would aspire to be a world leader, i.e. among the first ten tourist destinations in numbers of arrivals and, more importantly, in tourist receipts. This would certainly require a thorough understanding of the changes occurring in international tourism trends and a valid compatibility with the globalisation process. This needs professional articulation of its tourist goals having in mind relevant exigencies of the new tourism and globalisation trends. Some details of this strategy are as follows:

1   Important consumer changes in tastes and expectations. Such changes are constantly occurring owing to rapid technological, socio-cultural and economic aspects which exert a strong impact on lifestyles. This would necessarily be the outcome of serious search in market conditions

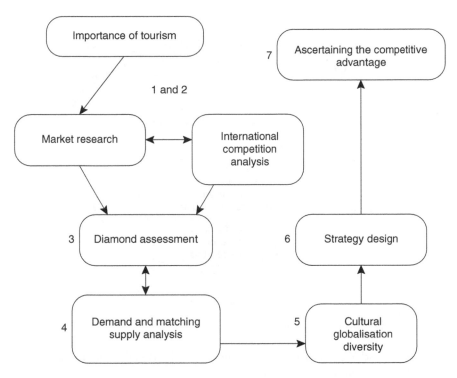

Figure 1.2 Globalisation and competitive strategy.

in both the generators and destinations of tourism. Thus continuous marketing research has to be engaged in to unveil new and changing global market communication systems on the one hand and new information and distribution systems on the other hand.

2   The tourism product's points of strength and weakness should be carefully established. Attention should be focused on tourist supply components, market conditions, competitive destinations, management standards, prices, production conditions and relevant manpower training.

3   The third aspect of globalisation is the development and adjustment of new integrated corporate structures as a result of alliances, mergers and acquisitions and which on the global level have dichotomous implications.

4   The fourth aspect of globalisation relates to culture in its global or national identification in the context of developing the 'global village'. Such environmental conditions are likewise important aspects that should help develop sustainable tourism development.

All these aspects constitute the destination's tourism policy in the new era of globalisation. Figure 1.3 is illustrative.

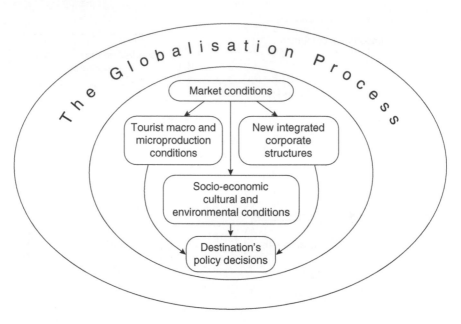

Figure 1.3 The process of globalisation.

Ritchie and Zins (1978) among others, identified eight general attraction factors of a regional destination:

- natural beauty and climate;
- cultural and social characteristics;
- accessibility of the destination;
- attitudes towards tourists;
- general infrastructure and tourist infrastructure;
- price levels;
- shopping and commercial facilities; and
- sportive, recreational and educational facilities.

Such a list is not exhaustive, as it lacks at least five factors, namely:

- management and service quality;
- market and organisational structures that make the trip a quite pleasant experience for the tourist;
- convenient factors of production;
- safety and security; and
- the successful and innovative management of change through the cherished partnership between government and the private sector.

Smeral (1996) prescribed a list of guidelines for improving the global competitiveness of a tourist destination. These guidelines include factor conditions, structure and quality of suppliers, market structures, organisation and strategies, demand conditions and government. The last three groups of guidelines are of particular importance to the tourism sector and therefore should be strictly focused on:

1 Market structures, organisation and strategies
   - image building within the context of global competition;
   - aggressive and innovative marketing to faster growth and expansion of tourism's value-added through special interest motivations; and
   - information coordination and intensification of knowledge pertaining to a destination's strengths and weaknesses within a competitive environment at the international, national and regional levels.
2 Demand conditions
   - expanding the destination's share of quality tourism movement from primary, secondary and opportunity markets offering them quality facilities and services;
   - reducing demand seasonality through strategies aiming at guaranteeing a steady flow of tourist traffic from various markets around the globe;
   - enhancing tourist receipts by concentrating mostly on high spending tourist arrivals;
   - encouraging repeat visitors through offering them diversified attractions separately presented or in combined forms; and
   - holistically-oriented local, regional and national policy.
3 Government
   - encouragement of systematic and continuous research into tourism market trends, demand changes and innovations in the leisure and tourism activities;
   - serious and systematic control of and guidance to the travel and tourism industries to keep total quality at its best suited for global competition;
   - improving academic and professional education and intensifying quality training in tourism to meet industry requirements;
   - eliminating red tape and avoiding all administrative hurdles including any conflict or overlapping of jurisdictions;
   - ameliorating environmental quality; and
   - proactive management of change and better usage of state commitments under the Free Trade in Services Agreement.

See Figure 1.4 for Egypt's tourist competitive advantage as an example of determining the competitive advantage of nations (ECES 1998).

Factor Conditions Moderate/High

Physical Resources
+ Rich and diverse Pharaonic, Greek, Roman, Christian and Islamic attractions.
+ Extensive coastline for leisure tourism.
+ Warm climate for almost nine months a year.
+ A number of venues in Cairo that can accommodate conference tourism.
+ The Nile River offers vast potential for promoting cruise tourism

Human Resources
+ Ample labour force and relatively low cost of labour.
+ Increasing number of tourism facilities and technical institutions.
–/+ Moderate guide skills.
– Limited local hotel management skills.
– Limited specialised training centres for hotel staff, guide training and customer services.

Capital Resources
+ Growing capital market.
+ Tourists are satisfied with banking services.
– Difficulty in accessing finance.
– High cost of finance.

Infrastructure
+ Improvements in telecommunication services.
– Inadequate transportation services (public buses and taxis).
– Underdeveloped information system for tourists.
– Unclean cities, major pollution.
– Overcrowded airports and poor quality of roads.

Knowledge Resources
– Lack of statistical information on international markets and domestic tourism.
– Weak marketing skills and poor advertising material.
– Inaccurate statistics on tourism's contribution to the economy.
+ All the cluster elements exist (albeit at different levels of efficiency).
+ Accommodations have been an important force in cluster development.
+ High linkages between H&R and the rest of the cluster.
+ High occupancy rate of accommodations.
–/+ Moderate guide language skills.
–/+ Unsatisfactory domestic means of transportation.
–/+ Moderate level of entertainment services.
–/+ Increasing privatisation and deregulation of cluster activities.

Firm Strategy, Structure & Rivalry Moderate
+ Strong foreign competition (regional and international).
+ Large number of firms in every component of the cluster, generating healthy domestic rivalry.
–/+ Slowly building alliances between local and international management companies.
–/+ Some public/private cooperation for tourism, cluster development.
–/+ Some firms are adopting diversified strategies.
–/+ Some firms are adopting niche strategies.
– Most business is run on family, rather than corporate basis.
– Limited private expenditure on promotion and marketing.
– Weak local management and promotion skills.
– Lack of accurate data which impedes strategic planning.
– Promotion efforts neglect potential domestic tourists.

Demand Conditions Weak
–/+ There is a growing trend of domestic tourism, especially to Sinai & Red Sea resorts.
–/+ Future prospects indicate a possibility of increasing home demand as a result of recent upward trends in per capita income, home demand sophistication/ and outbound tourism expenditures.
– Limited home demand does not compensate for fluctuations in foreign demand due to low level of per capita income.
– Low level of home demand sophistication does not encourage firms to innovate.

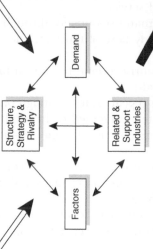

Egypt competes on the basis of *comparative* advantage rather than *competitive* advantage

–/+ Increasing promotion budget, from US$ 3 billion annually at the beginning of the 1990s to US$ 14 billion in 1996/97.
–/+ Growing coordination with private sector.
–/+ Shift in strategy toward diversification of destination: Red Sea, Sahl Hashish, Abu Soma, etc.
–/+ More concern about environmental protection and wealth conservation.
– Inefficient coordination among ministries related to the tourism service.
– Limited tourism information and accurate statistics.
– Inadequate quality control services (especially for food).
– Inadequate public transportation.
– Inadequate police force.
– Poor signs in many tourist areas.
– Uneven level of health, food, information services in different tourist destinations.
– Repeated events threatening security of tourists.

Government   Moderate

Related & Supporting Industries Moderate
–/+ Some travel agencies are internationally competitive.
–/+ Moderate level of marketing and promotion services.
– Weak support services (education, health, information, police).
– Accommodations at mid-range (2- and 3- stars) are below international standards.
– Weak linkages among other cluster components (excluding accommodation).
– Lack of coordination among support organisations.

Structure, Strategy & Rivalry

Demand

Related & Support Industries

Factors

Figure 1.4 The competitive environment of the Egyptian tourism cluster.[1]

Concluding remarks

The world is under strong winds of change. The world economy is becoming more competitive, more global, and increasingly dominated by information and communications technology (Drucker 1993). Globalisation has become a term that everybody talks about without a true agreement on what it means and on how it could influence tourism in the world. Globalisation becomes even less apprehensive when we look at it from a competitive advantage of nations' viewpoint which is mainly dynamic and primarily based on local distinctiveness and supremacy in certain qualities.

As Porter (1990: 19) noted, 'Competitive advantage is created and sustained through a highly localised process. Differences in national economic structures, values, cultures, institutions and histories contribute profoundly to competitive process.' This view is even more confirmed in the socio-political field where some writers (Jaques 1989) observed: 'As power moves upwards from the nation-state toward international units (global firms) ... so there is a countervailing pressure, whose roots are various, for it to move downwards. There is a new search for identity and difference in the face of impersonal global forces, which is leading to the emergence of new national and ethnic demands.'

As Michael Hall puts it (Hall 1996), 'Tourism is a part of the search for identity and a desire for economic positioning in contemporary globalisation.' And 'Because tourism helps in promoting and developing geographical entities, it is of particular significance for third world countries which are in greater need of enhancing their socio-economic and political wealth to earn and upgrade in international relations. The instrumental reasoning of today's new global capitalism and its "social imaginary" provide not only the initial impetus for sound tourism development but also sets the trajectory.'

Finally, tourism, as a new force of competitive advantage between nations, is an effective way of bridging the gap between globalisation and localisation particularly in culture. This would be the way to reach a full recognition of the context in which tourism operates and the complex network of relationships, connections and interfaces between various states and markets that it creates. Consequently, we could identify various differences in socio-cultural and economic systems between generators and destinations in this new globalised world.

In this context, we may ask ourselves whether globalisation and convergence mark the end of national ambitions and strife and whether the very idea of international economic competitiveness no longer makes sense. The answer in the various minds might differ but through tourism, in our opinion, the answer would still be 'No'.

Note

1   Acknowledgement of source of citation of Figure 1.4 is made to Fawzy, Samiha (March 1998) *Egypt: The Tourism Cluster*, The Egyptian Center for Economic Studies (unpublished).

References

Axford, B. (1995) *The Global System*, New York: St Martin's Press.
Barnet, R. and Cavanagh, J. (1995) *Global Dreams, A Touchstone Book*, New York: Simon & Schuster.
Bauer, P. T. (1973) *Dissent on Development*, India: Vikas.
Berger, S. and Dore, R. (1996), *National Diversity and Global Capitalism*, Ithaca, New York: Cornell University Press.
Butler, R. (1980) 'The concept of a tourist area cycle of evolution: implications for management of resources', *Canadian Geographer* 24, 5–12.
Castells, M. (1993) 'Information economy and the new international division of labour' in R. Heilbroner, *Visions and its Future* (1995), Oxford: Oxford University Press/New York: Public Library.
Cleverdon, R. (1998) *Vision 2020*, Madrid: World Tourism Organisation.
Corten, D. (1996) 'The failures of Bretton Woods', in J. Mander and E. Goldsmith, *The Case Against the Global Economy*, San Francisco: Sierra Club Books.
Daremus, P., Keller, W., Pauly, L. and Reich, S. (1998) *The Myth of the Global Corporation*, Princeton, NJ: Princeton University Press.
Drucker, P. (1993) *Post Capitalist Society*, New York: Harper Business.
ECES, Egyptian Center for Economic Studies (1998) *Egypt: The Tourism Cluster*, Final Report, Phase 1.
Dunning, J. and Hamdani, K. (1997) *The New Globalism and Developing Countries*, Tokyo: United Nations University Press.
Euromic (1999) Unpublished report, May. Egyptian Travel Agencies Association.
Gartner, W. (1996) *Tourism Development*, New York: Van Nostrand Reinhold.
Grant, R. M. (1991) '"Porter's competitive advantage of nations", an assessment', *Strategic Management Journal*, 12, Table 1: 540.
Hall, M. (1996) 'Globalisation and tourism, connecting and contextualising culture, environment, economy and place' in *Globalisation and Tourism*, Editions AIEST, Switzerland: St Gallen.
Jaques, M. (1989) 'Britain and Europe', in S. Hall and M. Jaques (eds) *New Times, The Changing Face of Politics in the 1990s*, London: Lawrence and Wishart.
Keller, P. (1996) *Globalisation and Tourism*, AIEST Tourist Review, 4. and AIEST Editions, 38, Switzerland: St Gallen.
Kotler, P. Jatusripitale, S. and Maesincee, S. (1997) *The Marketing of Nations*, New York: The Free Press.
MacCannell, D. (1992) *Empty Meeting Grounds, The Tourist Papers*, London: Routledge.
McGrew, A. (1992) 'The state in advanced capitalist countries', in A. P. Braham and P. Lewis (eds) *Political and Economic Forms of Modernity*, Oxford: Polity Press.
Ohmae, K. (1991) *The Borderless World*, New York: Harper Perennial.
Poon, A. (1993) *Tourism, Technology and Competitive Strategies*, Wallingford: CAB International.

Porter, M. (1990) *The Competitive Advantage of Nations*, Hampshire and London: Macmillan.

Ritchie, J. B. R. and Zins, M. (1978) 'Culture as a determinant of the attractiveness of a tourist region', *Annals of Tourism Research*, 5.

Sessa, R. (1996) 'Tourism production, tourism products', in *Globalisation and Tourism*, 38, Editions AIEST, Switzerland: St Gallen.

Smeral, E. (1996) 'Globalisation and changes in the competitiveness of tourism destinations', in *Globalisation and Tourism*, Editions AIEST, Switzerland: St Gallen,

Thurow, L. (1996) *The Future of Capitalism*, New York: William Marrow.

Wahab, S. and Pigram, J. (eds) (1997) *Tourism, Development, and Growth*, London: Routledge.

2 Territorial economic integration and globalisation

C. Michael Hall

Introduction

Globalisation is not just a fashionable idea, it is 'a concept with consequences' (Hirst 1997: 424). Globalisation has had the effect of changing the 'rules of the game' in the struggle for competitive advantage among firms, destinations and places within, as well as between, countries and regions (Hall 1997; Higgott 1999). Globalisation is a complex, chaotic, multiscalar, multitemporal and multicentric series of processes operating in specific structural and spatial contexts (Jessop 1999). It should be seen as an emergent, evolutionary phenomenon which results from economic, political, socio-cultural and technological processes on many scales rather than a distinctive causal mechanism in its own right. It is both a structural and a structuring phenomenon the nature of which depends critically on sub-global processes. According to Jessop (1999: 21) 'structurally, globalisation would exist in so far as co-variation of relevant activities becomes more global in extent and/or the speed of that covariation on a global scale increases'. Therefore, global interdependence typically results from processes which operate at various spatial scales, in different functional sub-systems, and involve complex and tangled causal hierarchies rather than being a simple, unilinear, bottom-up or top-down movement (Jessop 1999). Such an observation clearly suggests that globalisation is developing unevenly across space and time. Indeed, 'a key element in contemporary processes of globalisation is not the impact of "global" processes upon another clearly defined scale, but instead the relativisation of scale' (Kelly and Olds 1999: 2). Such relativities occur in relation to both 'space-time distantiation' and 'space-time compression'. The former refers to the stretching of social relations over time and space, e.g. through the utilisation of new technology such as the Internet, so that they can be coordinated or controlled over longer periods of time, greater distances, larger areas, and on more scales of activity. The latter involves the intensification of 'discrete' events in real time and/or increased velocity of material and non-material flows over a given distance; again this is related to technological change, including communication technologies and social technologies (Jessop 1999).

The discourse of globalisation therefore goes further than the simple description of contemporary social change, it also carries with it the power to shape material reality via the practical politics of policy formulation and implementation (Gibson-Graham 1996; Leyshon 1997; Kayatekin and Ruccio 1998; Kelly and Olds 1999). It can also construct a view of geographical space that implies the deferral of political options from the national to the supranational and global scales, and from the local to the national. In effect, globalisation 'itself has become a political force, helping to create the institutional realities it purportedly merely describes' (Piven 1995: 8).

In addition to the 'structural context' of globalisation noted above, authors such as Ohmae (1995), Jessop (1999) and Higgott (1999) point to a more strategic interpretation of globalisation, which refers to individual and institutional actors' attempts to promote the global coordination of activities on a continuing basis within different orders or functional systems. For example, interpersonal networking, inter-firm strategic alliances, the creation of international and supranational regimes to govern particular fields of action, and the broader development of modes of international and supranational systems of governance. Therefore, given the multiscale, multitemporal and multicentric nature of globalisation we can recognise that globalisation 'rarely, if ever, involves the full structural integration and strategic coordination across the globe' (Jessop 1999: 22). Instead, processes usually considered under the rubric of 'economic globalisation' include:

- the formation of regional economic and trading blocs – particularly in the triadic regions of North America (North American Free Trade Area (NAFTA)), Europe (European Union (EU)) and East Asia-Pacific (Asia Pacific Economic Cooperation) – and the development of formal links between those blocs (e.g. the Asia–Europe Meetings);
- the growth of 'local internationalisation', 'virtual regions', through the development of economic ties between contiguous, e.g. 'border regions', or non-contiguous local and regional state authorities, e.g. growth regions and triangles, in different national economies which often bypass the level of the nation-state but which still retain support at the national level;
- the widening and deepening of international and supranational regimes which cover economic and economically relevant issues and which may also provide for regional institutionalised governance;
- the internationalisation of national economic spaces through growing penetration (inward flows) and extraversion (outward flows);
- the extension and deepening of multinationalisation by multinational firms; and

- the 'emergence of globalisation proper through the introduction and acceptance of global norms and standards, the development of globally integrated markets together with globally oriented strategies, and "deracinated" firms with no evident national operational base'.

(Jessop 1999: 23)

Although all of the above processes have a spatial dimension to them, this chapter will focus on the two first mentioned processes as examples of territorial economic integration and the increased importance of regionalism in the world economy and for tourism in particular.

Tourism and regional economic and trading blocs

There has been a dramatic increase in the number of regional economic unions and trading blocs in the world. According to the Japanese External Trade Organisation (JETRO) as of July 1996, there were 101 such regional economic unions based on formal agreements (Table 2.1) (To distinguish them from substantive, but informal economic regions, regions with agreements on some sort of institutional framework are referred to here as 'regional economic unions' as per the approach of JETRO (1996)). Sixty-nine of these unions, corresponding to over 60 per cent, were established since 1990. The reason why so many regional economic unions were formed in the 1990s is related to the general reorganisation of the international economic order particularly with respect to the collapse of the planned economies of Eastern Europe with the fall of state communism. According to JETRO (1996) the idea that 'achievement of country's own economic prosperity required building closer ties with central countries in the market economic system spread'. This movement towards regional

Table 2.1 Number of regional economic unions by decade

	Before 1950	1950–9	1960–9	1970–9	1980–9	After 1990	Total
Africa	0	0	2	2	0	4	8
Asia/Oceania	0	0	0	0	1	2	3
Europe	0	0	1	2	0	36	39
Middle East	0	0	0	0	3	1	4
North America and Latin America	0	0	2	1	15	22	40
Others (spanning several regions)	0	0	1	1	1	4	7
Total of regions	0	0	6	6	20	69	101

Source: After JETRO (1996).

economic unions was global in scope and contributed to the enlargement of the EU, the establishment of the North American Free Trade Agreement (NAFTA) and Asia-Pacific Economic Cooperation (APEC), and the development of the Free Trade Area of the Americas (FTAA).

In classifying the regional economic unions by location, the most numerous are to be found in North America and Latin America, forty in all, followed by Europe with thirty-nine. As of the late 1990s, there were only three established unions in Asia and Oceania although more had been mooted (Table 2.2). The characteristics of the regional economic unions of these three regions in the 1990s was a mix of recycling existing agreements and the establishment of new ones in Latin America, acceleration of establishment of unions due to the collapse of state communism in Europe in the late 1980s and early 1990s, and the formation of substantive multi-lateral relationships before formal agreements in Asia (JETRO 1996; Hall and Samways 1997; Hall and Page 2000) (Table 2.3). In North America and Latin America, new unions include the G3 formed among Mexico, Colombia and Venezuela which was influenced by the establishment of NAFTA and the FTAA launched to compete with the formation of APEC in Asia and movements aimed at breathing new life into existing agreements such as CARICOM (Caribbean Common Market), ANCOM (Andes Common Market), and CACM (Central American Common Market) (Table 2.4). In Europe, the main features have been the increase

Table 2.2 Regional economic unions

Region	WTO notifications[a]	IMF[b]	JETRO[c]
Africa	7	14	8
Asia/Oceania[d]	10	6	3
Europe	73	15	39
Middle East	3	5	4
North America and Latin America	18	24	40
Others (spanning several regions)[e]	33	4	7
Total[f]	144	68	101

Source: After JETRO (1996).
Notes:
a Indicates agreements notified to the World Trade Organisation (previously GATT) from 1948 to July 1996.
b Indicates major regional economic unions determined by the International Monetary Fund after the conclusion of the Uruguay Round.
c Indicates totals of the World Trade Organisation and IMF minus overlapping agreements, agreements losing effect due to absorption by other economic unions, and non-reciprocal agreements plus some agreements independently confirmed by JETRO.
d APEC is not an economic union established by a treaty, so is not included.
e Indicates unions spanning several of the five regions.
f Unions which have been established by agreements, but which have not yet taken effect and unions currently being considered are not included.

Table 2.3 Regional economic zones in the Asia-Pacific

Name	Time of formation	Region	Comment
Japan Sea Rim Economic Zone (Northeast Asia Economic Zone)	Early 1990s	Japan (Japan sea side), Russian Far East, Mongolia, Republic of Korea, North Korea, China (Jilin, Heilongjiang and Liaoning provinces)	Centred around mouth of the Tumen River of China and principally including Jiln province and cities of Nakhodka Vladivostok and Chogjin (North Korea). This economic zone includes the Tumen Triangle linking the Russian Far East (Nakhodka and Vladivostok), Jiln province and the northern part of North Korea. Limited tourism development is occurring in the Russian Far East.
Yellow Sea Rim Economic Zone	Late 1980s	China (Liaoning and Shandong provinces), part of Northeast China, west coast of Republic of Korea	Incorporating the three Northeast provinces of China and the Dalian free economic zone. The area has received numerous investments from Japanese companies. There is also the East China economic zone linking Japan; the southern part of the Republic of Korea and various regions of China. Tourism development is occurring in south-western Korea while increased travel flows between China and Korea is also beginning to develop.
South China Economic Zone	Early 1980s	China (Fujian and Guangdong provinces), Hong Kong, Taiwan	Based on ties between Hong Kong and Guongdong province. Ethnic Chinese capital from Taiwan and Singapore is a major component in the development of the region. The region is currently subject to massive aviation and transport infrastructure development and the development of resorts along the South China coast, geared

			currently for the Chinese domestic market but increasingly taking international Chinese visitors. The west coast of Taiwan also holds great potential for tourism development once relations between China and Taiwan are normalised.
Cross-Straits Economic Zone	Early 1980s	China (Fujian), Taiwan	In broad sense, a subregion of the South China Economic Zone. Investment in China from Taiwan is increasing. There is also the Chinese Economic Zone based on ties between China and Hong Kong and Taiwan and extending to Macao and even Singapore.
Greater Mekong Subregion (Indochina Economic Zone)	Late 1980s	Thailand, Cambodia, Laos, Myanmar, Vietnam	Centred on the complementary relationship between Thailand which is growing economically at a speed bringing it close to the NIEs and surrounding countries. Ethnic Chinese capital sustains the industrial activity while tourism development is being financed from a number of sources including Australia, France, Malaysia and Singapore. Thailand is the only country with significant domestic investment in tourism. Thailand is currently attempting to make itself the transport and tourism hub of the region.
Singapore-Johor-Riau Growth Triangle	Late 1980s	Malaysia, Singapore, Indonesia	Aimed initially at a fusion of Singapore's capital, technical know-how, and information resources and the inexpensive labour forces of Malaysia and Indonesia. Has led to the development of substantial leisure and tourism infrastructure in Malaysia and Indonesia in particular for the rapidly growing Singapore middle class. The triangle contributed almost 25 per cent of inbound travel in Indonesia prior to the Asian economic crisis.

Table 2.3 (continued)

Name	Time of formation	Region	Comment
Brunei Darussalam-Indonesia-Malaysia-Philippines East ASEAN Growth Area	Early 1990s	Brunei, Indonesia, Malaysia, Philippines	Objective is construction of complementary relationship between economically underdeveloped regions in order to promote trade. Large-scale tourism development is presently focused on the Philippines although smaller-scale cultural tourism developments are occurring throughout Borneo/Kalimantan.
Indonesia-Malaysia-Thailand Growth Triangle	Early 1990s	Indonesia, Malaysia, Thailand	Aims at various tourist, agrobusiness and electronics related projects for economic cooperation. The establishment of new railway, road and shipping linkages is also regarded as having a positive spin-off in terms of intra-ASEAN and domestic tourism. Political instability in the region has slowed economic development.
Indonesia-Northern Territory-Papua New Guinea Growth Triangle	Mid-1990s	Eastern Indonesia, tropical Northern Territory, Papua New Guinea	Aims to develop better transport and tourism links between the Northern Territory and Papua New Guinea and to provide better links with Indonesia's eastern attractions. Although affected by political instability in Eastern Indonesia the Australian federal government is still interested in encouraging economic linkages.

Source: After JETRO (1996); Hall and Samways (1997); Hall and Page (2000).

Table 2.4 Regional agreements in North America and Latin America
(as at 1997)

Name	Year of inception	Main participating countries and regions	Remarks
CACM (Central American Common Market)	1960	Guatemala, Honduras, Nicaragua, El Salvador Costa Rica	Tariff union. Agreement on start of common market.
ANCOM (Andes Common Market)	1969	Colombia, Ecuador, Venezuela, Bolivia, Peru	Changed from free trade area to take common tariffs.
CARICOM (Caribbean Common Market)	1973	13 countries and one region in Caribbean Sea	Tariff union. Aims at common market. Significant for tourism because of relation to Caribbean regional promotion.
ALADI (Latin American Integration Association)	1981	Mexico, Colombia, Venezuela, Argentina, Brazil, Chile, Peru, Paraguay, Bolivia, Ecuador	Loose union using intraregional trade.
Israel-US Free Trade Agreement	1985	US, Israel	Tariffs eliminated for a number of products by end of 1994.
NAFTA (North American Free Trade Agreement)	1994	US, Canada, Mexico	Aims at free trade area. Surpasses EU in size. Mobility for tourism still being determined because of US fears of illegal immigrants.
Israel-Canada Free Trade Agreement	1996	Canada, Israel	Tariffs eliminated for majority of its agricultural and fishery products as well as textiles.
MEROOSUR (Southern Zone Common Market)	1991	Brazil, Argentina, Paraguay, Uruguay (Chile from 1997)	Tariff union since January 1995. Aim of free trade agreement with Chile.
G3	1995	Mexico, Colombia, Venezuela	Aims to create a free trade area.
ACS (Alliance of Caribbean States)	1995	G3, six Central American nations and 16 Caribbean countries	Aims at integrating economies starting from greater economic and cultural cooperation. Tourism significant.

Source: After JETRO 1996.
Notes: The forty agreements in North America and Latin America mentioned in Table 2.2 was determined from the above ten agreements and the thirty-two supplementary agreements of ALADI minus the free trade area agreements between the USA and Israel and between Canada and Israel.

in the number of agreements between the EU and neighbouring countries and the numerous agreements concluded among the countries of the former Soviet Union and countries of Eastern Europe (Table 2.5).

Tourism is often a significant component in the establishment of regional economic unions, and the establishment of more formal economic relations, such as free trade agreements which aim to enhance the flow of goods and services, also tend to provide for increased mobility of people. However, while increased personal mobility between countries is regarded as important for leisure travel other considerations include improving accessibility for business travellers and providing for regional labour mobility. For example, the Japanese Ministry of Foreign Affairs (1999) *Report of the Mission for Revitalization of [the] Asian Economy* specifically recommended the promotion of two-way interaction between people in Japan and Asian countries and called for the formulation and implementation of 'long-term, large-scale programs for human exchange' through relaxing the constraints on human exchange, particularly through improved international air links and international airport hub development, and 'simplification and acceleration of immigration, quarantine, and customs procedures'.

In the case of Asia, three sets of inter-related factors underpinned the extension of production networks which provided much of the region's economic growth in the early 1990s and which contributed to greater economic integration at the supranational level but which also left it vulnerable to the problems of the Asian financial crisis in the late 1990s:

- The first set of factors was the change in relative factor costs within the region. For companies in north-east Asia (Japan, Korea and Taiwan) seeking to increase their production of relatively mature products, the costs of undertaking the investment necessary to increase domestic capacity were far in excess of those of establishing new facilities elsewhere in the region where labour, infrastructural and land costs were much lower (Garnaut 1990).
- The second group of factors was political. The original tensions over trade imbalances and market access between Japan and the United States, and later Japan and Europe and Australasia, were extended to Korea and Taiwan in the late 1980s. International tensions in turn generated domestic political forces that interacted with and reinforced underlying economic change. For instance, the Japanese government encouraged corporations to increase both their foreign direct investment and their sourcing from their overseas subsidiaries. In tourism terms it is notable that Japan utilised outbound tourism as a means of improving trade relations and providing its trading nations with a source of funds to purchase Japanese manufactured goods (Hall 2000). Indeed, economic interdependence between Japan and Asia has been deepening steadily since the mid-1980s when the first wave in Japanese offshore investment began with Japanese companies establishing

Table 2.5 Regional agreements in Europe (as of 1997)

Region	Name of agreement	Year of inception	Main participating countries and regions	Remarks
EU	Association Agreements	1964	EU-Turkey (1964), EU-Malta (1971), EU-Cyprus (1973)	The EU-Turkey Association Agreement became a tariff union from January 1996. The three countries have applied for membership in the EU.
	European Union	1993	Belgium, Denmark, France, Germany, Greece, Ireland, Italy, Luxembourg, Netherlands, Portugal, Spain, UK, Sweden, Austria, Finland	Economic and currency union commenced in 1999, although certain countries have opted out of currency union as at the time of writing. Travel and tourism increasingly significant in EU for achievement of cultural and economic objectives.
	EA (European Agreement)	Starting 1994	EU-Poland (1994), EU-Hungary (1994) EU-Czech Republic (1995), EU-Slovakia (1995), EU-Romania (1995), EU-Bulgaria (1995)	Aims at realisation of EU membership of Eastern European countries in future from free trade area. Signed in 1995 by Estonia, Latvia, Lithuania and Slovenia. Albania is engaged in advance preparations. Further, separate free trade agreements were concluded with the Baltic States with these taking effect in 1995.
	Partnership and Cooperation Agreements	1994	EU-Russia (1994), EU-Moldavia (1994), EU-Ukraine (1994), EU-Belarus (1995), EU-Kirghizia (1995), EU-Kazakhstan (1995)	As of 1996 Georgia, Armenia and Azerbaijan were in negotiations. Uzbekistan was engaged in advance preparations.
	EU-Mediterranean Association	1996	EU-Tunisia	Aims at free trade area. Signed by Israel (1995) and Morocco (1996). Egypt, Jordan and Lebanon were in negotiations. Algeria was engaged in advance preparations. Syria was studying membership. Further, the EU and Israel concluded a free trade co-operation agreement in 1995 and the EU and Mediterranean countries concluded a trade coopera-tion agreement in the 1970s. Will be significant for tourism once Middle East political tensions decrease.

Table 2.5 (continued)

Region	Name of agreement	Year of inception	Main participating countries and regions	Remarks
EFTA	EFTA (European Free Trade Association)	1960	Norway, Switzerland, Iceland, Liechtenstein	Launched as a free trade area to counter the then EEC (European Economic Community) – the forerunner to the EU.
	EFTA, Central Europe, and Eastern Europe and EFTA and Mediterranean Free Trade Agreements	1992 (1993)	EFTA-Turkey (1992), EFTA-Israel EFTA-Romania (1993), EFTA-Poland (1993), EFTA-Czech Republic (1993) EFTA-Slovakia (1993), EFTA-Bulgaria (1993), EFTA-Hungary (1993), EFTA-Slovenia (1995), EFTA-Estonia (1996), EFTA-Latvia (1996), EFTA-Lithuania (1996)	As of 1996 Albania was in negotiations. Egypt, Morocco and Tunisia were engaged in advance preparations.
EU-EFTA	EEA (European Economic Area)	1994	EU and EFTA countries other than Switzerland	World's largest unified market.
Eastern Europe	CEFTA (Central European Free Trade Agreement)	1993	Poland, Czech Republic, Slovakia, Hungary, Slovenia	Aims at free trade area. Exceptions in agricultural products, etc. Slovenia joined in 1996. Romania, Bulgaria and the Baltic States are expected to join. Significant for increased intra-regional travel.
	Czech Republic and Slovak Republic Customs Union Agreement	1993	Czech Republic, Slovakia	Tariff alliance. Aims at maintaining past trade in goods.
	Baltic Free Trade Agreement	1994	Estonia, Latvia, Lithuania	Aims at free trade area. Exceptions in agricultural products.
	Czech Republic-Romania Free Trade Agreement	1995	Czech Republic, Romania	Aims at complete abolition of tariffs, border taxes, etc. on industrial products.

	Slovak Republic-Romania Free Trade Agreement	1995	Slovakia, Romania	Aims at complete abolition of tariffs, border taxes, etc. on industrial products.
	Romania-Moldavia Free Trade Agreement	1995	Romania, Moldavia	Abolishes tariffs on products originating from two countries.
CIS	Commonwealth of Independent States (CIS) Economic Union	1993	Armenia, Azerbaijan, Belarus, Georgia, Kazakhstan, Kirghizia, Moldavia, Russia, Tadzhikistan, Turkmenistan, Ukraine, Uzbekistan	Aims at gradually broadening union from free trade area to tariff union, common market, and currency coordination and in future economic union.
	Agreement between Republic of Belarus, Republic of Kazakhstan, Kirghiz Republic and Russian Federation on Deepening Integration in Economic and Humanitarian Fields	1996	Russia, Belarus, Kazakhstan, Kirghizia	Aims at free movement of goods, services, capital and labour and development of economic relations. Important for intra-regional travel.
	Agreement on the Foundation of Commonwealth	1996	Russia, Belarus	Aims at single economic zone and currency union.

Source: Based on JETRO (1996).

Notes :

1 In Table 2.2, the association agreements, European agreements, partnership and cooperation agreements, and free trade agreements of the EFTA, Central Europe and Eastern Europe and the Mediterranean countries are counted separately since they are concluded with individual countries.

2 The EU-Turkey Association Agreement, EU-Tunisia Association Agreement, EU-Israel Free Trade Agreement, EFTA-Turkey Free Trade Agreement and EFTA-Israel Free Trade Agreement are counted as 'Others' (spanning several regions) in Table 2.2.

3 The number of European agreements of Table 2.2 include the free trade agreements between the EU and Estonia, Latvia and Lithuania (See remarks for EA).

offshore operations and Japanese imports from the region increasingly substantially, providing new markets for Asian economies. As the economies of the ASEAN nations and China grew at unprecedented rates, Japan and the NIEs continued to increase their consumption of imports and their industries continued to move offshore. As a result, Asia now has an extremely high degree of intraregional trade and investment interdependence (JETRO 1999). The NIEs have been the major investors in ASEAN and China, followed by Japan. Trade inter-dependence within the region – including Japan and the nine East Asian countries (the four NIEs, four major ASEAN economies and China) – has risen as high as 52 per cent. This is almost on par with intraregional interdependence in the European Union, which is 61 per cent, and exceeds that of NAFTA, at 48 per cent (Arai 1998). Around 40 per cent of Japan's trade is now with Asia, which is also the destination for around 30 per cent of Japanese total investment (Yosano 1998).

- The third set of factors is related to changes in the production process which have facilitated flexible production techniques through the advent of the microelectronics and communications revolution. These, in turn, decreased the significance of economies of scale thereby opening up markets for a whole range of non-standardised products. Smaller companies have therefore been been able to gain a footing in increasingly regionalised production chains.

Many areas of the Asian regional political economy, as with many regions in Europe and North America, now consist of clusters of inter-related industrial sectors that are better described as networks rather than as unconnected 'industries'; the basic organisational 'unit' being the inter-action between firms linked together in chains of production, exchange and distribution. Firms, or even decentralised divisions within firms, main-tain a degree of autonomy in the chain, but all significant activity is, in some way, coordinated with other organisations in the network. Such network relations have been facilitated by the willingness of government to encourage overseas investment, often in partnership with local industry, and by the development of policy settings which have sought to reinforce freer trade and investment in the region.

Although not formally classified as a regional economic union by JETRO (1996), APEC is a major force for free trade in the Pacific Rim and in cre-ating improved dialogue and relations between the major economic pow-ers in the region. APEC has no charter which forces contractual obligations on its members, instead it works by a consensus driven approach (Hall and Samways 1997). APEC was established in 1989 to provide a forum for the management of the effects of the growing interdependence of the Pacific Rim and to help sustain economic growth. Originally established as an informal group of twelve Asia-Pacific economies, APEC has since grown with the admission of the People's Republic of China, Hong Kong and Chinese Taipei (Taiwan) in 1991, Mexico and Papua New Guinea in 1993 and Chile in 1994. A permanent secretariat is based in Singapore, with the APEC Chair and ministerial meetings being rotated annually among members. The first ministerial meeting was held in Canberra, Australia, in 1989; since then meetings have been held in Singapore, Seoul, Bangkok, Seattle, Jakarta, Osaka, Manila, Vancouver and Kuala Lumpur, with New Zealand hosting APEC in 1999 (Hall and Page 2000).

The declarations and recommendations of the annual ministerial conference carry significant political and diplomatic weight and provide the framework for the development of an action agenda which serves as the blueprint for meeting the commitment to 'free and open trade and investment in the Asia-Pacific', with all barriers to trade and investment to be dismantled before 2010 or 2020 by developed and developing national members, respectively (Bureau of East Asian and Pacific Affairs 1996). As part of the APEC framework, ten working groups have been established, covering broad areas of economic, educational and environmental cooperation, including tourism 'on the basis that the tourism industry is of growing importance in promoting economic growth and social development in the Asia-Pacific region' (APEC Tourism Working Group (ATWG) 1995). The ATWG was formed in 1991 based on the recognition that the tourism industry is of growing importance in promoting economic growth and social development in the Asia-Pacific region. The ATWG brings together tourism administrators to share information, exchange views and develop areas of cooperation on trade and policies. Participation by the business and private sector has been channelled through the involvement of representative travel organisations such as the Pacific Asia Travel Association (PATA), the World Travel and Tourism Council (WTTC), and the World Tourism Organization (WTO). The ATWG plans to expand participation of the business and private sector by enlarging the number of representatives invited to attend as observers to ATWG meetings (ATWG 2000). The vision statement of the ATWG says that it 'will foster economic development in the Asia-Pacific region through sustainable tourism growth that is consistent with the enhancement of the natural, social and cultural environment', recognising that:

- tourism is one of the region's fastest growing industries and is of significant importance to the economic development of the APEC economies;
- tourism is important in fostering regional understanding and cooperation;
- the tourism industry in member economies is at different levels of development; and
- member economies share the common goal of quality development and services.

(ATWG 1999)

Under its objectives, the Tourism Working Group will endeavour to achieve long-term environmental and social sustainability of the tourism industry by giving priority to the following:

- removing barriers to tourism movements and investment and liberalising trade in services and tourism;
- developing and implementing the concepts of environmental and social sustainability in tourism development;
- facilitating and promoting human resource development;
- enlarging the role of the business/private sector;
- developing cooperation and programmes in the fields of information-based services related to trade and tourism; and
- sharing information among APEC economies.

(APEC Tourism Working Group 1995)

During its two meetings held in 1999 (Manzanilo, Mexico and Lima, Peru), the ATWG's work programme included activities in the areas of trade and investment liberalisation and facilitation, and economic and technical cooperation including the development of an APEC Tourism Charter. A tourism task force, comprising Brunei Darussalam, Korea, Mexico, New Zealand and Singapore, with inputs from PATA and the WTTC, identified four main APEC Tourism Charter goals. These are:

1   Sustainable management of tourism impacts and outcomes – environmental, economic, social and cultural.
2   Increase mobility of visitors and demand for tourism goods and services in the APEC region.
3   Remove impediments to tourism business and investment.
4   Enhance recognition and understanding of tourism as a vehicle for social and economic development – harmonisation and sharing of tourism information, expanding the knowledge base.

(ATWG 2000)

In principle, the APEC Tourism Charter, proposed to be adopted by APEC Tourism Ministers in June 2000, is to include measures that will:

- contribute to the minimisation of the regulatory impediments to tourism;
- promote environmentally and socially sustainable tourism;
- reduce congestion and improve passenger processing facilitation;
- identify emerging issues in tourism;
- improve the understanding of tourism; and
- enhance visitor services and tourism infrastructure.

(ATWG 2000)

Undoubtedly, the actions of a group such as the APEC Tourism Working Group are not spectacular in terms of the appointment of media coverage which they receive. However, they are extremely influential in terms of the direction of tourism development and serve to strengthen the horizontal and vertical integration of the tourism industry within the Asia-Pacific region. In addition, it should be noted that organisations such as APEC while seeking greater freedom of trade and investment in the Asia-Pacific are part of a hierarchy of organisations which are seeking to develop new regimes of international tourism trade with corresponding increased supranational economic integration (Hall 2000). However, these actions reflect the policy settings of nation-states. The next section examines the development of economic globalisation at a regional level in terms of growth triangles and border agreements which involve sub-national government.

Tourism and 'local internationalisation'

A major development in the world trade system in the 1990s has been the emergence of regional cooperation with one of the principal factors affecting the competitiveness of economies is the emergence of region-alised networks of production. Regional cooperation is now considered a substantial means of enhancing economic development and providing economic security within regions. Transnational economic zones utilise the different resources of the various countries and exploit cooperative trade and development opportunities. The private sector provides capital for investment while the public sector at the sub-government and national level provides infrastructure, fiscal incentives, and the administrative frame-work to attract industry. Generalisations about the new production networks which develop in such regions are difficult as they take various forms and differ, for instance, according to the origins of the lead firms in the network (Bernard 1994). Nevertheless, it has been apparent that growth triangles or polygons (depending on the number of national actors involved) have been a substantial driving force for growth in a number of regional economies throughout the 1990s, although in the Asian context much of the initial driving force for such integration has come from the private sector rather than government (Toh Mun Heng and Low 1993; Amsden 1994).

The regionalisation of production in East Asia might appear to be sub-stantially explained merely by reference to changing patterns of compara-tive advantage as indicated in the movement in relative factor prices. Such an account, however, would be incomplete. As noted above, the changing geo-political context and technological developments also played an impor-tant role, particularly with respect to the level of economic integration of Japanese firms in the region. Regional cooperation provides a competitive model to attract investment and technology. In East Asia a number of subregional growth areas have been established to maximise cross-border movements of goods, services, investment and human resources, and to exploit comparative advantages of geographical areas divided by political boundaries (see Table 2.3). In several of these areas, the business/private sector has established business councils to link business entities and other economic instutions, and to influence government policy. According to the Secretary General of ASEAN, Ajit Singh, 'These growth areas will have to be flexible to change where necessary, innovative, and always attentive to the needs of the investors and the businessmen. They also have to be aware that they are competing with much larger countries such as China and India, whose capacities for attracting investors are much greater than their own' (Kruger 1996: 17). For example, the Southern Growth Triangle in South-East Asia, also known as SIJORI (Singapore, the State of Johore in Malaysia, and Riau Province of Indonesia), was formed in 1989 and covers a population of about six million people. It attracted US$10 billion in private sector investments during its first five years. Such regional economic cooperation has occurred in other Asian regions as well, spurring economic development. The Indonesia, Malaysia, Thailand Growth Triangle (IMT-GT) is an economic network of three regions in Indonesia (Medan, Aceh and West Sumatra), four northern Malaysian states (Kedah, Penang, Perak, and Perlis), and all the southern provinces in Thailand (the fourteen southern Thai provinces in South Thailand comprise Ranong, Chumphon, Surat Thani, Phangna, Krabi, Phuket, Nokho. Si Thammarat, Trang, Phattalung, Satun, Songkhla, Pattani, Yala and Narathiwal). The overall goal of the IMT-GT is to accelerate private sector-led economic growth through the following initiatives (Ministry of Agriculture nd):

• promote direct investment and facilitate economic development of the subregions by exploiting underlying economic complementarities and investment opportunities;
• enhance subregional international competitiveness for direct investment and export production;
• lower transportation and transaction costs arising from geographic proximity; and
• reduce production and distribution costs through economies of scale.

The IMG-GT activities are divided into eight working groups: (1) Tourism, (2) Investment, Trade, and Industry, (3) Agriculture and Fisheries, (4) Services, (5) Infrastructure, (6) Human Resources Development, (7) Transportation and Telecommunication, and (8) Energy. Five industrial sectors have been identified as critical for the development of the IMT-GT area with the goals for each sector being designed to support the overall IMT-GT development objective. The sectors are divided according to activity. The productive sectors (i.e. investment, trade, industry, agriculture and fisheries, and tourism) are intended to generate the production, employment, trade, investment and other activities contributing to IMT-GT area economic growth. The facilitating sectors (i.e. labour mobility, human resource development, transportation, communications, and energy), are comprised of activities that are driven by, and are designed to facilitate and accommodate economic growth (Ministry of Agriculture nd).

A comprehensive set of policy, project and institutional actions is recommended to facilitate the achievement of the overall IMT-GT goal and various sector objectives. A total of 97 initiatives (including major sub-components) are proposed for implementation, comprising 24 policy, 30 programme, 32 physical project, and 11 institutional initiatives. Of these, 55 of the initiatives are proposed for Indonesia, 58 for Malaysia, and 57 for Thailand: 10 would be implemented at the national level and 32 among all three subregions. Three would be implemented between the Indonesian and Malaysian subregions of the IMT-GT and thirteen between the Thai and Malaysian subregions. None of the initiatives would be directed specifically for action between the Indonesian and Thai subregions. Eighteen of the initiatives are in the tourism sector. Table 2.6 identifies tourism projects that had been developed as part of the IMT-GT by 1998.

The development of local internationalisation is clearly not isolated to Asia. For instance, Europe has many such transnational developments, often with EU funding, as part of the processes of European integration (Hall 2000). Examples also exist in North America particularly with respect to cross-border relations between Canada and the United States and Mexico and the United States. For example, the Pacific NorthWest Economic Region (PNWER) consists of the American states of Alaska, Idaho, Montana, Oregon and Washington plus the Canadian provinces of Alberta, British Columbia and Yukon Territory, as well as numerous private sector members. The combined Gross Domestic Product (GDP) of the region is in excess of US$350 billion annually. If it were a single country, the Pacific NorthWest Economic Region would rank tenth among the world's industrial economies (PNWER 2000a).

Initially established in 1989 as the Pacific NorthWest Legislative Leadership Forum, PNWER was created in 1991 by statutes in all seven states and provinces. PNWER brings together legislative, government and private sector leaders to work towards the development of public policies that promote the economies of the Pacific Northwest region in the global marketplace: 'The objective of PNWER is to build the necessary critical mass for the region to become a major player in the new global economy' (PNWER 2000a). According to PNWER (2000b): 'Increasing competition requires economic policies that build on the strengths of the region, increase efficiency through cooperation and collaboration, and create ongoing opportunities for policy development beyond the state, provincial, and federal level.'

As with APEC, policy development and implementation is done through the Working Groups. The six original Working Groups of PNWER reflected the region's strengths including Environmental Technology,

Table 2.6 IMT-GT tourism projects (as of 1998)

Companies	Nature of MOU	Project amount	Project location
ALFRA Destination Management/ P.T. Shahiba Coral Resort	Tourism	RM0.4m	Medan
Malayan Banking Berhad/ P.T. Shahiba Coral Resort	Hotel – Loan – Agreement	RM15.6m	Medan
P.T. Shahiba Coral Resort / P.T. Lyduma Intermas	Consultancy	RM0.15m	Medan
Sri ASEAN Eko Resort	Time share resort	–	Medan
Kedah State Development Corporation	Southern Border Administrative Centre	–	Penang
P.T. Shahiba Coral Resort/P.T. Kasih Cintra Timur	IMT Hotel	–	Medan
Impian Samudera Sdn Bhd/ P.T. Adenan Harue Sulaiman	Hotel/Golf	–	Sabang
Bayview Hotel Penang / Perak Tourist Association	Human resource development and training at the Sultan Hotel Aceh	–	Aceh
Bayview Hotel Penang	Human resource development and training at the Siantar Hotel	–	Aceh

Source: Derived from SIBEXLINK (1998).

Tourism, Recycling, Value-Added Timber, Workforce Training and Tele-communications, with further working groups being established in Transportation (1993), Export (1994), Government Procurement (1994) and Agriculture (1995). The Tourism Working Group has undertaken work in a number of areas including creating regional tourism partnerships; developing a list of barriers to tourism in region; successfully lobbying for an open skies policy between Canada and the USA; recommended expansion of visa waver, CANPASS and other barrier reducing programmes; and sponsoring a Regional Tourism Marketing Summit (PNWER 2000c). The activities of PNWER are similar to other local internationalisation initiatives around the world. However, the full impact of such activities are still to be adequately examined.

Although models of international relations are well developed to deal with the activities of nation-states, the growth of subgovernmental governments as international actors provides significant challenges to our understanding of economic globalisation and its spatial and policy implications. Moreover, in an era of economic globalisation the rise of 'new' supranational policy issues such as the environment, trade and labour mobility has become profoundly different in scope from the traditional strategic and security issues of international relations. Such issues may be described as 'intermestic' in nature, that is they are simultaneously both domestic and international policy issues which, while being of substantial domestic concern, cross international boundaries thereby creating international interest in the setting of policy (Cohn and Smith 1996). Tourism is increasingly becoming recognised as such an intermestic policy issue particularly as subnational actors such as states, provinces, regions and cities respond to an increasingly globalised business environment and seek to attract investment, employment and tourists. Moreover, the role of tourism in establishing regional images through place-marketing processes may see it being given closer attention as a factor in supranational policy development (Hall 1997). Nevertheless, the understanding of the policy and territorial dimensions of tourism's functions in economic integration remains a poorly developed though potentially highly significant subject of study.

Conclusions

International regionalism is a manifestation of globalisation but it is both a set of processes and an ideology of economic and political management. *De facto* economic regionalism and integration has not been policy driven by government. Instead, it has been firm and network-led within markets and has therefore been essentially private led (Toh Mun Heng and Low 1993; Amsden 1994). At the *de jure* level regionalism is an example of national and subnational led institutional cooperation which can take a number of forms from trade commitments through to binding international agreements which serve to reinforce and expand the networks that have already been developed by the private sector. National and subnational governments have tended to introduce policies which support international regionalism in order to enhance the credibility of the state and of members with external actors, especially in relation to sources of Foreign Direct Investment. Furthermore, it has been argued that entering regional agreements may also provide for domestic policy discipline, often in the face of domestic or even international opposition (as witnessed for example during the Seattle meetings of the World Trade Organisation in 1999), the aim of which may be to make up collectively in the region for diminished domestic political and economic autonomy at the government and subgovernment level (Higgott 1999). However, these processes clearly do not take place in a uniform fashion.

This chapter has examined two specific examples of economic integration that have significant spatial dimensions. The development of regional economic and trading blocs and the growth of local internationalisation is intimately related to tourism in that tourism and associated accessibility and mobility concerns are usually a key component of the creation of vertical and horizontal linkages with regional economic networks. Historically, tourism has been ignored as an economic component in research on globalisation. However, government and subgovernment recognition of the economic role of tourism, not only in generating employment and attracting investment but also in contributing to regional promotion, provide a firm basis for further research on the significance of tourism as a factor in territorial economic integration. Thereby, reinforcing Jessop's observation that if adequately addressed 'trends towards globalisation can certainly help situate and interpret current changes in the spatial scale of economic (and other) institutions, organisations, and strategies' (1999: 21). The relationship of economic integration and interdependence at various transnational spatial scales to globalisation is not unproblematic. The regional crisis in East Asia at the end of the 1990s reinforces the consequences of globalisation in terms of its economic, political and social impacts while simultaneously reflecting that globalisation processes operate at various spatial scales and involve complex hierarchies rather than being a simple, unilinear, bottom-up or top-down movement which can be easily contained within a territorial framework. Nevertheless, such frameworks will continue to act as the spatial setting within which economic globalisation will be negotiated within the foreseeable future.

References

Amsden, A. H. (1994) 'The World Bank's the East Asian miracle: economic growth and public policy', *World Development*, 22(4): 613–70.

APEC Tourism Working Group (1995) *Asia-Pacific Economic Cooperation (APEC) Tourism Working Group*, APEC Secretariat/Singapore Trade Development Board, Singapore.

APEC Tourism Working Group (1999) *Terms of Reference* (updated 3 December 1999), APEC Secretariat/Singapore Trade Development Board, Singapore. (http://apec-tourism.org/terms-of-reference/)

APEC Tourism Working Group (2000) *Activities by Groups: Tourism* (updated 9 February 2000), APEC Secretariat/Singapore Trade Development Board, Singapore. (http://www.apecsec.org.sg/workgroup/tourism_upd.html)

Arai, H. (1998) *A Scenario for Dynamic Recovery from the Asian Economic Crisis* Thai-Japanese Association and JETRO Bangkok, Bangkok, Thailand, 21 August 1998 Hisamitsu Arai, Vice-Minister for International Affairs, MITI, Tokyo: Ministry of International Trade and Industry.

Bernard, M. (1994) 'Post-Fordism, transnational production and the changing global political economy', in R. Stubbs and G. R. D. Underhill (eds) *Political Economy and the Changing Global Order,* London: Macmillan.

Bureau of East Asian and Pacific Affairs (1996) *Fact Sheet: Asia-Pacific Economic Cooperation,* Bureau of Public Affairs, Washington.

Cohn, T. H. and Smith, P. J. (1996) Subnational governments as international actors: constituent diplomacy in British Columbia and the Pacific Northwest. BC Studies: *The British Columbian Quarterly* 110, Summer: 25–59.

Garnaut, R. (1990) *Australia and the Northeast Asian Ascendancy,* Canberra: Australian Government Publishing Service.

Gibson-Graham, J. K. (1996) *The End of Capitalism (as we knew it): A Feminist Critique of Political Economy,* Oxford: Blackwell.

Hall, C. M. (1997) 'Geography, marketing and the selling of places'. *Journal of Travel and Tourism Marketing* 6(3/4): 61–84.

Hall, C. M. (2000) *Tourism Planning.* Harlow: Prentice-Hall.

Hall, C. M. and Page, S. (eds) (2000) *Tourism in South and South-East Asia,* Oxford: Butterworth-Heinemann.

Hall, C. M. and Samways, R. (1997) 'Tourism and regionalism in the Pacific Rim: An Overview', in M. Oppermann (ed.) *Pacific Rim Tourism,* Wallingford: CAB International.

Higgott, R. (1999) 'The political economy of globalisation in East Asia: the salience of region building', in K. Olds, P. Dicken, P. F. Kelly, L. Kong and H. W. Yeung (eds) *Globalisation and the Asia-Pacific: Contested Territories,* Warwickshire Studies in Globalisation Series, London: Routledge.

Hirst, P. (1997) 'The global economy – myths and realities, *International Affairs* 73(3): 409–25.

Japan External Trade Organisation (JETRO) (1996) *White Paper on International Trade* 1996, Tokyo: Japan External Trade Organisation.

Japan External Trade Organisation (JETRO) (1999) *White Paper on International Trade Fall in Prices Causes Slowdown in World Trade (Summary),* Tokyo: Japan External Trade Organisation.

Jessop, B. (1999) 'Reflections on globalisation and its (il)logic(s)', in K. Olds, P. Dicken, P. F. Kelly, L. Kong and H. W. Yeung (eds) *Globalisation and the Asia-Pacific: Contested Territories,* Warwickshire Studies in Globalisation Series, 19–38, London: Routledge.

Kayatekin, S. and Ruccio, D. (1998) 'Global fragments: subjectivity and class politics in discourses of globalisation', *Economy and Society* 27(1): 74–96.

Kelly, P. F. and Olds, K. (1999) 'Questions in a crisis: the contested meanings of globalisation in the Asia-Pacific', in K. Olds, P. Dicken, P. F. Kelly, L. Kong and H. W. Yeung (eds) *Globalisation and the Asia-Pacific: Contested Territories,* Warwickshire Studies in Globalisation Series, 1–15, London: Routledge.

Kruger, D. (1996) 'Asians form "growth polygons" to end poverty', *The Japan Times* 29 March: 17.

Leyshon, A. (1997) 'True stories? Global dreams, global nightmares, and writing globalisation', in R. Lee and J. Wills (eds) *Geographies of Economies,* London: Arnold.

Ministry of Agriculture (Malaysia) (nd) *Indonesia, Malaysia and Thailand – Growth Triangle (IMT-GT)* (http://agrolink.moa.my/moa1/imt_gt/) (Accessed 9 February 2000)

Ministry of Foreign Affairs (MOFA) (1999) *Report of the Mission for Revitalization of Asian Economy: Living in Harmony with Asia in the Twenty-first Century*, November 1999, Tokyo: Ministry of Foreign Affairs.

Ohmae, K. (1995) *The End of the Nation State: The Rise of Regional Economies*, New York: HarperCollins and The Free Press.

Pacific NorthWest Economic Region (PNWER) (2000a) *PNWER Profile, Seattle: PNWER* (http://www.pnwer.org/PNWER Profile) (Accessed 18 February 2000).

Pacific NorthWest Economic Region (PNWER) (2000b) *Background, Seattle: PNWER* (http://www.pnwer.org/background/backgrou.htm) (Accessed 18 February 2000).

Pacific NorthWest Economic Region (PNWER) (2000c) *Accomplishments, Seattle: PNWER* (http://www.pnwer.org/background/accompli.htm) (Accessed 18 February 2000).

Piven, F. (1995) 'Is it global economics or neo-laissez-faire?' *New Left Review* 213: 107–14.

SIBEXLINK (1998) *Services* (http://www.sibexlink.com.my/imtgt) (Accessed 9 February 2000).

Toh Mun Heng and Low, L. (eds) (1993) *Regional Cooperation and Growth Triangles in ASEAN*, Singapore: Times Academic Press.

Yosano, K. (1998) *Revitalizing Japanese and ASEAN Economies*, Speech by Kaoru Yosano Minister of International Trade and Industry, 23 September, 1998, Singapore. Tokyo: Ministry of International Trade and Industry.

3 Globalization, national tourism policy and international organizations

Eduardo Fayos-Solà
and Aurora Pedro Bueno

Introduction

So much has been said about globalization and the need for corrective policies at this crossroads in history, that the demand for solid institutions to design, apply and develop these policies has all too often been overlooked. It was only in the late 1990s, with the Asia-Pacific economic crisis, the situation in Russia and other countries of the former communist bloc – Bosnia, Kosovo or Chechnya – that the mainstream of thought was admitting to the need for second generation reforms, that is to say, policies aimed at reinforcing institutions.

Given this situation, theoretical thinking in tourism and actual tourism policies are still somewhat behind the times. The need for an explicit national tourism policy is at times questioned: such tourism powers as the United States have dismantled their central tourism agencies; others, like the European Union, assign their tourism policy to small departments with absurd budgets, inadequate staff resources and incoherent programmes. In some countries, especially in those which have recently become active in tourism, interventionism is practised, while others employ the discredited if well-intentioned formula of leaving business to the businessmen, while the public administrations carry out the promotion.

The panorama is even more baffling at the international level. While all main agents of tourism – airlines, hotel chains, tour operators, tourism administrations, etc. – discuss and/or defend globalization, little is done to analyse its contents and implications. In this regard, much less has been done to propose frames and instruments of action for international tourism policy and, beyond that, global institutions committed to the development and coordination of world tourism. In some cases, it seems that the concept of globalization is introduced to justify non-action or, more specifically, the dismantling of national and international *public* tourism policy. Somehow, the argument goes, the market forces will find a way, and it will be to the benefit of consumers.

In spite of that business philosophy, this chapter deals with public sector intervention, through a specific sectoral policy, tourism policy, which responds to at least two crucial issues:

- the significant contribution of tourism activity to the general aims of economic policy – development, stability, efficiency, etc.; and
- the large component of public goods in tourism activity.

An overview of explicit tourism policy implemented in the last decade shows an evolution in its content, from interventionist attitudes and a focus on promotional mechanisms, to the creation of frameworks that foster the competitiveness of tourism clusters and the use of broad scope instruments, quite similar to those present in industrial policy.

Globalization – or rather, the growing trend towards globalization – increases the need for national and international tourism policies. But the question of who are to be the decision-makers of such policies, and what will be their substance, needs to be examined. Multinational/transnational enterprises will obviously play a role, and it will be naturally concerned with the efficiency and profitability of international tourism. Non-governmental organizations will also want to intervene to ensure that tourism is compatible with their own agendas, with specific cultural, environmental or social objectives in mind.

Thus, it is vital to analyse what is the role of governments and inter-governmental organizations in global tourism and what may be the adequate substance of international global tourism policy, regarding both the development of tourism and its contribution to global society.

Globalization and tourism: the new age

In recent years, reference to globalization in the academic and professional world is constant (e.g. Waters 1996; Scott 1997; Burda and Dluhosch 1998; Gummet 1999; Cordella and Grilo 1998; Spybey 1996; Oxley and Yeung 1998; Richardson 1997). The notable interdependence between economies (e.g. OECD 1996) and the trend towards greater similarity of lifestyles (e.g. Yearley 1996; and Swarbroke and Horner 1998) are two conventional points of reference in the globalization concept.

Nevertheless, globalization, under other names, is not a new concept but rather an acceleration of trends that have been active for decades and even centuries (Foreman-Peck 1998). In fact, in the twentieth century, apart from technological advances and the political and social transformations of the past few decades, there had not been a great advance of this trend. It can even be said that financial and economic institutions in the second half of the nineteenth century were more internationalized than at the beginning of the twentieth century.

What is often understood as globalization comprises diverse economic, social and political phenomena. The intensification of commercial exchanges, marked by the progressive dismantling of protectionist barriers, the growing integration of financial markets, the presence of new industrialized countries and technological developments, especially in the area of know-how and information, are affecting both the national economies and the lifestyles of societies (e.g. Amin 1997).

All of this is creating the basis for a global system or organization, distinctly characterized by a high level of economic, socio-cultural and environmental interdependence. In the latter area, some authors have pointed to the problems caused by the global warming, pollution and the danger of nuclear war as factors which accelerate globalization (e.g. Kilminster 1997). The development of these and other phenomena on a global scale clashes with the entrenched policies and institutions designed for national frameworks, as public policies, debated and occasionally agreed upon in international fora as they may be, still lack the global dimension.

In this context, it is not remarkable that tourism activity is both the cause and effect of accelerated globalization. It is useful to point out three essential elements of contemporary tourism (e.g. Fayos-Solà 1993 and 1996; Gee and Fayos-Solà 1997; Go and Pine 1995; Go 1998; Poon 1993; Vanhove 1998; and WTO 1998, 1998a and 1999):

1   The extension of tourism demand throughout the world: the increase in intra- and inter-regional travel – although many strata of the population are still travelling only locally or are strangers to tourism.
2   Similarity of tourism demand: convergence of consumer preferences, tastes and lifestyles – although the type of travel is segmented.
3   Concentration and similarity of tourism supply: expansion of distribution systems, business mergers, etc. – although new specialists agents are appearing on the scene.

To all of this it must added without doubt the impact of new technologies on tourism, which is even more significant than that of changes in consumer taste or in institutional structures. The traditional tourism resources, the comparative advantages (climate, landscape, culture, etc.), are becoming less and less important compared with other factors in tourism competitiveness. Information (or rather, the strategic management of information), intelligence (innovative capacity of teams within an organization) and knowledge (know-how, or a combination of technological skills, technology and organizational culture – *humanology*) now constitute new tourism resources and key factors in the competitiveness of tourism organizations (enterprises, destinations and institutions). The major (most-visited) tourism destinations of the world are no longer the famous beaches or traditional cultural capitals, but rather man-made products, such as Orlando or Las Vegas. In fact, the greatest foreseeable competition in the medium term for the present tourism activity is not the appearance of new exotic resorts, but instead the massive use of the increasingly accessible and efficient information and communication technologies for new leisure products: virtual travel and experiences.

Tourism thus finds itself in a situation, which Kuhn (1962) would clearly define as a paradigm shift, and which is not casually related to or far from the globalization process of economy and society in general. The concept of a business paradigm, understood as a set of theories, values, attitudes, methods and instruments, rules and practices, is useful when analysing business strategy in given framework conditions.

Thus, it is increasingly seen that, in recent decades, mass tourism business strategies (the Fordian Era of Tourism) – and especially profit-making through economies of scale and the consequent standardization of rigid tourism packages – are giving way to a new paradigm shaped by the segmentation of the new consumer demands, new technologies, new forms of business production and management and new framework conditions (Poon 1993).

This new post-Fordian business paradigm in tourism, which Fayos-Solà (1994) called the *New Age of Tourism*, has repercussions on business strategy, and also very profound ones on the policy and even the organization of tourism administrations.

The main objective of tourism policy is to improve the conditions under which tourism activity is carried out. In the Fordian era, increase in tourism activity required a quantitative type of action – which goal was maximizing the number of visitors. In this paradigm, the emphasis on attracting demand corresponded well to the then traditional type of economic policy, based on Keynesian mainstream thinking. Tourism receipts enhanced foreign currency earnings and the creation of employment, so that the success of this sectoral tourism policy allowed other economic policy objectives to be reached, especially those related to economic growth and full employment.

The Keynesian-style policies are based on handling the components of aggregate demand: consumption, investment, public expenditure and net exports (exports minus imports). In this intervention framework, tourism expenditure is considered as an item within exports. The income multiplier – i.e. the mechanism which explains how an increase of the variables of aggregated demand (investment, public expenditure and exports) creates an increase in income which exceeds the initial effort – has constituted the central explanatory and justifying mechanism of demand policies.

The transmissions and leakages in the multiplier chain depend on the marginal propensity to consume (the part of each increment in income destined to consumption), on the marginal propensity to import, and the average tax rate. These last two elements constitute leakages in the Keynesian model: the greater these two elements are, the less the multiplier effect in the domestic economy is. Interior tourism expenditure (domestic plus incoming – traditionally considered under exports) favours the initial effort in the chain. Outgoing tourism expenditure (national less domestic – usually considered under imports) diminish, cancel or make this initial effort negative. The capacity of the national economy to supply the needs of tourism activity affects the marginal propensity to import, increasing or diminishing the final effect of income creation.

Within this theoretical framework we should emphasize how Keynesian-style economic policy makes full sense within a framework of nation-states, where the effects of these policies are highly predictable and controllable. However, this model lost validity after the financial and economic crises at the end of the 1970s, its theoretical base weakened by the difficulty of applying Keynesian formulas when facing wage-price spiralling inflation – stagflation – and a growing internationalization of the economy.

The dismantling of international trade restrictions (tariffs, import duties, etc.) stimulated by the GATT strongly modified the preconceptions on marginal propensity to import. Imports have become less expensive and the previously described chain of induced consumption was easily channelled towards 'foreign' products. 'National' economic areas are now more receptive to foreign products. The leakages of the model described have increased considerably.

In addition, in international tourism markets the emergence of new destinations and products with competitive prices has been constant since these years, and is a strong threat to traditional destinations. With growing competition, it has been necessary to undertake the restructuring of traditional offers.

Within this setting, it is logical that tourism administrations since the 1980s have switched their emphasis to supply policies. The main aim of aggregated supply policies is to increase and improve the productive capacity of a country. Without abandoning supply policies, it is necessary to point out the change in perspective caused by this shift, since it was no longer only a question of creating internal or external demand, of improving demand conditions, of fostering their increase or moderation in accordance with the current economic cycle; or of tourism administrations concentrating on promotion. From this time, the need to improve tourism production to meet with an ever growing competition was felt. This implied the dismantling of sub-sectors, enterprises and unproductive products, a greater research and development (R&D) effort in education and training, in the business quality clusters, in the infrastructure, public services and goods for the sector, etc.

This change, from a tourism policy based on demand to supply models, takes into consideration that the basic problem is not tourism demand, which will continue to grow according to all forecasts (WTO 1998a). Globalization and the increased competition in tourism markets after the 1980s has required a consistent improvement in the price/product-characteristics ratio, that is to say, a continuous striving towards quality and efficiency. The Spanish case provides a good example of the new approach to tourism policies (DGT.Spain 1992), which prioritizes action on the quantity and, especially, quality of tourism supply, making available to businessmen the necessary mechanisms to increase competitiveness:

- improving know-how by fostering R&D, tourism education and training, and information management;
- diversifying the supply, with new products and destinations;
- physical modernization of installations and infrastructure;
- improving business clusters, encouraging action by ancillary businesses, associations and the *coopetition* (cooperation-competition) between private agents and the public sector;
- improving promotion, with greater quality (responding to promotion needs of the actual, existing supply) and efficiency;
- conservation and regeneration of tourism areas; and
- improving horizontal (interdepartmental) and vertical (local-regional-state-intergovernmental) coordination of public administrations for tourism policy.

Globalization, governance and the nation-state: implications for national tourism policies

As several authors have pointed out (Hirst and Thompson 1996; Amin 1997; Scott 1997), the discussion on globalization often covers up highly ideological visions of the future; there is no true evidence that globalization has gone beyond the acceleration of political, economic and cultural internationalization processes which began in many cases centuries ago, and in any case, it does not appear to be totally just to use this concept to defend a radically anti-political vision of the world in the twenty-first century.

It is obvious that the debate on globalization has rekindled extreme right and left ideological points of view. For the former, globalization will offer new hope (after the failure of the monetary experiences of the 1970s and 1980s) for a world where free trade, world capital markets and transnational organizations can fully use productive resources without the clumsy interference of governments. For the radical left (also affected by the fall of state socialism and the anti-imperialist movements of past decades), globalization of capitalism indicates the uselessness of social-democratic style 'welfare' initiatives carried out at a national level.

If we examine the general role of the state at present before defining its tourism policy functions, it is quite obvious that its capacities have been redefined. The *sovereignty*, exclusive control (excluding other authorities) of a territory no longer exists. The capacity to 'defend' its citizens from macro-conflicts has been questionable since the era of nuclear arms. The claim to standardize and control culture within borders is also no longer justifiable; citizens of the world establish cultural affinities and links through means of communications that escape the control of nation-states and almost any censoring attempt. Finally, and this has been remarked previously, economic internationalization makes it practically impossible to carry out autonomous economic policies.

But all this does not signify either the disappearance of the political role of nation-states, or a great change in international relations. Perhaps it should be recalled that in the first place the very sovereignty of states, as defined in the seventeenth and eighteenth centuries, has always depended on their international recognition. The guarantee of non-intervention by other states allowed the consolidation of sovereignty in the state itself. Also, although nation-states have seen their capacity for exclusive control of a *territory* enormously diminished, what is true is that they still have a central role in the control of the *population* of this territory, taking into account that it is much less mobile than either information or economic flows.

For this reason, it seems that the issue under debate is not the avowed disappearance or reduction to a minimum of the role of the nation-state in a 'globalized' world, but rather the question of governance in a more integrated society at world level and the role of governments in this society. There is no doubt that the nation-states have a vital role in this process: they still possess a great deal of the power and, in any case, maintain the legitimacy of representing the populations that live within their frontiers; and beyond having an unquestionable role as 'local' suppliers of certain public goods in the world context, they are the natural interlocutors in intergovernmental organizations that can possibly make advances in the task of designing, proposing and maintaining standards (voluntary, agreed or legislated) for the functioning of international and/or global systems. Although there may be other protagonists in the creation of framework conditions – multinational companies for example – it does not seem that they can claim a greater representation of the world's citizens, and the authentic multinational nature of many companies – which in reality are strongly established in one of the developed regions of the world and also have operations and subsidiaries in other countries and regions – is questionable. In any case, it seems evident that the *second generation* tasks, to which the introduction of this chapter refers – the creation and strengthening of responsible international and global institutions in charge of these framework conditions – will in great part depend on the collaboration between nation-states and other major protagonists on a world level.

Within the area of tourism activity, and, as has been previously indicated, the transition from national policies almost exclusively quantitative in dimension (maximizing the number of tourists through promotion), to others stimulating competitiveness (quality and efficiency) in an international context, entails a change of traditional functions in the tourism public sector. This change can be summarized as follows:

•   First, the transition from a situation where the public sector owns and operates all types of tourism facilities and intervenes customarily in the direct provision of goods and services, to a role of coordinator of private and public actions in tourism.

- Second, the opening up of the goals and means of tourism policy, from an almost exclusive promotional content (generic advertising, trade missions and exhibits, publications, etc.), to a broad range of instruments to foster and facilitate the activities of tourism decision-makers.
- Finally, the evolution, from a philosophy of rigid regimentation of entrepreneurial activities, to deregulation and privatization of tourism.

This ties in well with the new role of the nation-state: international representation of populations (and enterprises) located within its frontiers, concordance of interests which are not always in agreement – through the stimulation of associative and cooperative activity – and the improvement of the quality of life within its territorial limits.

The implications for the formulation of national tourism policies are clear:

1   The objectives of these policies must refer to the creation of competitive frameworks on a local–regional–national scale which, by improving the conditions of the economic, social and environmental framework, achieve contributions of the tourism sector to the well-being of the citizens.
2   Although the use of promotional instruments by tourism administrations (communication, publicity, etc.) continues to be requested by decision-makers in tourism, its importance is decreasing. On the other hand, the need to coordinate promotion with a wider range of instruments in tourism policy has become evident.
3   The new public instruments for tourism development and management are not fundamentally different from those used in sectoral industrial policy. Essentially, they foster the competitiveness of existing tourism clusters and the adoption of strategies for success in international markets of emerging destinations or those which are in the process of restructuring.

Although a study of the budgets of national tourism administrations (WTO 1996) indicates that public expenditure in tourism is still greatly concentrated in promotional instruments, in the area of competitiveness and strategy there is a growing dedication to Porterian-type instruments (Porter 1990; Ritchie and Crouch 1993; DGT.Spain 1992, 1996 and 1999; CDT.Australia, 1992; DEAT.South Africa 1996). Specifically, the new tourism policies of countries such as France, Spain, Italy, Germany, Canada, Australia and South Africa concur in the use of the following instruments:

Strengthening supply conditions

- Human resource development in tourism. Education and training which is more in line with the short, medium and long-term needs of the tourism employers. Awareness of the need to anticipate the upsurge of new tourism professions and the continuous training of professionals in the sector.

- Fostering innovation and development (R&D) specifically for use by tourism enterprises. Awareness of the need to give priority to the R&D of tourism processes over the R&D directly applicable to products and services.
- Modernization of the productive plant, installations and infrastructure in tourism. Awareness of the need to make compatible the modernization of the supply of public and private goods and services, and to create lasting mechanisms to permanently (and not on a one-time basis) carry out this task.
- Impetus of diversification and specialization in tourism destinations, products and services, although taking into account that the competitive advantages are more easily achieved in the realm of tourism processes.
- Stimulus to the conservation of natural, cultural, urban or rural areas in which tourism is carried out, and of heritage sites sensitive to tourism use. Awareness that these areas form part of the consumer *tourism experience* and that there is no tourism product and destination quality without eco-tourism quality.
- Fostering a more even geographical distribution of tourism supply and demand, in support of other more generic objectives of public policy (e.g. an incomes policy for farming in disadvantaged areas). This has to be tightly coordinated with the responsible regional and local governments.

Strengthening the business fabric

- Fostering of associative and cooperative initiatives between tourism enterprises and destinations. Awareness of the need to cooperate and not only compete at the intra-cluster and inter-cluster level in the context of a national tourism policy.
- Stimulus to give tourism activities an adequate dimensional scope. Awareness that this dimensional scope depends on the nature of niche-markets and the (changing) state of available technology. New information and communication technologies allow for new solutions in this respect.
- Public and cooperative contributions to sectoral and sub-sectoral information management, useful for decision-making at the enterprise, sectoral association, *cluster*, tourism destination level or even at a macro-national level.
- Adaptation of the judicial and institutional framework to give confidence and greater efficiency to business decision-making. Awareness of the need to make this framework flexible so that it continuously adapts to rapidly changing circumstances.

- Contribution of the tourism administration and sectoral cooperation institutions to *strategic* decision-making, beyond considerations of short- or medium-term quality and efficiency and the search for excellence in established market niches.

Strengthening demand conditions

- In the context of the Porterian paradox that a better informed, exigent and sophisticated demand favours competitiveness and strategic positioning of enterprises (Porter 1990 and 1998).
- Obtaining and disseminating market information on consumer groups typology and communication channels for this purpose.
- Improving the promotion policy, from an instrument based on passive information on the positive characteristics of products, destinations and cultural and environmental amenities, to a means of forming and modulating the expectations and even the perceptions of the clients.
- Support to marketing efforts of tourism enterprises and destinations. Awareness of the role of new technologies and stimulus to innovation in tourism marketing processes.
- Improving the tourism information milieu in which consumers, workers, enterprises and tourism administrations move. Awareness that the cost of obtaining this information for the individual decision-maker may be high and that it is therefore preferable to approach this matter as the provision of a public good.
- To strengthen tourism training and qualifications, not only within the business context, but also in that of the consumer and host societies.
- Protection of the consumer-tourist, improving the applicable standards and the inter-administrative coordination. Awareness that today's tourist demands a high degree of confidence in the quality of the product and in his personal safety as an essential condition to increasing his loyalty to tourism products, services and destinations.
- Integral management of tourism quality to increase the level of consumer satisfaction and the well-being of receiving societies. Awareness that it is essential to have existing client loyalty to be competitive in tourism destinations, and that this cannot be achieved only through aggressive commercial promotion aimed at new clients.

Strengthening of linked industries and services

- Stimulus to the creation and adequate functioning of related industries and services in tourism clusters and destinations. Awareness of their relevance, both in the horizontal (complementary) and vertical (suppliers, sub-contractors and client companies) sense to achieve competitiveness.

- Coordination of public administrations concerned with tourism, both in the horizontal (departments within an administration) and vertical (local, regional and national administrations) sense. Awareness that the public administration in general (and not only the *tourism* departments of the same) constitutes part of this institutional and business milieu essential for competitiveness in tourism.

- Stimulus to the re-engineering of the macro-processes within the tourism clusters and destinations. Awareness that it is possible to achieve the necessary quality objectives by improving the efficiency of the optional useable processes. Re-engineering of public administrative processes and their coordination with the private sector, improving their quality and dedication to service, usually is an important part of this instrument.

The role of intergovernmental organizations in tourism

In spite of abundant references in professional and academic literature (Frangialli 1999; Gee and Fayos-Solà 1997; Go and Pine 1995, Hanlon 1996; Harris *et al.* 1998; Jones and Pizam 1993; Laws *et al.* 1998; Oppermann and Chon 1997; Pearce *et al.* 1998; Pizam and Mansfield 1995; Swarbroke and Horner 1998; Teare *et al.* 1997; Theobald 1998; and Vellas and Becherel 1995) to globalization in tourism, the fact is that the tourism business fabric is mainly made of sub-sectors with a large number of medium, small and micro industries, often of a marked local character.

The most notable exceptions are the air transport sub-sector in itself and the existence of some large enterprises in the hospitality (hotel chains), travel agency (certain tour operators) and entertainment (macro theme parks) sub-sectors. In fact, though, one of the best-known business lobbies in the sector, the World Tourism and Travel Council (WTTC), which defines itself as comprising the chairpersons and highest executives (CEOs) of the largest companies in the world, has only seventy-five members. In addition, even when considering these large tourism enterprises, there is reason to question their status as *global* enterprises. Most frequently, they are strongly identified – by origin, business culture, major operations and decision-making strategies – with one of the countries of the G3 triad (North America, Europe and Japan), with their presence in other countries being as subsidiaries, franchises, etc.

For these reasons, it is difficult to agree with the statement that 'tourism is one of the most globalized industries'. The fact that many of the industry's clients have to cross borders to travel and that there are suppliers with products in several countries can grant it, at the most, a partially *international* character. According to WTO estimates, only one in ten tourist movements is international, while the rest are domestic. Thus, although international tourism demand is already more than 650 million trips annually and is growing at an accumulative yearly rate of 4.3 per cent, most travel takes place within world-regions (Europe, North America-Caribbean, East Asia-Pacific) and within national borders. Additionally, tourism, unlike financial transactions or information flows, requires the physical transport of people, a characteristic which makes it highly controllable by the sovereignty, albeit residual, of nation-states.

Thus, it seems reasonable to defend the premise that the tourism industry is still in a phase of international activity, although it is also true that tourism is, on the other hand, contributing to the worldwide dissemination of cultural and social habits, and is therefore in this regard, a *factor* in the globalization process.

However, the importance and growing expansion of international tourism, its contribution to the development of regions and countries, to income and employment creation, its status, which has already been mentioned, as a transmitter of cultural identity images, all justify the attention it has merited and still merits from institutions on a worldwide scale as well as the existence of an intergovernmental organization (WTO) specifically dedicated to international tourism policy.

Neither an analysis, nor an exhaustive inventory of the international institutions, which have had or do have influence, either direct or indirect on tourism flows, is attempted here; the list is too long and the analysis complex. Many international organizations and agencies, from the OECD, with a tourism committee whose existence is now at risk, to the World Bank, have given attention to some of the most relevant functions of tourism activity. The European Union, in 1989 for the first time, granted responsibility for tourism policy to a specific department – its General Directorate XXIII – although the budgetary provisions given to the tourism unit were always minimal and the main part of the European budget had a much stronger impact on tourism activity through the structural funds, programmes for innovations and training or, in the case of third world countries, through development aid programmes.

Perhaps the most relevant issue, in the context of this chapter, is the substance of supranational intergovernmental action in tourism, i.e. *international tourism policy*, and the viability and pertinence of hypothetical global tourism policy.

Justification of an international tourism policy, apart from the arguments already indicated, is also based on the growing importance of the knowledge factor in the production of tourism services and *experiences*. Perhaps, somewhat paradoxically, the tourism industry, the origins of which tied it geographically to nature resources and/or historical or cultural heritage, has freed itself from these conditioning factors, to the extent where the most sought after tourism destinations at present are often totally artificial (man-made). The use of communication in tourism, often tied to the leisure-entertainment industry, has shaped new consumer needs and created a demand for new tourism destinations in a process where communication-entertainment (the film industry, computerized games, the internet, television, publishing, etc.) has created expectations in potential tourists which are later satisfied (theme parks, theme hotels, dramatized tourism experiences, etc.) with major contributions from information-communication technology and the entertainment industry,

It is in this context, where one can see a rapid tendency towards globalization in the tourism industry, standardizing supply and demand and freeing them from the confines of stationary cultural or natural realities, and making it almost indistinguishable from the leisure industry, advancing towards a future of *virtual experiences* that could easily escape the control of the sovereignty of states.

What should the substance of a contemporary international tourism policy be, and in what direction should this policy move, taking account of the previously mentioned tendencies towards future globalization?

The first issue should, without doubt, be to identify the players in tourism policy. If the globalization of tourism is not considered to have already happened, there is still time and the opportunity to identify those players who are more desirable and those who are less so. There is also time to favour the most sensible future scenarios, seen from the perspective of the contribution of tourism to the well-being of citizens, and their participation in the decisions as to what type of well-being they truly desire.

Far from accepting extreme positions on globalization – i.e. that it is already determined, that the decision-making power of multinationals/transnationals is above that of the traditional sources of governance – it is possible to determine explicitly the current players, and possibly future ones, of governance in general and tourism policy in particular:

1 **Regional and local administrations.** Although their area of competence falls within the framework of higher level administrations, they have the effective advantage of being close to the citizen and the entrepreneurial units. They constitute the ideal public players to implement sectoral policies, which can be decided on occasion within the local and regional context or coordinated with the administrations having a wider scope of action. In the democratic context they are validated by the vote of their citizens.

2 **National administrations.** They still have a wide magnitude of sovereignty. On occasion, they have devolved part of this to regional and local administrations and/or relinquished part to institutions or administrations with an international mandate. In the democratic context, they are endorsed by the vote of their citizens and frequently discharge this representation within international institutions. The consequent limitations (free circulation of capital, elimination of tariffs, etc.) are accepted in view of the benefits expected from a better distribution of resources on an international/world scale, but other

objectives of national economic policies can be in contradiction with this self-limitation.

3   **Supranational administrations.** Their historical origin lies in commercial agreements. The European Union is the most significant experience in this sense. The size of this type of administration enables economic, social, environmental and sectoral policy objectives to be set, which are out of the reach of national administrations, and they can more successfully confront the undesirable aspects of the globalization process.

4   **Agreements between countries or even between blocks of countries (G3 or G7 type).** These are established to confront specific problems (financial speculation, international crime, etc.) and sometimes lead to the creation of international legislation.

5   **International agencies and organizations created by a group of states to permanently handle specific issues arising from economic, social or environmental activities.** In the field of tourism, the paradigmatic player in this category is the World Tourism Organization (WTO).

This final type of player is the one whose decisions on tourism policy are considered here, although the substance of the tourism policy which can evolve should be analysed within the context of other players in international governance. Non-governmental organizations (NGOs) and private sector businesses and institutions are excluded here as principal players in governance, since they lack democratic representation, although their important role as partners or associates in governance by the previously mentioned players is obvious.

The instruments which are useful to the international organizations usually belong to one of the following categories:

1   **Legislation.** Agreements with a judicial scope to standardize the laws of member states and even of other states which may join the initiative. These agreements are directed at remedying the non-extraterritoriality of national laws and/or the lack of international legislation and/or the lack of enforcing bodies.

2   **Agreements without a judicial scope.** Aimed at eliminating or lessening the repercussions of frontier restrictions by creating a framework for greater security in international transactions. These agreements have a technological, economic, social and/or ethical content. They are generally enforced through specific mechanisms to penalize infractions.

3   **Voluntary quality standards.** These are proposed without the need of previous consensus, at the initiative of the organization in question or by a group of member states. They propose a model for conduct with regard to technological, economic, sociological and/or

ethical matters. They do not usually have authority to penalize, but are intended rather to oversee or coordinate. They are accepted voluntarily due to the added value they give in terms of promotional image, facilitation in communication with other players in the market, interspatial and inter-temporal measurements, etc.

Given the demonstrated difficulty in establishing and developing the tourism policy instruments indicated in the first two categories, it can be said that international organizations specialized in tourism, and concretely WTO, are showing a growing tendency to use voluntary quality standard instruments. This signifies, without doubt, an advance over the previous situation, where the insistence to establish legal agreements, or even simply enforceable agreements, led to a general impasse given the inability to achieve consensus or wide majorities because of:

- the diverse economic, social, cultural and political situations of the member states;
- the frequently heterogeneous nature of the member state's representation in the organization: Departments of Foreign Affairs, of Commerce and Tourism, of Culture and Tourism, of the Economy (Tourism Department), of Industry and Tourism, of Tourism, etc.;
- the variable importance of departments of tourism and of tourism affairs within governments of member states, where on occasion they play a minor role;
- with regard to legislative instruments, the difficulty that departments with competence in tourism have to influence sufficiently the deliberations and decisions of the national legislative powers.

The instruments in the voluntary quality standards category can point to models, of varying types, for flexible and rapid action, which can be gradually adopted by member states and even as a global voluntary standard. These models can successfully bring added value to international markets, which are potentially global, and can be adapted to different national situations. Although the explicit adoption of a voluntary standard by a sovereign state facilitates its global establishment, this can be expedited, in the case of delay or a lack of will, if the standard is *de facto* adopted by the industry and/or citizens of the country in question. Furthermore, the standards proposed can also fail when their format is rejected or ignored by potential users.

Thus, in this context, it should be noted that the role of international organizations in tourism, and specifically that of WTO, is rapidly evolving, from the traditional one of a forum where countries meet, to that of serving as an information broker between the countries and of being responsible for carrying out economic development projects and giving specific assistance to countries, up to the present role of also serving as an institution where voluntary quality standards are created and implemented in such key areas of tourism as (a) the development of human resources for tourism – education, training, strategy, management and labour conditions; (b) statistical information; (c) market intelligence; (d) know-how in products, services and processes; (e) infrastructure, collective services and urban environments; (f) cultural and environmental aspects of tourism; (g) economic and social effects of tourism; (h) facilitation of international movements of travellers and tourists; (i) financing of tourism; (j) quality of products, services and tourism environments; (k) communication in tourism; (l) ethical aspects of tourism activity; (m) legislative processes and contents in tourism; and (n) coordination of administrations with competence in tourism – intra-administration, inter-administrations, and with the private sector.

The content of the tourism policy instruments being used in contemporary action is justified by the two major reasons for the existence of international organizations specialized in tourism (and that of the WTO itself):

1   International and global public goods, externalities, market imperfections and merit and demerit goods.

To analyse the contemporary scope of this justification it must be realized that the foundations – i.e. the so-called resources – of tourism activity are rapidly changing. In principle, as has already been indicated, cultural and natural resources were those backing tourism development. The addition of financial capital and work efforts to these resources created tourism products. The comparative advantages of tourism destinations were based on the abundance and correct combination of these elements.

At present, the relevance of natural and cultural resources has diminished – except for world-class resources and in specialized niche markets – while the importance of financial capital and above all that of information, intelligence-creativity and know-how – used by human teams in business and organizational cultures prepared for competiveness and strategic success – has increased.

Given these circumstances, the role of international organizations specializing in tourism is clear: (a) The provision of public goods which previously were the domain of the states, such as quality education and training, strategic information and basic know-how; (b) The internalization of externalities in the planetary context, such as the costs of pollution or possible climate changes; (c) The correction of imperfections in international markets such as the costs of information or the appearance of highly monopolized tourism operations; and (d) the introduction of ethical criteria in carrying out tourism activity on an international scale (working conditions in tourism, sexual exploitation in tourism and the like).

This first justification of the activity of WTO or other international organizations clearly shows the differences between intergovernmental agencies or organizations – created by and responsible to a (large) group of states, and occupied with international governance – and other international organizations (such as NGOs, motivated by more specific aims), international business lobbies or large enterprises, whose objectives differ from, and on occasion are in conflict with, those mentioned above.

2 The benefit to member states of exercising their sovereignty in optimal conditions and of supplementing it when it is questioned or proves inefficient in the globalization process.

This is where the traditional role of international organizations is evolving towards greater technical contents, which give depth and relevance to the member states' fora of discussion. Without assuming a conceptual breach with this traditional role, it is obvious that the use of a voluntary standard type of instruments, already mentioned above, gives greater flexibility and scope to the resolutions discussed and adopted in these fora, which are later subjected to a validation process – through an appraisal of their value in the market and society and their acceptance or not by businesses, institutions and the citizen.

The acceptance and establishment of these standards, when it takes place, represents a real step towards international and global governance for: (a) it makes it possible to have truly global rules (standards), backed by states, which are more representative than businesses or other types of organizations; (b) this acceptance happens only when the standards create added value for a large enough number of social and economic players; (c) it reinforces the role of the organizations creating and overseeing these standards as well as the capacity of such organizations to adapt them to changing circumstances with greater democratic legitimacy.

Conclusions

Although the concept of globalization is broadly used in academic and professional literature, its exact definition, its measurement and its effects are far from being clear. The mere reference to phenomena affecting the world is still conceptually weak and cannot be used as evidence of globalization without being qualified. Economic, technological or cultural internationalization processes are not new, and their present acceleration does not imply that globalization is inexorable. Historically these processes have stopped and reversed several times.

It is also not evident that these processes and their future culmination in globalization imply the disappearance or impotence of the sovereignty of nation-states. This assumption on occasion responds to a highly ideological view of world society. States still have mechanisms to control large enterprises and to create new instruments for governance, guaranteeing democratic control of future scenarios compatible with the well-being of a majority of the citizens. Multinational/transnational enterprises are not necessarily global, since their cultural and strategic bases and the greater part of their business volume are generally found in only a few countries, normally located in the area of the G3 (North America, Europe or Japan).

The phenomena related to globalization affect tourism differently, depending on whether demand or supply are being considered. Demand shows clearer globalization tendencies as consumer preferences and expectatives converge, even though the *type* of holidays sought is becoming more diverse. On the other hand, tourism supply is still far from being global; thus, multinationals in tourism have not permeated the markets, with a few exceptions such as airlines, and hotel chains. The business fabric in almost all tourism sub-sectors is formed by hundreds of thousands of small and micro enterprises. In addition, international tourism, although significant and rapidly growing, only represents a minor part of the total volume of the tourism business, in the most part domestic. As tourism implies the physical movement of people, the capacity of the states to exercise their sovereignty in this activity is obvious.

National tourism policy will remain a key factor in the development of tourism in a majority of countries for at least the next decade, although devolution to regional and local governments may change its role in some areas. The importance of tourism and its economic, social and environmental implications, which affect governance and broader scope economic policy, speak in favour of establishing explicit national tourism policy frameworks. This sectoral policy may then be implemented by regional and local administration, which is closer to concrete tourism destinations and business clusters. The question is therefore one of reassigning tasks and it does not imply the automatic weakening of national tourism administrations.

The substance of tourism policy in key countries has been broadly in line with other economic sectoral interventions, particularly industrial policy. Emphasis has shifted, from almost exclusive concern with promotion, to a wider range of instruments acting on productive conditions as well. However, the specific characteristics of tourism supply ask for special attention being put in certain elements of competitiveness. The comparative advantages (natural and cultural resources) which used to be the base for the success of tourism destinations are giving way to *competitive* advantages in a new business paradigm (the New Age of Tourism) where information, inteligence and know-how play a vital role.

These national tourism policies increasingly have a central theme: the use of Porterian style instruments to foster the competitiveness of tourism clusters (destinations, sub-sectors and/or groups of enterprises). These instruments belong to one or several of the following types: (a) Strengthening the supply conditions; (b) Strengthening the business fabric; (c) Strengthening the demand conditions; (d) Strengthening of linked industries and services.

Even though it may be premature thinking of tourism as an already globalized activity, the importance and expansion of international tourism does justify the treatment of tourism matters in international-scope organizations and the existence of an intergovernmental institution (the WTO) specifically dedicated to international tourism policy.

The work programme of any intergovernmental institution committed to tourism, and, in particular, that of WTO, must respond to two types of rationale: (a) the importance in tourism of international and global public goods, externalities, market imperfections and merit and demerit goods; and (b) the benefit to member states of exercising their sovereignty in optimal conditions and of supplementing it when it is questioned or proves inefficient in the globalization process. When in this context, tourism policy implemented by an intergovernmental organization represents a real step towards international and global governance.

International legislation or enforceable agreements are rather rigid instruments for international tourism policy. The evolution of national tourism policies – towards deregulation, privatization and a role coordinating public–private partnerships – leads the way to a more participative and less coercive kind of tourism policy. The preferred type of instruments of such a policy is found in the realm of voluntary standards of quality; they can be very flexibly adopted by countries, destinations or the industry. These types of instruments adapt best to the difficulties found in developing international tourism policy given: (a) the heterogeneous nature of government departments competent in tourism; (b) the diverse economic, social, cultural and political conditions in nation-states; (c) the variable importance of departments of tourism within governments of nation-states, where on occasion they play a minor role; and (d) the difficulty that departments with competence in tourism have to sufficiently influence the deliberations and decisions of the national legislative powers.

The threats and opportunities characterizing the internationalization and globalization processes in contemporary society require, in tourism as well, responses beyond *ad hoc* legislation, treaties or agreements. International and global matters need international and global *institutions*. The liberalization of trade and tourism, the removal of obstacles and the consequent improvement in the allocation of resources make for big improvements in the well-being of the peoples of the world. However, it is important to pay attention to the *actors* of the globalization processes. Multinational/transnational companies and non-governmental organizations are without doubt very relevant decision-makers in the new realities – but they cannot play the leading role in representative governance, which is a question of increasing concern at world level. Tourism, because of its importance in the development of regions and countries and its capacity to convey images of cultural identity – so deeply needed in the configuration of global society – requires international and global *representative* organizations, to play a key role in world governance.

References

Amin, S. (1997) *Capitalism in the Age of Globalization: The Management of Contemporary Society*, London: Zed Books.

Burda, M. and Dluhosch, B. (1998) *Globalization and European Labour Markets*, London: Centre for Economic Policy Research.

CDT.Australia (1992) *Tourism. Australia's Passport to Growth: A National Tourism Strategy*, Canberra: Commonwealth Department of Tourism.

Cordella, T. and Grilo, I. (1998) *Globalization and Relocation in a Vertically Differentiated Industry*, London: Centre for Economic Policy Research.

DEAT.South Africa (1996) *The Development and Promotion of Tourism in South Africa: White Paper*, Government of South Africa, Pretoria: Department for Environmental Affairs and Tourism.

DGT.Spain (1992) *FUTURES, Plan Marco de Competitividad del Turismo Español* (1992–5), Madrid: Dirección General de Turismo, Government of Spain.

DGT.Spain (1996) *FUTURES, Plan Marco de Competitividad del Turismo Español* (1996–9), Madrid: Dirección General de Turismo, Government of Spain.

DGT.Spain (1999) *PICTE 2000: Plan Integral de Calidad del Turismo Español (2000–06)*, Madrid: Dirección General de Turismo, Government of Spain.

Fayos-Solà, E. (1993) 'El Sector Turístico como Sector Industrial', *Revista de Economía Industrial*, July.

Fayos-Solà, E. (1994), 'Competitividad y Calidad en La Nueva Era del Turismo', *Revista de Estudios Turísticos*, 123: 5–10

Fayos-Solà, E. (1996) 'Tourism policy: a midsummer night's vision', *Tourism Management*, Sept. 405–12

Foreman-Peck, J. (ed.) (1998) *Historical Foundations of Globalization*, Cheltenham: Edward Elgar.

Frangialli, F. (1999) *Observations on International Tourism*, Madrid: World Tourism Organization.

Gee, C. Y. and Fayos-Solà, E. (eds) (1997) *International Tourism: A Global Perspective*, Madrid: World Tourism Organization.

Go, F. and Pine R. (1995) *Globalization Strategy in the Hotel Industry*, London: Routledge.

Go, F. (1998) 'El Turismo en el Contexto de la Globalización', *Papers de Turisme* 23: 6–47.

Gummet, P. (ed.) (1999) *Globalization and Public Policy*, Cheltenham: Edward Elgar.

Hanlon, P. (1996) *Global Airlines: Competition in a Transnational Industry*, Oxford: Butterworth-Heinemann.

Harris, R., Heath, E., Toepper, L. and Williams, P. (1998) *Sustainable Tourism: A Global Perspective*, Oxford: Butterworth-Heinemann.

Hirst, P. and Thompson, G. (1996) *Globalization in Question*, Cambridge: Polity Press.

Jones, P. and Pizam, A. (eds) (1993) *The International Hospitality Industry: Organisational and Operational Issues*, Harlow: Longman.

Kilminster, R. (1997) 'Globalization as an emergent concept', in A. Scott (ed.) *The Limits of Globalization: Cases and Arguments*, London: Routledge.

Kuhn. T. S. (1962) *The Structure of Scientific Revolutions*, University of Chicago Press.

Laws, E., Moscardo, G. and Faulkner, B. (eds) (1998) *Embracing and Managing Change in Tourism*, London: Routledge.

Oxley, J. and Yeung, B. (eds) (1998) *Structural Change, Industrial Location and Competitiveness*, Cheltenham: Edward Elgar.

OECD (1996) *Globalisation and Linkages to 2020: Challenges and Opportunities for OECD Countries*, Paris: OECD.

Oppermann, M. and Chon, K. S. (1997) *Tourism in Developing Countries*, London: ITBP.

Pearce, P., Morrison, A. and Rutledge, J. (1998) *Tourism: Bridges across Continents*, Australia: McGraw-Hill.

Pizam, A. and Mansfield, M. (eds) (1995) *Tourism, Crime and International Security Issues*, Chichester: Wiley.

Poon, A. (1993) *Tourism Technology and Competitive Strategies*, Wallingford: CAB International.

Porter, M. E. (1990) *The Competitive Advantage of Nations*, London: Macmillan.

Porter, M. E. (1998) 'Clusters and the economics of competition', *Harvard Business Review*, Nov.–Dec.: 77–90

Richardson, P. (1997) *Globalization and Linkages: Macro-Structural Challenges and Opportunities*, Paris: OECD.

Ritchie, J. R. B. and Crouch, G. I. (1993) 'Competitiveness in international tourism: a framework for understanding and analysis', unpublished research paper, World Tourism Education and Research Centre, University of Calgary.

Scott, A. (ed.) (1997) *The Limits of Globalization: Cases and Arguments*, London: Routledge.

Spybey, T. (1996) *Globalization and World Society*, Cambridge: Polity Press.

Swarbroke, J. and Horner, S. (1998) *Consumer Behaviour in Tourism: An International Perspective*, Oxford: Butterworth-Heinemann.

Teare, R., Canziani, B. F. and Brown, G. (eds) (1997) *Global Directions: New Strategies for Hospitality in Tourism*, London: Cassell.

Theobald, W. (ed.) (1998) *Global Tourism: The Next Decade*, Oxford: Butterworth-Heinemann.

Vanhove, N. (1998) 'La Globalización de la Demanda Turística y el Impacto sobre la Estrategia de Mercado', *Papers de Turisme* 23: 49–87

Vellas, F. and Becherel, L. (1995) *International Tourism*, London: Macmillan.

Waters, M. (1996) *Globalization*, London: Routledge.

WTO (1996) *Budgets of National Tourism Administrations*, Madrid: World Tourism Organization.

WTO (1998) *Yearbook of Tourism Statistics 1998*, Madrid: World Tourism Organization.

WTO (1998a), *Tourism: 2020 Vision – Executive Summary*, Madrid: World Tourism Organization.

WTO (1999) *Compendium of Tourism Statistics 1999 Edition*, Madrid: World Tourism Organization.

Yearley. S. (1996) *Sociology, Environmentalism, Globalization: Reinventing the Globe*, London: Sage.

Part II

The globalisation of tourism demand and marketing

4 The tourism phenomenon

The new tourist and consumer

Dimitrios Buhalis

Introduction: tourism demand and competitiveness in the globalisation era

Globalisation is already evident in most aspects of tourism activity. International tourism and hospitality enterprises have taken advantage of numerous factors to expand their operations globally. Globalisation has raised competitive pressures by bringing more entrants into the market and as a result enterprises have to compete within a much more complex environment. Emerging technologies enabled greater homogeneous control and operational systems as well as coordination with head office despite geographical location and distance. Changes in the political and legal environment introduced greater freedom of trade and more specifically in travel deregulation of transportation and more flexible and adaptive international investment and development systems. A wide range of forms and arrangements is followed, from direct ownership, partnerships with local operators and/or governments, to franchising and marketing consortia. Labour mobility also enabled people to travel to different countries to manage properties and systems. Perhaps more importantly the emerging multi-culturalisation of investors' employees generated through education and training, media reports, and extensive travelling experience developed a new breed of global enterprises which offer their products at a standard quality regardless of locality. As a result of the emerging globalisation new tools are required to manage processes, multi-ethnicity and culture and to support employees and enterprises in satisfying all their stakeholders. A whole range of changes in society and the global economy will need to be taken into consideration in planning and managing tourism destinations and enterprises in the era of globalisation (Parker 1998; Go 1996; Go and Pine 1995; Makridakis 1989; Dicken 1992; Smeral 1998; Cooper and Buhalis, 1998).

This chapter concentrates on leisure tourism and identifies the main trends influencing demand. It illustrates that four main factors propel changes in the international tourism demand and explains that globalisation magnifies the scale and scope of the implications emerging. The chapter illustrates that the most critical factors affecting demand are:

1   Proliferation of technology both on transportation and information technology.
2   Ecology and environmental concern.
3   An increase of multicultural societies.
4   A quest for edu- and enter-tainment, where education and entertainment merge to offer personal development opportunities.

As demonstrated in Figure 4.1 the chapter suggests that tourism demand is going through a transformation which can be explained through the change of the 4Ss framework for seaside tourism: Sea-Sun-Sand-Sex; and the 4Ss framework for urban tourism; Sightseeing-Shopping-Shows-Short breaks, to Segmentation-Specialisation-Sophistication-Satisfaction. This framework should facilitate the interpretation of major demand trends in the international tourism arena in order to assist tourism managers to develop suitable solutions, which will delight, rather than just satisfy, all tourism stakeholders.

Tourism demand trends and globalisation

Tourism demand evolved rapidly in the 1990s altering conventional wisdom and changing a whole range of factors influencing tourism planning and management. Attempting to interpret tourism phenomena and forecast the future of international activity is similar to reading the 'crystal ball'. Tourism has grown enormously in the last half century and become the world's largest 'industry'. It has also developed a multidimensional and multidisciplinary character making the analysis of both demand and supply a complex task. The globalisation experienced alters the competitiveness of destination regions and provokes a whole range of new activities and requirements from the demand side. Increasingly people are becoming more aware of their limited time and are looking for both value for time

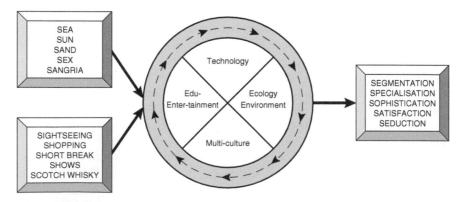

Figure 4.1 Dynamic tourism demand trends.

and value for money. Predicting international demand trends is therefore a very challenging task, as the dynamic nature of these developments clearly demonstrates that the only constant in tourism is continuous change. Nevertheless, successful tourism management and planning will increasingly need to identify the factors changing demand trends. The industry should therefore offer meaningful tourism products and also provide strategic and operational tools, which can delight consumers and enhance the competitiveness of destinations and enterprises within the global market.

Workers have established their right for leisure time, dedicated to their recreation. Paid annual holidays of about four weeks is nowadays a right for most people in Western developed countries, and the rest of the world is gradually heading in this direction. Leisure time is also increasing gradually, as discussions are in progress in the European Union to reduce the working week to 35 hours and to establish a maximum of 48 hours per week. O'Brien (1996) explains that 'The West European leisure travel market is undergoing structural and cultural changes. These changes are critical to the future demand for, and supply of, leisure products both to consumers and to intermediaries who distribute travel products.' The European market has experienced a certain level of maturity as the vast majority of North Europeans take annual holidays abroad. In contrast the majority of South European tourists as well as people in North America have traditionally consumed domestic tourism products for a variety of reasons.

A large proportion of these holidays is spent on international trips, especially during the summer season, when people from northern climates traditionally visit southern resorts in order to enjoy the warm weather and waters. These leisure products are widely referred to as the '4Ss', i.e. Sun, Sea, Sand and Sex (Lowry 1993: 183). Leisure 4Ss products are packaged together and consumers purchase a combination of transportation, accommodation and activities packaged together by tour operators. In addition, several other types of demand emerge, especially for short-break holidays, which tend to concentrate on sports and educational activity, hobbies and visiting cultural attractions. This kind of tourism is generally domestic and often takes advantage of resources located in urban environments (such as theatres, cultural centres) or rural areas (e.g. agriculture or heritage) in close proximity to the main residence of consumers.

In recent years, however, tourism demand started changing towards a new type of activity where the individuality and independence of travellers are placed at the heart of the leisure activities. An environmental awareness is evident and consumers are actively selecting destinations which manage their environmental resources properly (Middleton and Hawkins 1998). Moreover, 'a return to nature and its pace, the search for a measure of isolation, the concern for hygiene and health, the taste for do-it-yourself, home handicrafts and sport' can be observed along with an increasing interest in cultural issues. In this sense 'people prefer to live their holidays rather than to spend them' (WTO 1985: 9). As a result, Goodall (1988:

34) suggested that 'the days of 4S's holidays are numbered'. Buhalis (1994: 261) proposed that the traditional 4S's for tourism (sea-sun-sand-sex) be transformed in 'specialisation-sophistication-segmentation-satisfaction'. This process started in the late 1980s and it is expected to dominate the transformation of tourism demand as well as the re-engineering of the industry during the next century.

Table 4.1 demonstrates that by the year 2010 about 1 billion tourists will undertake international tourism activity, spending almost 9 billion nights away and almost $1 trillion at 1995 prices. Tables 4.1 and 4.2 illustrate that long-haul travel will be increasing at a higher rate and as a result traditional destinations will be challenged, losing their market share in the future (Edwards and Graham 1997). As a result, both tourism destinations and enterprises will need to appreciate demand trends as well as the factors that affect them in order to predict the needs and wants of their travellers and develop satisfactory tourism products.

Critical factors affecting tourism demand

As illustrated in Figure 4.1, a wide range of forces from the external environment propel the changes in tourism demand. The ones that are more critical are summarised in the following points.

Table 4.1 Forecast growth of worldwide travel 1995–2010

Volume of travel	Actual 1985	Actual 1995	Forecast 2000	Forecast 2005	Forecast 2010
Trips abroad excl. day trips (millions)	307	535	632	782	964
Short/medium haul (millions)	272	455	518	617	724
Long haul (millions)	35	79	114	165	240
Nights abroad (millions)	2,828	4,571	5,518	6,903	8,654
Spending abroad (US$bn at 1995 prices)	206	393	516	686	922
Travel characteristics					
Nights per trip	9.2	8.5	837	8.8	9.0
Spending per trip (constant 1995 US$ excl. fares)	671	735	816	877	956
Spending per night (constant 1995 US$-excl. fares)	73	86	93	99	107
Growth rates %					
Trips abroad excl. day trips		5.7%	3.4%	4.4%	4.3%
Short/medium haul		5.3%	2.6%	3.6%	3.3%
Long haul		8.5%	7.5%	7.7%	7.7%
Nights abroad		4.9%	3.8%	4.6%	4.4%
Spending abroad (US$ at 1995 prices)		6.7%	5.6%	5.9%	6.1%

Source: Adapted from Edwards and Graham (1997).

Table 4.2 Forecast trips abroad by destination region 1995–2010

Trips (mn) Excl. day trips	Actual 1995	Of world (%)	Forecast 2000	Forecast 2005	Forecast 2010	Of world (%)	Growth pa, actual 1985–95 (%)	Growth pa, forecast 1995–2000 (%)	Growth pa, forecast 2000–05 (%)	Growth pa, forecast 2005–10 (%)
Europe/Mediterranean	379.6	**71.0**	412.2	489.5	566.7	**58.8**	5.1	1.7	3.5	3.0
North America	58.3	**10.9**	75.0	95.2	121.8	**12.6**	5.1	5.2	4.9	5.1
Caribbean	8.0	**1.5**	12.8	18.4	27.0	**2.8**	7.1	9.8	7.5	7.9
Central/South America	29.7	**5.6**	47.2	63.1	84.5	**8.8**	6.6	9.7	6.0	6.0
Africa (excl. North)	5.0	**0.9**	5.8	7.4	9.7	**1.0**	8.0	3.2	4.8	5.6
Middle East	4.2	**0.8**	5.5	6.2	7.5	**0.8**	7.6	5.5	2.4	3.8
South Asia/Indian Ocean	4.0	**0.8**	6.0	8.9	13.8	**1.4**	12.4	8.2	8.4	9.0
South East Asia	18.8	**3.5**	30.9	45.5	66.4	**6.9**	13.8	10.4	8.0	7.9
Australia/New Zealand	4.5	**0.8**	7.5	12.7	22.7	**2.4**	13.0	10.7	11.3	12.3
Far East/Pacific	22.3	**4.2**	29.4	35.3	44.3	**4.6**	8.8	5.6	3.7	4.6
Total	534.4	100	632.3	782.2	964.4	100	5.7	3.4	4.4	4.3

Source: Adapted from Edwards and Graham (1997).

Technology in general and information technology in particular

Bradley, Hausman and Nolan (1993a: 33) illustrate the profound role of technology on the competitiveness of organisations in the global economy, as they claim that 'globalisation and technology are mutually reinforcing drivers of change'. In addition, Metakides (1994: 2) asserts that the global information revolution obliges enterprises to 'act local and think global', while transforming dramatically both production and consumption patterns.

The technological revolution since the 1970s has facilitated tourism activity and has enabled consumers to travel further afield at a fraction of the cost and time required earlier on. The proliferation of jet engines, the ubiquitous motor car and new technology vessels and trains have not only reduced money and time required but has also provided the infrastructure for more people to travel (Westlake and Buhalis 1998).

Consumers are also empowered by information technology (Poon 1993). They not only require value for money, but also value for time for the entire range of their dealings with organisations. This reflects people's shortage of time, evident in Western societies. The emerging Internet tools enable consumers to search on-line for information and to undertake reservations. Increasingly, IT and the Internet in particular, enable travellers to access reliable and accurate information as well as to undertake reservations in a fraction of the time, cost and inconvenience required by conventional methods. IT can also improve the service quality and contribute to higher guest/traveller satisfaction (Buhalis 1998). The availability of information on everything conceivable enables consumers to personalise their tourism bundles and to purchase only the most suitable products. The usage of IT on the one hand is driven by both the development of the volume and complexity of tourism demand, and on the other hand it alters their characteristics and enables individuals to select a much more personalised bundle of tourism products. Nobody really knows how many consumers are currently connected to the Internet and how many of them buy products electronically. It was estimated that 150 million people or 2 per cent of the global population used the Internet in the late 1990s (Taylor 1999). Most Internet users match the profile of the most desirable market segments: they are well-educated professionals who travel frequently and have a higher disposable income, as well as a higher propensity to spend on tourism products (Smith and Jenner 1998).

The proliferation of the Internet revolutionised communications as it enabled organisations to demonstrate their offerings globally using multimedia interfaces (Smith and Jenner 1990). Suppliers have an unprecedented opportunity to communicate with their target markets globally, to develop their global presence and to establish direct relationships with consumers. The WTO (1985) argues that 'the key to success lies in the quick identification of consumer needs and in reaching potential clients with comprehensive, personalised and up-to-date information'. The rapid growth rate and the expeditious increase of on-line revenue experienced in most industries, including tourism, illustrates that electronic commerce will dominate by the year 2005. This justifies massive investments by organisations to develop their electronic presence (Buhalis 1998; O'Connor 1999).

The Internet has revolutionised flexibility in both consumer choice and service delivery processes. Every tourist is different, carrying a unique blend of experiences, motivations and desires often as a result of previous experience, background and social status. Increasingly customers become much more *sophisticated and discerning*. Tourists become demanding, requesting high quality products and value for both their money and – perhaps more importantly – time. Having experienced several products the new/ experienced/sophisticated/demanding travellers rely heavily on electronic media to seek information about destinations and experiences, as well as to be able to communicate their needs and wishes to suppliers rapidly. Tourists are increasingly frequent travellers, linguistically and technologically skilled and can function in multicultural and demanding environments overseas. The Internet empowered the 'new' type of tourist to become more *knowledgeable* and to seek exceptional value for money and time. New consumers are more culturally and environmentally aware and they often would like a greater involvement with the local society.

Ecology and environmental concern

Ecology and environmental concerns are increasingly becoming more important and attract a higher degree of interest by consumers. 'Green consumers' especially in Scandinavia and North Europe lead a new movement where regions and products, which fail to demonstrate a certain degree of sustainability, are increasingly becoming unacceptable in the marketplace. In tourism, there is a gradual growth of an environmentally friendly tourist who is often referred to as 'green', 'responsible', 'eco', 'ethical', 'alternative', etc. (Swarbrooke and Horner 1999). A wide range of considerations are related to green tourism which can influence the selection of destination, transportation modes, activities undertaken and products consumed during the holiday. As a result, a growing number of consumers are attracted to natural areas and ecotourism has emerged as one of the more significant powers of change in the international tourism industry (Fennell 1999; Diamantis 1998; Hvenegaard 1994; Shackley 1996).

Different consumers have dissimilar tolerance levels. As a result, Swarbrooke and Horner (1999: 202) illustrate that there are shades of green tourism from 'very green' to 'not green at all'. Nevertheless, consumers are becoming less tolerant to environmental damage and actively seek unspoilt areas to spend their holidays. Escaping from environmentally unfriendly urban regions holidaymakers often require sustainable environments where they can relax and play. Middleton and Hawkins (1998: 12) explain 'there is overwhelming evidence of customer preference for product qualities that are unambiguously concerned with environmental quality at chosen destinations. Even more interesting is the clear evidence of growing preference among experienced travellers'. As a result, a new sector is emerging in the industry to offer green products and at the same time to preserve their sustainability. However, the ability of destinations and enterprises to restrict themselves and to avoid overexploiting resources is questionable and often it is only a matter of time before greedy entrepreneurs and unwise planning procedures push a destination through the different stages of its life cycle to overdevelopment and oversupply forcing mass tourism (Buhalis 1999). Nevertheless, environmental concern and preference will increasingly dominate consumer choice and it will also determine their willingness to pay as well-preserved destinations and facilities will be able to charge premium prices for the privilege.

Multi-cultural background

People increasingly live in a multi-cultural environment. A great labour mobility as well as immigration effectively means that societies are often composed of a multi-ethic population. Different cultural backgrounds often entail different customs and values which create dissimilar if not conflicting tourism needs and wants. Multi-culture is also promoted by the emerging global television channels, such as CNN, MTV, etc. which on the one hand broadcast global images and social behaviour paradigms, and on the other hand generate interest and curiosity for the 'global village'. As a consequence. Consumers become more aware of other places, their political situations and special conditions. In addition, the exposure of consumers to many cultures through previous travelling experiences provides plenty of examples for comparisons and a wealthy basis for building expectations. Globalisation effectively implies that increasingly tourists and the industry need to interact in a culturally diverse environment and to learn how to manage, negotiate and compromise with people from different cultural backgrounds and experiences. A whole range of new skills are therefore required by the industry to communicate with the entire range of customers as well as to interact with all stakeholders (Guirdham 1999; Hocklin 1995).

Edu- and enter-tainment

Consumers are also increasingly using their leisure time for personal development. Instead of lying by the swimming pool, there is evidence that a greater percentage of tourists use their time at destinations to learn about other cultures, history and customs. Special interest and activity holidays are attracting larger numbers of holidaymakers not only because people lack time to undertake these activities whilst at home, but also because they assume a more active and participative style of holidays where they take the opportunity for personal development and exploration (Weiler and Hall 1992; Ryan 1997). Several levels of activity can be identified. People may use the time to practise their favourite sport, such as skiing, tennis, etc.; explore an area for a specific interest, e.g. archaeology, architecture; learn a new skill, such as cooking, painting; or simply interact with local people to meet, understand and appreciate the local culture. Table 4.3 illustrates a wide range of tourism themes and specific interests catered for by the industry.

The transformation of tourism demand

The development of mass tourism, since the early 1950s, has been based on a combination of 'sea-sun-sand-sangria-sex' products for summer and sea-side holidays and on 'sightseeing, short breaks, shopping, shows, scotch whisky' for urban-based tourism. Holidaymakers from northern/cold regions traditionally 'escape' for a certain period to southern/warmer destinations in order to relax, 'recharge their batteries', restore their physical and mental strength for another heavy winter and hard work at home (Krippendorf 1987). Tourists largely enjoyed a 'mass, standardised and rigidly packaged' holiday product, which enabled them to consume tourism products at reasonable prices due to economies of scale (Poon 1993). The transformation is illustrated graphically in Figure 4.1.

Recently, however, a shift can be identified in the marketplace, which takes customers away from the traditional tourism demand prototypes to the new era of tourism (Poon 1989: 92 and 1993: 85). The inclusive tour sector in the UK, for example, is set to experience the first decline since 1980, not just simply due to short-run factors or airport congestion and increased prices. A structural shift in consumer preference towards independent or semi-independent trips, and away from perceived mass tourism destinations can be observed (Cooper 1992: 94). Other Europeans, and particularly Germans and Danes, also move away from traditional mass destinations and select new, environmentally-friendly and more authentic regions for their holidays. O'Brien (1996) illustrates that across the West European market the estimated ratio of independent to package-booked travel is in the order of 70:30, with the majority of the French, Spanish, Italian and Greek markets arranging their travels themselves. In addition, consumers take a larger number of short holidays. Consequently, they normally spend a couple of weeks away during the summer and also have two or three short breaks throughout the year (Aderhold 1992). Holidays are not only regarded as opportunities to escape from the daily routine, but also as a personal development opportunity, where tourists can explore cultures and develop new skills, interests and hobbies. WTO (1985) claims that 'this non-mainstream tourism presently accounts for no more than

Table 4.3 Special interest holidays

Abseiling	Carriage Drivers Hunting	Falconry
Adventure	Carriage Driving	Farm Holidays
Agricultural	Instructions	Fencing
Air Borne	Cartoon Drawing	Fishing
Air Sports	Castles	Fishing Deepsea
Aircraft	Ceramics	Fishing Game
Anti-smoking	Champagne	Fitness
Antiques	Chateau	Flower Arranging
Archaeology	Chess	Flower Festivals
Archery	China Painting	Fly-Drive
Architecture	Christianity	Flying
Aromatherapy	Churches	Flying Lessons
Art Tours	Cider and Wine	Folk
Astrology	Cinema	Food Holidays
Astronaut School	Circus Training	Four Wheel Drive
Astronomy	Clay Shooting	French Teaching
Badminton	Climbing	Furniture History
Ballet	Coach Tours	Gardening
Ballooning	Coarse Fishing	Gastronomy
Basket Making	Collage	Genealogy
Basketball	Computers	Geography
Battlefield Tours	Conferences	Geology
Beauty Care	Congresses	German Teaching
Bee Keeping	Cookery	Glacier Tours
Beer Festival	Crafts	Gliding
Bible Cruises	Cricket	Go Karts
Bible Study	Cruising	Gold Panning
Bicycle Touring	Culture	Golf
Biking	Cultural Heritage	Gorilla Viewing
Bird Watching	Curling	Gourmet Break
Boat Tours	Curry Tours	Grass Scootering
Bonfire Dance	Cycling	Gullet
Weekend	Dancing	Gundog Training
Bookbinding	Dancing on Ice	Hacking
Botany	Darts	Hang-Gliding
Bowling	Deer Shooting	Harpseal Viewing
Bowls	Desert Tours	Health & Fitness
Bridge	Diet	Health Farms
Bulb Fields	Dinner Dance	Helicopter Flights
Bungee Jumping	Do it Yourself	Hiking
Butterfly Study	Drama	Historic Houses
Calligraphy	Dress Making	History
Camel Riding	Driving Schools	Hogmanay
Camping	Duck Shooting	Holy Shrines and
Canadian Canoes	Ecological Tours	Mosques
Canals	Educational	Home Exchange
Canoeing	Eggcraft	Honeymoons
Car Rallying	Elephant Riding	Horse Racing
Caravans	Elgar Weekend	Horse Riding
Caring	Environmental	Horse Viewing
Carnival	Expeditions	Hunting
	Fabric Sculpture	Ice Skating

Indoor Bowls
Industrial Archaeology
Italian Teaching
Jazz
Jewellery Making
Jewish Tours
Judo
Karting
Kayaks
Kibbutz Holidays
Language Study
Learn to Drive
Learn to Swim
Leather Work
Leisure Rowing
Literature-Poetry
Management
Marathon
  Running
Mardi Gras
Marionettes
Microlighting
Military History
Military Tattoos
Motor Car
Motor Cross
Motor Racing
Motorcycle Rider
Mountain Climb
Mystery Tours
Murder Mystery
Museum Visits
Music Concerts
Music Country
Music Pop
Music Tours
Myths, Monks and
  Miracles
Narrow Boat
National Parks
Natural History
Nature Reserves
Nature Studies
Naturism
Needlecraft
Netball
Opera
Orienteering
Ornithology
Painting
Parachuting
Paragliding
Parascending
Partridge Shooting

Petanque
Pheasant Shooting
Photography
Pilgrimages
Polar Bear Viewing
Polo
Pony Trekking
Pool
Porcelain
Pottery
Powerboating
Powerchuting
Quad Bikes
Races
Raft Building
Rafting
Railways
Rally Karts
Ranch Holidays
Reindeer Driving
Religion
Religious Study
Riding & Stable
Rifle Shooting
River Adventure
River Boat Cruises
Romantic Breaks
Rough Shooting
Rowing
Rugby
Running of the Bulls
Rural Surprises
Russian Teaching
Safari
Snowmobile
Safaris
Safaris (Camel)
Sailing
Salmon Fishing
Schools
Scuba Diving
Sea Canoeing
Senior Citizens
Shooting
Shooting Rifle
Silk Painting
Simulator
Single Parent
Skating
Ski Safaris
Skiing
Skiing, Cross Country
Sledging, Dog
Sledging, Grass

Slimming
Snooker
Snorkelling
Snowboard
Soccer
Social History
Spanish Teaching
Spas
Speedboat Trip
Spinning
Sporting Events
Sporting Tours
Sports Breaks
Squash
Steam Trains
Steamboats
Students
Sub-Aqua
Sugarcraft
Superkarts
Surfing
Survival
Swedish
Swimming
Tea Dance
Television
Tennis
Thalasotherapy
Theatre
Tiger Viewing
Trade Fairs
Training Courses
Trail Riding
Treasure Hunts
Trekking
Trout Fishing
Valentine Breaks
Vegetarian
Video Skills
Vineyards
Wagon Train
Walking
Water Skiing
Water Sports
Wedding
Weekend Breaks
Whale Viewing
White Water Canoeing
White Water Rafting
Wild Flowers
Wild Life
Wildlife
Weekends
Windsurfing

Table 4.3 (continued)

Wine Tasting	Women's Group	Yacht Charter
Wine Tours	Expeditions	Yachting
Winter Sports	Workout	Yoga

Source: 1. Special Interest Holiday Tour Index, *Travel Trade Gazette*, Directory, October 1992. 2. Acorn Holidays Brochure.

5% of the total tourism demand, but it is growing much more rapidly than traditional resort-based or round-tour tourism. A ceiling of 10% of total tourism is forecasted by the travel trade for this new type of tourism, though this may rise over time as alternative becomes standard'.

Traditional destinations have been victims of their own success. They grew to attract a large amount of people and inevitably have become over-crowded. The development of facilities and services, which cater for the mass markets, has forced them to lose parts of their character and has reduced their appeal. Operating on low margins also prevented principals from reinvesting and regenerating their products. Hence, mass tourism products are often regarded as responsible for both the aesthetic and environmental degradation of various destinations. Failure to control tourism development and practices has had disastrous impacts on well-known resorts. Thus, traditional tourism products and destinations have become outdated and have lost their ability to attract their intended market segments. Instead, they can attract consumers by reducing their prices and by developing a volume based product. Hence, they jeopardise their resources further and are unable to generate the positive impacts attributed to tourism. Marketing can therefore assist the management of tourism behaviour at the destination. Marketing should encourage a responsible attitude towards local resources (Buhalis 2000).

From Sea-Sun-Sand-Sex-Sangria

The transformation of tourism demand follows a wide range of trends and developments, which propel several differences in consumer behaviour. The summer/sea-side holiday is changing due to a wide range of environmental and climatic reasons. Firstly, the *sea* in well-established resorts has often been polluted by sewage, waste leaked by leisure boats and litter left behind by holidaymakers (Poon 1993: 64). In addition, pollution caused by other industrial sectors as well as accidents in oil tankers, has also degraded the quality of the water environment. A number of diseases and viruses can also be transformed through the sea, causing serious health problems. Examples include the algae in the Adriatic in 1989 and the pollution of the sea in several British seaside resorts. As modern tourists are reluctant to tolerate environmental pollution, the sea becomes less attractive at destinations that have failed to protect the natural environment.

Although the *sun* has been a prime motivation for sun-lust tourists, it becomes under attack as the Green House Effect and the Ozone Layer Loss have increased temperatures to uncomfortable levels. In addition, skin cancer from sun overexposure reduces its appeal as a tourism motivator, while Poon (1993: 124) claims that 'the sun sets on tourism'. Wall (1992) suggests that 'the Greenhouse Effect and the likely climate changes to which it will give rise will likely impinge upon tourism at a global scale and may lead to diverse and profound consequences of global climate change for tourism'. Consequently, these phenomena will probably change tourism demand patterns in the next century and lead tourists towards northern and cooler climates.

Sand has similarly been degraded since masses of tourists are normally packed into a limited space on beaches, and spoil the environment. Poon (1993) claims that 'degradation of beaches and soil erosion from construction too close to the shore line, for example, some hotels and resorts in the Caribbean region are experiencing a loss of sand and bathing area as the sea begins to reclaim some of the area, exposing ugly building foundations'. In addition, sand has been identified as responsible for the transformation of various diseases and hence future tourists may avoid being exposed to the sand. Thus, excessive numbers of holidaymakers and tourism overdevelopment eventually destroy both the environment and the 'escape' element for consumers (Krippendorf 1987; Poon 1993: 126). In addition, the transformation of various diseases from sand starts being another negative factor against attracting tourists.

Sangria symbolises the tendency of traditional holidaymakers to consume great quantities of alcohol. New consumers use holidays as a form of personal development and hence alcohol will increasingly be less important. Excessive alcohol consumption is not only unhealthy and often responsible for accidents and injuries, but it also creates a range of social problems at destination areas. Frequently hooliganism and conflicts between locals and tourists are also attributed to excessive consumption of alcohol. New tourists will therefore need less alcohol in order to enjoy themselves.

Sex and romance have always been an important, but often untold, element of leisure and tourism activities. It may be one of the major motivations for 'sex-tourists', especially for some 'specialised' destinations (e.g. Bangkok, Las Vegas, etc.) or part of the sightseeing experience (e.g. Amsterdam, Paris) (Hall 1992; Burns 1993: 6; O'Malley 1988; Wickens 1994: 820; Truong 1989; Crick 1989: 324). Eroticism accompanies almost all tourism activities, and it has been extensively used for advertising purposes (Lowry 1993: 183; Burns 1993: 3). Specific products have been developed to accommodate this type of demand. For example holidays in particular clubs (e.g. the 18–30 Club or the Club Med.) traditionally had the image of the wild, care-free bachelors looking for companions while on holiday; various destinations (e.g. Amsterdam, Mykonos) are prime destinations for homosexuals; some Far East and African destinations have developed a prostitution industry which caters for wealthy Western tourists; whilst 'romantic' destinations (e.g. Venice, Mauritius, Rome and Paris) fiercely target the 'couple/honeymooners' market. However, modern diseases (i.e. Aids) have increased public concern and have reduced sexual activity with non-regular partners during holidays. This is influencing the tourism demand patterns. Destinations where sex is a primary activity or areas with a large Aids-sufferers population might face a decline in arrivals (Kurent 1991: 80; Cohen 1988). In contrast, 'romance' destinations might gain a greater market share; tourism enterprises like the Club Med. have altered their image significantly and target different segments (i.e. family or sportive markets); while other clubs (e.g. Sandals) emerged to cater exclusively for couples. It is quite apparent therefore, that the sex element in the traditional holiday patterns is also changing radically and it will influence the decision-making processes of future holidaymakers.

As a consequence of the above trends, Mediterranean destinations lose some of their market share and appeal. Trips from main European countries to this region are projected to fall from 45 per cent in 1995 to 38 per cent in 2010 (Edwards and Graham 1997).

From short breaks, sightseeing, shopping, shows, Scotch whisky

A combination of short breaks, sightseeing, shopping, shows and Scotch whisky has dominated urban tourism in the past (Davidson 1994; Law 1993, 1996; Page 1994; O'Brien 1998). Often urban tourism is closely related with business travel (characterised by MICE [Meetings, Incentives, Conferences, Exhibitions]) as business travellers usually consume some leisure tourism products or they may stay for a few more days to enjoy the local resources. Hence, urban tourism is often characterised by *short breaks*, sometimes combined or as an extension of business trips. Leisure tourists also tend to be curious to visit metropolitan centres which they are familiar with through the media. They also take advantage of the large amount of cultural and heritage resources often found in urban destinations, such as museums, galleries, theatres, cathedrals, monuments, etc. A wide range of facilities such as hospitals, centralised government agencies and educational establishments also act as attractions for consumers.

Often urban tourism is consumed through organised *sightseeing*, which aims to pack as many attractions as possible into the limited time available at the destination. Sightseeing is facilitated through transfers and guided tours. However, sightseeing programmes are often characterised by their rapidity and inflexibility. Consumers take specific interest in fewer but more personalised attractions and use their leisure time to concentrate on their interests. They will be more interested in experiencing elements of the destination, rather than tick them off their 'must see' items. This will be particularly the case for repeat visitors to destinations who become familiar with local resources and people. Hence future tourists will need a greater flexibility and control over their time at the destination in order to explore in detail and experience resources of their choice. Thus they will need tailor-made sightseeing programmes which will enable them to increase their flexibility.

Shopping has dominated urban tourism as people from peripheral areas were lured by a great variety and often cheaper prices in urban centres. The proliferation of shopping malls out of city centres and the globalisation of manufacturing and retailing, as well as the distribution of products through the electronic media are expected to reduce the appeal of shopping as an attraction to urban destinations. Instead, shopping will be integrated with other attractions and experiences at the destination. Perhaps tourism shopping will be themed in relation to the local tourism product.

Another element of urban tourism has traditionally been shows of any kind, such as theatre, opera, cinema, circus, etc. Only cities have the infrastructure as well as the critical mass of consumers required in order to stage performances. However, the growth of electronic media and the development of facilities in peripheral regions as well as the maturity of the market demonstrate that it will be difficult to impress consumers in the future. Even shows and entertainment activities will need to be customised to suit the feelings of consumers during their holiday.

Similarly with sea-side tourism, *Scotch whisky* symbolises the contribution of alcohol to the urban tourism product. It can be observed that, although alcohol has been playing a significant role in urban tourism, health considerations as well as other life style influences discourage tourists from consuming large quantities of alcohol. As a result, alcohol will be themed with the overall experience rather than being a stand-alone product. Examples of that can be demonstrated by the expansion of Irish pubs where Irish food and drinks can be consumed within a themed environment.

Towards sophistication-specialisation-segmentation-satisfaction-seduction

The shift of demand towards quality and value-for-money products is increasing rapidly in the tourism industry. Tourists demand higher quality products and services and real experiences during their holidays. The traditional annual family holiday in a seaside resort will play a less dominant role in the future. Multi-interest travel is therefore replacing part of the present bread and butter products of the industry. Future products will probably combine beach holidays with pleasure and special interest of some kind or culture. Future tourists will 'prefer to live their holidays rather than to spend them', and they do so by engaging in cultural, physical, educational and spiritual activities (WTO 1985).

Rigidly packaged tours are not in line with trends towards individual expression. As a result, the independently organised tourism segment emerges rapidly whilst there is a decline of the relative importance of packaged tours. O'Brien (1996) suggests that 'Growth in the inclusive tour market will continue, though at a much slower rate, but in the larger summersun markets, particularly Germany and the UK, the major IT operators will develop a wider product range to ensure that sales remain buoyant.' One of the most important obstacles in organising individual tourism packages hitherto is the lack of economies of scale and bargaining power, which will enable the reduction of the individual package prices to affordable levels. WTO (1991: 18) suggests that packaged tours are not in line with trends towards individual expression. As a result, the decline of the relative importance of packaged tours is expected in favour of independently organised tourism (Moutinho 1992: 10; Poon 1989: 93; Cooper 1992: 94; Mintel 1991: 4, *Economist* 1992; Ryan 1989: 67). Bentley (1991: 57) states that 'tour operators in the 1990s will seek to combine inclusive tour elements with individual variations, in order to satisfy the desire for an individual experience, but at a cost lower than an individually arranged holiday'. Meanwhile, the tourism industry is moving towards the accommodation of activity/adventure/wildlife/culture/independent/special interest holidays (Weiler and Hall 1992; Ryan 1997; Mintel 1991; Ashworth 1993; Shackley 1996; Smith and Jenner 1990; Ogilvie and Dickinson 1992; Cockerell 1991). The new generation of tourists is more educated, experienced, sophisticated, knowledgeable and demanding. This is reflected in the kind of travel experiences they seek, their behaviour and preferences whilst at the destination and also in the information they require in travel decision-making (Poon 1993: 114; Ritchie 1991: 153; Buhalis 1993; WTO 1991: 32). The WTO (1985) estimates that 'ecotourism (or nature based tourism) is growing by 25–30 per cent per year while culture-based tourism is recording annual expansion of 10–15 per cent. Endemic tourism (based on the individual character of the locality or community) is expected to become an important means of differentiating tourist destinations and appealing to the 'new' types of tourism'.

Sophistication

Tourists in the post-industrial era are better prepared for living in an international world. Modern people are able to work and function in a demanding environment. They are often familiar and capable to cope with foreign languages, customs and cultures and as a result a 'new global lifestyle is emerging', often as a result of the globalisation of the media as well as the extensive travelling experiences of consumers (Bentley 1991: 55 and WTO 1991: 10). The enhancement of the media, and especially television, has reduced the distances and increased the eagerness of modern people to approach and experience remote cultures and foreign areas (Coates and Coates 1991: 68). Bennett (1992: 87) states that:

educational improvements, together with enhanced communications have led to more sophisticated requirements from holidaymakers who are now looking for new activities to fill their leisure time and satisfy their cultural, intellectual and sporting interests. Increased linguistic ability among the younger generations, communication and financial services have made travel easier.

Hence, sophistication is a major element in the 'new types' of holidays and the products emerging to satisfy modern demand. Holidaymakers take advantage of their education as well as the availability of information through the new media and plan their holidays in advance. They are eager to approach and experience remote cultures and foreign areas in order to 'live their leisure time' and satisfy their cultural, natural, intellectual and sporting interests. The linguistic abilities of younger generations (both hosts and tourists), as well as the prevalence of communication and financial services at a global level have made travel more accessible and easier.

Extensive travelling has also increased the required sophistication of the tourist product. As many modern travellers have been in several countries and treated by several tourism enterprises, they have developed a set of assessment criteria, which they utilise in order to compare their tourism experiences. Moutinho (1992) illustrates that the 'sophistication of the customer will have an impact on all product development throughout the industry. There will be an increased requirement for high standards of product design, efficiency and safety'. Thus, people will seek more varied, personal and authentic experiences, while a wide range of new, imaginative, tourism products will be demanded. In addition, personal development and special interest travel are expected to provide rewarding, enriching, adventuresome experiences. Not only is mass tourism less able to satisfy a large proportion of the marketplace, it is also regarded as environmentally unacceptable and thus 'politically incorrect'. Moreover, Cooper and Ozdil (1992: 378) suggest that:

> In particular the realisation of the negative impacts of tourism on host environments and societies has prompted a search for alternative forms of tourism and a move away from 'mass' tourism. Indeed, this movement is largely consumer rather than industry driven and may lead to 'politically correct' or acceptable forms of tourism which will be chosen by the consumer in preference to more damaging forms.

Thus, adventure and green tourism are driven by consumer requirements at the expense of mass and environmentally threatening tourism (Middleton and Hawkins 1998).

Tourist product sophistication should probably aim to deliver the appropriate product, at the appropriate time, at the appropriate price. The amalgamation of tourism products and the delivery of seamless travel experiences are increasingly important. Information technology facilitates the development of suitable businesses and communication networks for achieving these purposes. Modern travellers demand customer convenience in all aspects, while 'total consumer satisfaction' and enrichment of the holiday experience are required; all these at a competitive and fair cost. The anticipated sophistication of the consumer will have an impact upon product development as well as on customer retention throughout the industry. Not only will there be an increased requirement for high standards of product design, efficiency and safety, but also the tourist will be more critical of the product and will have the experience to compare offerings.

Specialisation

Tourists' requirement for sophistication has recently initiated the demand for specialised products. Tourism motivation is a very complicated set of needs and desires, which differ for various people. Several motivators are consistently rated in a variety of consumer behaviour surveys (Table 4.4). However, tourism researchers are generally unable to identify the motivators and determinants of tourism activity with accuracy, as a result of the diversified needs, desires and decision-making criteria used by each individual consumer each time s/he selects a tourism product. Hitherto, tourism products used to be general, unspecialised, with almost identical characteristics and have been traded as commodities rather than services for the satisfaction of specific needs. The mass tourism philosophy, where tourism products should appeal to all different tastes and should be as cheap as possible, in order to attract customers of all purchasing abilities and achieve economies of scale, used to dominate the marketplace. These products become less attractive as consumers are exceedingly conscious and keen to explore the cultural, social, gastronomical, political, and environmental aspects of destinations. (Smith and Jenner 1990; Ogilvie and Dickinson 1992; Cockerell 1991; Ryan 1989: 67; WTO 1985; Weiler and Hall 1992).

They are expected therefore to organise more independent activities, adventure and sport holidays and devote their holidays to special interest

Table 4.4 Generic motivators for tourism activity

Change and escape from routine
Social interaction and recognition
Rest and relaxation
Enjoyment and adventure
Special activity often related to special interests
Weather: experience different climates
Geography: enjoy the sea, landscapes, mountains
Culture and gastronomy
Educational and personal development

Source: Gilbert (1992): 235.

activities. As a result tailor-made travel arrangements are expected to grow at a faster pace than pre-packaged holidays over the next decade (WTO 1991). Kotler (1988) suggests that 'each buyer is potentially a separate market because of unique needs and wants'. This statement could not be more valuable in any other industry than tourism. Technology makes possible more tailored products to meet individual tastes and hence we turn from mass production of all kinds to customisation. One-to-one marketing initiatives emerge gradually to take advantage of expressed consumer requirements and to develop individualised product solutions. Long-term customer segmentation will be based on the feeling of individual consumers at each particular moment, rather than on broad segmentation variables. This will enable enterprises to offer instantaneous tourism products to satisfy the needs of consumers at each moment. Hence, the competitiveness of tourism organisations and destinations will depend on their ability to differentiate their product and serve individual consumer needs.

New specialised tourism products are marketed and distributed differently from ordinary offerings. Advertising will primarily be carried out in specialised media, while direct selling and relationship marketing will be utilised for understanding the consumer needs and promoting specialised products. The Internet and technology in general will facilitate one-to-one marketing, as specialised products will be distributed directly to the right market segments. Offering specialised products on-line will not only improve the specialisation of the industry, but it will also enable the reduction of the brochure used for the promotion of tourism products, reducing both the printing cost as well as the environmental damage caused from the production and distribution.

Apart from including special interest activities in traditional holiday brochures, a wide range of new programmes emerge to cover these markets. New programmes such as 'Battlefields', 'Italian Cooking', 'Dutch Bulbfields', etc. address specific interest markets and aim to provide a thematic tour. These new products are based extensively on activities that are undertaken during holidays, while accommodation and transportation arrangements are given less importance. Themed leisure activities are also expected to increase in order to satisfy the demand for specialisation. In the catering industry, for example, restaurants used to be specialised only in national cuisines (e.g. French, Chinese, Greek restaurants). New themes are rapidly emerging based on the food served (e.g. Steaks: Aberdeen Steak House; Hamburger: McDonald's; Chicken: Kentucky; Pizza: Pizza Hut); on the entertainment provided (e.g. Hard Rock Café or Planet Hollywood); the surrounding environment (e.g. Country pub: Bass Pubs or Rainforest Café); or on the life style and preferences (e.g. vegetarian or game restaurants) are continually emerging.

As a result, Lickorish (1990) predicts that the importance of the 'mini market segments' will increase rapidly. Independent holidays increase their market share in the international arena. In the UK for example they increased their contribution to 45 per cent of the total holidays abroad in 1994, from 37 per cent in 1983 (O'Brien 1996). Thus, Moutinho (1992) suggests that 'tourist innovation is more likely to be about un-packaging rather than packaging, providing more individual attention within a number of price bands'. Eventually tourism marketing will move even further from one-to-one relationship marketing to the level of marketing towards how a person is feeling at a particular moment. This will be facilitated with the development of intelligent agents and push technology which will assess the situation and mood of a person and will promote the most appropriate products for that particular moment.

As a result, specialisation is expected to be a dominant element of holidays in the future. People will be able to select numerous specialised holidays all over the world. Both destinations and tourism enterprises should therefore identify their competitive advantage in offering specialised products and develop integrated and themed tourist experiences. For example, France could develop gastronomic themed activities, while Greece could emphasise themes based on archaeology-philosophy-culture. Orlando and Las Vegas in the USA, which are strongly themed towards Disneyland visitors and gamblers respectively, are good examples of specialised/themed tourism products which attract a large volume of visitors based on these attractions (Eadington and Smith 1992: 7).

Segmentation

Specialisation of tourism products entails a need for segmentation of tourism markets. Since tourists no longer have single, standardised and rigidly packaged wants, segmentation offers the opportunity to provide appealing tourism products to well-defined markets. Market segmentation can be defined as the grouping of individuals according to their preference or reaction to specific elements of the marketing mix, i.e. product, price, distribution and promotion or according to their characteristics (Kotler 1988: 280). WTO (1991) claims that 'segmentation is the process by which a travel vendor, whether an airline, hotel or destination, identifies and attracts consumers who will be satisfied by the product or service the vendor offers'. Figure 4.2 illustrates the ten broad categories of criteria which are often used in tourism marketing for segmentation (Swarbrooke

Geography	Behaviour
Demographics	Product related
Life cycle	Life style
Purpose of trip	Distribution channels
Psychographics	Price elasticity

Figure 4.2 Categories used in tourism marketing for segmentation.

and Horner 1999; Morrison 1989: 145; Fotis 1992: 22; Middleton 1988: 70; Mill and Morrison 1985: 362).

Traditionally, tourism marketers have been using geographic and demographic criteria in order to describe their markets, probably because these categories offer objective tangible and measurable variables. However, as a number of phenomena could not be explained and interpreted, additional segmentation categories and methods have been added. Consequently, psychographics and behavioural criteria are used nowadays, in tourist segmentation, in order to provide detailed customer profiles, identify tourist motivations, needs and determinants, and offer an appropriate tourist product mix. Thus, life-style segmentation has gained ground in modern tourism marketing (Lowyck *et al.* 1992: 13; Mazanec 1995). 'Lifestyle is a way of living, characterised by the manner in which people spend their time (activities), what things they consider important (interests) and how they feel about themselves and the world around them (opinions)' (Morrison 1989). Although life-style segmentation is probably the most difficult and subjective method, it provides the best prediction and understanding of tourist activities. Table 4.5 illustrates some of the main tourism typologies developed.

Tourism destinations, enterprises and organisations will need to undertake thorough segmentation in order to ensure that they design suitable tourist products, use proper communication media and charge acceptable prices. Moreover, segmentation is essential for treating seasonality problems, as well as for mitigating tourism impacts at destinations. Segmentation is also critical in bringing together the right types of consumers/tourists, as their interaction during the travel experience is largely responsible for the delivery and perception of tourism products. For example, the demographic changes worldwide demonstrate the development of a new market segment of older/retired but active people who have plenty of time and often money to spend. Several operators currently develop suitable products for this market segment. The UK company Saga Holidays specialises with holidays for the over 55s, while a new style of club is currently being developed by Mr Trigano, the ex Club Med. Chairman. The required sophistication and specialisation of tourism can only be offered to small segments of the market with very similar needs and motivations. Hence, customer satisfaction will largely depend on proper segmentation and achievement of a suitable customer mix.

Satisfaction

Consumer satisfaction is the essence of most developments in tourism demand. Satisfaction 'occurs when consumer expectations are met or exceeded' (Assael 1987) and can be defined 'as an evaluation that the chosen alternative is consistent with prior beliefs with respect to that alternative' (Moutinho 1987). Increasingly customer satisfaction will be inadequate for

Table 4.5 Major life-style typologies of tourists

Cohen 1972	Plog 1973	Perreault, Darden and Darden 1977
Organised mass tourists	Allocentric	Budget travellers
Individual mass tourists	Near allocentric	Adventures
Explorer	Mid allocentric	Homebodies
Drifter	Near Psychocentric	Vacationers
	Psychocentric	Moderates
Westvlaam Economisch Studiebureau 1986	Plog 1989	Dalen 1989
Active sea lovers	Venturesomeness	Modern materialists
Contact-minded holiday makers	Pleasure seeking	Modern idealist
	Impassivity	Traditional idealists
Nature viewers	Self-confidence	Traditional materialists
Rest seekers	Plainfulness	
Discoverers	Masculinity	
Family oriented sun and sea lovers	Intellectualism	
	People orientation	
Traditionalists		
American Express Travel Related Services 1989	Hall 1989	Fotis 1992
Adventures	New enthusiasts	Tranquillers
Worriers	Big spenders	Culturers
Dreamers	Antitourists	Budgeters
Economisers	Stay-at-home tourists	Nightlifers
Indulgers	New indulgers	
	Dedicated Aussies	

Source: Adapted from: Lowyck *et al.* (1992); Hall (1989); and Fotis (1992).

the fiercely competitive environment and thus 'delighting the customer' should become the target for all enterprises (Kotler 1992: 24).

Tourism satisfaction should be one of the strategic directions of every tourist enterprise, destination or organisation. This can be achieved by offering:

- at least the quality promised;
- undertaking consumer research and formulating innovative tourism products;
- improving services constantly;
- adjusting tourism products to customer needs and feelings;
- enriching the tourist experience; and
- offering value for money.

Satisfying consumers, as well as providing sufficient value for both money and time, are also ethical obligations and responsibilities of tourist destinations and organisations. Satisfying consumers also makes financial/business sense.

- First, satisfied tourists are normally very loyal. The increasing sophistication of travellers, and the fierce competition in the international market make repeat business difficult to maintain. Therefore, it should be extremely welcome and appreciated by the tourism industry.
- Second, satisfied customers are always the best, most reliable and cheapest promotional medium, as they usually recommend tourist destinations/enterprises to friends through word of mouth. In contrast, dissatisfied travellers spread their complaints to potential customers. Bearing in mind that most people get advice for their travel plans from friends and relatives, the image of the destination depends heavily on the description of the previous visitors.
- Third, providing adequate services and satisfying the tourist will eliminate enterprises and destinations from potential legal actions and suits against them. Consumerism and consumer protection are international movements forced by various private and governmental organisations. The European Community Package Travel Directive, especially on the package and timeshare holidays, is expected to have major impacts on tourism enterprises in the near future.

Due to the fact that the tourism product is an amalgam of many products and services, which formulate the 'tourist consumption chain', each trip is assessed as a total experience. Hence, the satisfaction of tourists normally depends on the harmonic delivery of the product throughout the chain. As destinations represent the 'raison d'être' for tourism, travel experience tends to be appraised at a destination level. Consequently, an integrated approach should be employed, by all enterprises and organisations involved at a destination level, in order to ensure that consumers are satisfied by the whole range of products consumed during their travelling experience. Thus, all tourist enterprises should consider the needs and wants of their customers before, during and after the delivery of their own products to ensure that everybody involved in the tourism product delivery chain offers satisfactory services. The ultimate measurement of tourist satisfaction is probably hidden in the answer of two critical questions:

- Would the customer come back?
- Would the customer recommend the product/destination to his/her friends?

Seduction

Tourism providers need to 'seduce' tourists through the development and delivery of offerings and marketing mixes which reflect consumers' feelings at a particular moment and satisfy their entire range of needs and wants. The inconsistency and unpredictability of tourism consumers, which is emerging due to the dynamic nature of modern life and the overexposure of consumers to the media, illustrates that tourism marketing will become much more difficult for the future.

Tourism destinations and organisations that manage to seduce consumers will need to make them feel special at each stage of their travelling experience. Offering tailor-made tourism products and caring for the individual needs and wants of consumers will be one of the most critical attributes tourism organisations will need to have in order to attract and satisfy tourists in the future. Extensive marketing research, using psychology methodologies will enable tourism organisations to explore consumer feelings and requirements whilst on holidays. Data mining on the information available through loyalty clubs, Passenger Name Records (PNRs), guest history, and other sources of information should be exploited in order to assess the needs and wants of particular consumers. Complex models of behaviour can then be developed in order to predict consumer requirements and develop instantaneous products before consumers require them. Developing relationship marketing will enable tourism organisations to establish a closer partnership with consumers. Destinations and organisations who manage to seduce their consumers will be able to increase the value added they offer and achieve high levels of customer retention and loyalty.

A word of conclusion

This chapter attempts to explore the trend of the tourism demand by providing a framework of analysis. It is argued that the traditional tourism products will no longer be adequate for the recreation of the new generation of tourists emerging. Instead, a more individualised product is expected to dominate demand in the near future. Tourism demand trends can be illustrated in a framework of five new S's, namely sophistication-specialisation-segmentation-satisfaction-seduction. Emerging tourism products need to be sophisticated in order to delight the new, experienced, and demanding consumer. Moreover, a certain degree of specialisation is required in order to cater for the individual needs and wants. This can only be achieved by detailed segmentation of the market where all cluster segments are identified and offered tailor-made products. The ultimate aim should be the total customer satisfaction before, during and after the consumption of the tourist product, which is underlined by both ethical and business motives. Tourism organisations and destinations which achieve the above will 'seduce' their clientele and achieve sustainable competitive advantages. A thorough understanding of this transformation framework as well as the utilisation of marketing and information technology tools will be essential for all players involved in the tourism industry. Enterprises and destinations which fail to appreciate these developments and modernise their offerings will be marginalised in this century.

References

Aderhold, P. (1992) 'Trends in German outbound tourism', in B. Ritchie, D. Hawkins, F. Go, D. Frechtling (eds) *World Travel And Tourism Review: Indicators, Trends, and Issues*, Vol. 2, Oxford: CAB International.

Ashworth, G. (1993) 'Culture and tourism: conflict or symbiosis in Europe', in W. Pompl and P. Lavery (eds) *Tourism in Europe: Structures and Development*, Oxford: CAB International.

Assael, H. (1987) *Consumer Behaviour and Marketing Action*, third edn, California: PWS-Kent.

Bennett, O. (1992) 'Is tourism going upmarket?' *Travel and Tourism Analyst*, 2: 84–98.

Bentley, R. (1991) 'World tourism outlook for the 1990s', in B. Ritchie, D. Hawkins, F. Go, D. Frechtling, (eds) *World Travel and Tourism Review: Indicators, Trends, and Issues*, Vol. 1, Oxford: CAB International.

Bradley, S., Hausman, J. and Nolan, R. (eds) (1993) *Globalisation, Technology and Competition. The Fusion of Computers and Telecommunications in the 1990s* Boston, MA: Harvard Business School Press.

Buhalis, D. (1993) 'Regional integrated computer information reservation management systems as a strategic tool for the small and medium tourism enterprises', *Tourism Management* 14(5): 366–78.

Buhalis, D. (1994) 'Information and telecommunications technologies as a strategic tool for small and medium tourism enterprises in the contemporary business environment', in A. Seaton, *et al.* (eds) *Tourism – the State of the Art: The Strathclyde Symposium*, Wiley, UK.

Buhalis, D. (1998) 'Strategic use of information technologies in the tourism industry', *Tourism Management* 19(3): 409–23.

Buhalis, D. (1999) 'Limits of tourism development in peripheral destinations: problems and challenges', *Tourism Management* 20(2): 183–7.

Buhalis, D. (2000) 'Marketing the competitive destination of the future', *Tourism Management* 21(1): 97–116.

Burns, P. (1993) 'Sun, sand, sea and sex: the problematique of tourism and the north–south debate', unpublished presentation at University of Surrey, 24 June, Guildford.

Coates, J. and Coates, J. F. (1991) 'Tourism and the environment: realities of the 1990s', in B. Ritchie, D. Hawkins, F. Go, D. Frechtling, (eds) *World Travel and Tourism Review: Indicators, Trends, and Issues*, Vol. 1, Oxford: CAB International.

Cockerell, N. (1991) 'European independent travel', *Travel and Tourism Analyst* 4: 38–48.

Cohen, E. (1988) 'Tourism and Aids in Thailand', *Annals of Tourism Research* 15: 476–86.

Cooper, C. (1992) 'United Kingdom outbound travel', in B. Ritchie, D. Hawkins, F. Go, D. Frechtling, (eds) *World Travel and Tourism Review: Indicators, Trends, and Issues*, Vol. 2, Oxford: CAB International.

Cooper, C. and Buhalis, D. (1998) 'The future of tourism', in C. Cooper, J. Fletcher, D. Gilbert, R. Shepherd, and S. Wanhill (eds), second edn, *Tourism: Principles and Practice*, London: Addison Wesley Longman.

Cooper, C. and Ozdil, I. (1992) 'From mass to "responsible" tourism: the Turkish experience', *Tourism Management* 13(4): 377–86.

Crick, M. (1989) 'Representations of international tourism in the social science: sun, sex, sights, savings and servility', *Annual Review Of Anthropology* 18: 307–44.

Davidson, R. (1994) *Business Travel,* London: Pitman.

Diamantis, D. (1998) 'Consumer behaviour and ecotourism products', *Annals of Tourism Research* 25(2): 515–28.

Dicken, P. (1992) *Global Shift: The Internationalisation of Economic Activity*, second edn, London: Paul Chapman Publishing.

Eadington, W. and Smith, V. (1992) 'Introduction: The emergence of alternative forms of tourism', in V. Smith, and W. Eadington (eds) *Tourism Alternatives: Potentials and Problems in the Development Of Tourism*, Philadelphia: University of Pennsylvania Press.

Economist (1992) 'Holidays: Unpackaged', 18 July, 29.

Edwards, A. and Graham, A. (1997) *International Tourism Forecasts to 2010, Research Report*, London: Travel and Tourism Intelligence.

Fennell, D. (1999) *Ecotourism: An Introduction*, London: Routledge.

Fotis, J. (1992) *Vacational Lifestyle Segmentation of Tourists on the Island of Rhodes, Greece*, MSc. Dissertation, University of Surrey.

Gilbert, D. (1992) *A Study of the Factors of Consumer Behaviour Related to Overseas Holidays from the UK*, PhD thesis, University of Surrey.

Go, F. (1996) 'A conceptual framework for managing global tourism and hospitality marketing', *Tourism Recreation* Research 21(2): 37–43.

Go, F. and Pine E. (1995) *Globalisation Strategy in the Hotel Industry*, London: Routledge.

Goodall, B. (1988) 'Changing patterns and structure of European tourism', in B. Goodall and G. Ashworth (eds) *Marketing in the Tourism Industry: The Promotion of Destination Regions*, London: Routledge.

Guirdham, M. (1999) *Communicating Across Cultures*, London: Macmillan Business.

Hall, M. (1989) 'Special interest tourism: A prime force in the expansion of tourism', in R. Welch (ed.) *Geography in Action:* Proceedings of Fifteen New Zealand Geography Conference, New Zealand Geographical Society Conference, Series No. 15, Dunedin, New Zealand, pp. 81–9.

Hall, M. (1992) 'Sex tourism in South East Asia', in D. Harrison (ed.) *Tourism and the Less Developed Countries*, London: Belhaven Press.

Hocklin, L. (1995) *Managing Cultural Differences: Strategies for Competitive Advantage*, London: Addison Wesley.

Hvenegaard, G. (1994) 'Ecotourism: a status report and conceptual framework', *Journal of Tourism Studies* 5(2): 24–35.

Kotler, P. (1988) *Marketing Management: Analysis, Planning and Control*, sixth edn, New Jersey: Prentice-Hall.

Kotler, P. (1992) 'Silent satisfaction', *Marketing Business*, December-January: 24–7.

Krippendorf, J. (1987) *The Holiday Makers: Understanding the Impact of Leisure and Travel*, London: Heinemann.

Kurent, H. (1991) 'Tourism in the 1990s: threats and opportunities', in B. Ritchie, D. Hawkins, F. Go and D. Frechtling (eds) *World Travel and Tourism Review: Indicators, Trends, and Issues*, Vol. 1, Oxford: CAB International.

Law, C. (1993) *Urban Tourism: Attracting Visitors to Large Cities*, London: Mansell.

Law, C. (1996) (ed.) *Tourism in Major Cities*, London: Thomson Business Press.

Lickorish, L. (1990) 'Tourism facing change', in M. Quest (ed.), *Horwarth Book of Tourism*, London: Macmillan.

Lowry, L. (1993) 'Sun, sand, sea & sex: a look at tourism advertising through the decoding and interpretation of four typical tourism advertisements', in K. Chon, (ed.), *Proceedings of Research and Academic Papers*, Vol. V, The Society of Travel and Tourism Educators, Annual Conference, 14–17 October, Miami.

Lowyck, E., Langenhove, L.V. and Bollaert, L. (1992) 'Typologies of tourist roles', in P. Johnson, and B. Thomas, (eds) *Choice and Demand in Tourism*, London: Mansell.

Makridakis, S. (1989) 'Management in the 21st century', *Long Range Planning*, 22(2): 37–53.

Mazanec, J. (1995) 'Constructing traveller types: new methodology for all concepts', in R. Butler, and D. Pearce, (eds) *Change in Tourism: People, Places, Processes*, London: Routledge.

Middleton, V. (1988) *Marketing in Travel and Tourism*, London: Heinemann.

Middleton, V. and Hawkins, R. (1998) *Sustainable Tourism: A Marketing Perspective*, Oxford: Butterworth-Heinemann.

Metakides, G. (1994) 'Opening address', Paper presented at the European Information Technology Conference '94, Brussels, 6–8 June.

Mill, P. and Morrison, A. (1985) *The Tourism System: An Introductory Text*, New Jersey: Prentice-Hall.

Mintel (1991) 'Independent travel: A bias towards youth, *Leisure Intelligence* 1: 1–32.

Morrison, A. (1989) *Hospitality and Travel Marketing*, New York: Delmar.

Moutinho, L. (1987) 'Consumer behaviour in tourism', *European Journal of Marketing*, 21(10): 3–44.

Moutinho, L. (1992) 'Tourism: The Near and Future Future from 1990s to 2030s or from Sensavision TV to Skycycles', Conference Proceedings: *Tourism in Europe*, Newcastle Polytechnic, 8–10 July.

O'Brien, K. (1996) '*The West European leisure travel market: forecasts for opportunities into the next century*', London: Financial Times Newsletters and Management Reports.

O'Brien, K. (1998) 'The European business travel market', *Travel and Tourism Analyst*, 4: 37–54.

O'Connor, P. (1999) *Electronic Information Distribution in Tourism & Hospitality*, Oxford: CAB International.

Ogilvie, J. and Dickinson, C. (1992) 'The UK adventure holiday market', *Travel and Tourism Analyst*, 3: 37–51.

O'Malley, J. (1988) 'Sex tourism and women's status in Thailand', *Society and Leisure*, 11(1): 99–114.

Page, S. (1994) *Urban Tourism*, London: Routledge.

Parker, B. (1998) *Globalisation and Business Practice: Managing Across Boundaries*, London: Sage.

Poon, A. (1989) 'Competitive strategies for new tourism', in C. Cooper (ed.) *Progress in Tourism Recreation and Hospitality Management*, Vol. 1, London: Belhaven Press.

Poon, A. (1993) *Tourism, Technology and Competitive Strategies*, Oxford: CAB International.

Ritchie, B. (1991) 'Global tourism policy issues: an agenda for the 1990s', in B. Ritchie, D. Hawkins, F. Go, and D. Frechtling (eds) *World Travel and Tourism Review: Indicators, Trends, and Issues*, Vol.1, Oxford: CAB International.

Ryan, C. (1989) 'Trends past and present in the package holiday industry', *The Service Industry Journal* 9(1): 61–78.

Ryan, C. (1997) *The Tourist Experience: A New Introduction* London: Cassell.

Shackley, M. (1996) *Wildlife Tourism,* London: International Thomson Business Press.

Smeral, E. (1998) 'The impact of globalisation on small and medium enterprises: new challenges for tourism policies in European countries', *Tourism Management,* 19(4): 371–80.

Smith, C. and Jenner, P. (1990) 'Activity holidays in Europe', *Travel and Tourism Analyst* 5: 58–78.

Smith, C. and Jenner, P. (1998) 'Travel agents in Europe', *Travel and Tourism Analyst* 4: 5–15.

Swarbrooke, J. and Horner, S. (1999) *Consumer Behaviour In Tourism,* London: Butterworth-Heinemann.

Taylor, P. (1999) 'Business urged to get a connection as web turns out to be more than a fad', *Financial Times,* 6 July: 14.

Truong, T. (1989) 'The dynamics of sex tourism: The case of Southeast Asia', in T. Singh, H. Theuns, and F. Go (eds) *Towards Appropriate Tourism: The Case of Developing Countries,* Frankfurt: Peter Lang.

Wall, G. (1992) 'Tourism alternatives in an era of global climatic change', in V. Smith, and W. Eadington (eds) *Tourism Alternatives: Potentials and Problems in the Development of Tourism,* Philadelphia: University of Pennsylvania Press.

Weiler, B. and Hall, G. (1992) *Special Interest Tourism,* London: Belhaven Press.

Westlake, J. and Buhalis, D. (1998) 'Transportation', in C. Cooper, J. Fletcher, D. Gilbert, R. Shepherd, and S. Wanhill (eds) second edn, *Tourism: Principles and Practice,* London: Addison Wesley Longman.

Wickens, E. (1994) 'Consumption of the authentic: The Hedonistic tourist in Greece', in A. Seaton, R. Wood, P. Dieke and C. Jenkins (eds) *Tourism – The State of the Art: The Strathclyde Symposium,* England: Wiley.

World Tourism Organisation (1985) *The Role of Recreation Management in the Development of Active Holidays and Special Interest Tourism and Consequent Enrichment of the Holiday Experience,* Madrid: WTO.

World Tourism Organisation (1991) *Tourism to the 2000. Qualitative Aspects Affecting Global Growth: A Discussion Paper, Executive Summary,* Madrid: WTO.

5 The effects of globalisation on tourism promotion

A. V. Seaton and Philip Alford

Introduction

Although the concept of globalisation is part of the linguistic currency of contemporary business, it is neither a precise construct with an agreed meaning, nor one that can be empirically pinned down with ease. In the few books which address globalisation in tourism there has been a general failure to define it, except as a general recognition that international tourist numbers are increasing across the world (Theobald 1994), or as an implicit synonym for the expanding presence of multinational enterprises in the hospitality market (Go and Pine 1995), and airlines (Hanlon 1996).

Globalisation is a particularly problematic concept in the context of discussion of promotion since it can be argued that the term is as much a part of the rhetoric of promotion as a social process distinct from it. The word 'global' is now regularly used as a promotional auxiliary, a strap line attached to corporate advertising as in, 'LBG – Global leaders', 'Corporation X sponsors the global game' or brand names like 'Global Holidays' where, arguably, its use is commercial hyperbole for 'international' (it would be interesting, for example, to attempt to put precise definitions and dates on when hotel groups and airlines ceased to be international and became global).

In general terms, globalisation has been interpreted as the increasing expansion of international transactions, made possible by modern communications, and the ubiquitous productive and commercial reach of modern transnational corporations, developing and utilising those communications to sell their products in many countries. Examples are the car industry, the oil industry, financial services.

This general orientation has been augmented in a critical overview (Warhurst *et al.*, 1998) which characterises globalisation, not as a single process, but as a number of emerging tendencies, each of which may be evolving at a different pace, and each of which may be seen as *problematic* or one that *demands qualification*. These tendencies, with some of the problematics and qualifications attached to them (in parentheses), include:

- The development of a stage beyond internationalism (Is globalisation a *qualitative* stage beyond internationalisation or simply a *quantitative* extension? This debate is very similar to the one about the relationship of modernism to postmodernism).

- Emergence of transnational organisations operating across borders outside the control of states (To what extent do they operate across borders, and to what extent are they really beyond the control of states?).
- The emergence of stateless executives who think outside national boundaries (Do executives jettison their nationality and become transnational 'new' men and women?).
- Emergence of a global consumer culture in which tastes and consumption patterns across the world are converging (To what extent is there a global consumer culture and to what degree are indigenous tastes affected by it?).
- The evolution of a supply of standardised products across borders to meet global consumer tastes (What evidence is there that different countries are consuming identical products?).
- Homogenisation of managerial techniques and promotional methods used to sell goods (To what extent have managerial techniques been exported across frontiers?).
- The standardisation of promotional messages to sell global goods across frontiers (What evidence is there of homogeneous promotional campaigns running in several or many countries?).
- The emergence of competition from many countries (How much increase has there been in competition and what impact has it had?).
- Convergence of media through IT led innovations that now penetrate every corner of the globe (Is the world now a 'global village' which is everywhere penetrated by the same communication media?).
- An awareness of the ease of access and communication to other countries: a frame of mind which makes the world everyone's oyster (How international in outlook does everyone feel?).

Within a single chapter it is impossible to examine, in the context of tourism and tourism promotion, all these issues, let alone the questions and qualifications that can be set against them. Instead this analysis will concentrate on just five issues and explore them in the context of tourism promotion, within the primary, though not exclusive, contexts of destination and attraction marketing. The five issues relate to the following questions:

- To what extent are tourism organisations global?
- Is there a global consumer for tourism and tourism promotion?
- Is there global convergence and homogenisation of tourism promotional practices?
- What are the distinctive competitive implications of globalisation?
- To what extent is tourism promotion part of a global consumer culture?

The final section will present an SME case study, reporting research into a promotional consortium of small destination organisations in five EU countries, which suggests that globalisation and global response is not simply an issue for transnational enterprises.

Are tourism organisations global in operation?

In the paradigm industries of cars, oil and finance one of the identifiers of globalisation is locational, the corporate establishment of subsidiaries or satellite operations (productive, distributive or administrative) in countries outside the parent country, each of which may be supported by regional promotional activities. There is only a partial equivalence to this situation in tourism. At the top end of the market large hotel and restaurant organisations, operating as chains and franchises, may have a physical presence, through their product portfolios, in many countries, but the vast majority of them do not. In the 'HoReCa' sector of the European Union (covering hotels and other accommodation, restaurants, canteens and catering), 95.5 per cent of the enterprises in the 15 countries are very small (0–9 employees). Half of the persons employed in this sector work in very small businesses (1 to 9 employees); a further 15 per cent are 'one-man' enterprises. On average, four persons work in a HoReCa business in the EU. About 10 per cent of persons employed work in large enterprises of more than 250 employees. The HoReCa sector accounts for more than 1.3 million enterprises in the EU; this is about 8.5 per cent of the total number of enterprises (EC, DGXXIII, 1999). From this data it is clear that most European hospitality organisations are mainly small ones, domestically located with no physical presence abroad at all.

The story is similar in the travel agency sector where it is unusual for many agents, except Thomas Cook, to have their own dedicated operations abroad, though there may be strong linkages in travel agency chains like British Travel International (BTI) which now has fifty-five partners in sixty-seven countries and generated over US$22 billion in 1997. Destination agencies also have a limited locational presence internationally. The norm tends to be for the big countries to have satellite offices in a few, often capital, cities of high generating countries, and none elsewhere. International airlines also have a limited presence, usually confined to offices in or near the main hub airports from which they operate.

Finally, in the attractions sector there is little incidence of organisations with satellite outposts except in the case of the bigger theme parks, notably Disney.

In summary, in tourism there is a quite small incidence of transnational operation at the locational level. Tourism is not another Shell, Esso, Sony, Laura Ashley, Benetton, Body Shop, except in important but limited parts of the hotel and restaurant sector.

Client globalisation: Is there a global tourism consumer?

The absence of an overseas presence in a bricks-and-mortar sense does not, of course, mean that tourism organisations are not global in their markets. The fact that they do not have offices or premises in other countries does not mean that they do not draw customers from them. In some ways tourism has *always* been global in that, throughout this century and before, a significant proportion of tourists to many destinations, particularly European capitals, have come from abroad. National destinations sought and served international markets long before the car industry, the oil industry and the financial houses.

Superficial evidence suggests that the global market in tourist numbers and nationalities has never been greater. Impressionistic evidence of this internationalisation can be seen in:

- the guest profiles of restaurants and hotels;
- the multinational clientele for large airlines, many of which have entered into code sharing agreements to service global demand better;
- the babble of languages at major, cultural tourism attractions such as the Uffizi, the Prado, the Louvre, the National Gallery; and
- the visual evidence that many destinations attract visitors from every corner of the globe.

To this impressionistic data could be added a library of international tourist trends, and visitor profiles showing how individual countries and cities attract their share of the 500,000,000+ tourists thought to travel each year.

However, underneath the surface of these obvious developments all is not as it appears to the globalisation spotter. Though world trends suggest dynamic growth in international travel few individual destinations or attractions have either a large or even spread of international visitors. The typical pattern is for both destinations and attractions to derive their main demand from a few, often longstanding and traditional markets. This Pareto effect – the phenomenon whereby the majority of tourism generation for most countries is not equally spread across many, but concentrated among a few, nations (commonly four or less), holds good in the era of globalisation as it did in the past. This is not surprising since the main factors which commonly create tourist generation are:

- geographical proximity and ease of access, often with shared borders (which is why Austria attracts German tourists; Norway attracts Scandinavians; Switzerland attracts Germans; and why Finland's three main generating nations in 1998 were Sweden, Russia and Estonia);
- cultural homogeneity (Ireland attracts Americans; New Zealand attracts the English, the Irish and the Americans); and

- population size (which is why more Germans go to Austria than vice versa; more Americans go to Canada and so on).

For most destinations, therefore, the major generating areas tend to be the most populated countries with which they share borders and/or cultural (particularly linguistic) links. The Pareto effect means that the past is often the most reliable indicator of the main shape of the future. For example, between 1985 and 1995 main international visitors to the State of Victoria in Australia were consistently from four main country groups: New Zealand, Japan, UK/Ireland and USA/Canada (Victoria 1996) and between 1980 and 1997 Scotland's main generating countries were England, America, Germany and France. Destination planners should thus be cautious in diversion of funds and effort away from their old-established markets because of the much-vaunted glamour of new 'global' opportunities. In the 1990s a number of NTOs rushed to divert marketing and promotional funds into special activities designed to attack new markets perceived as lucrative, particularly the Japanese, with results that rarely justified the outlay. The importance of retaining one's 'heartland' franchise is the first necessity for tourism planners. If a destination begins to lose its prime market(s) it will almost certainly be difficult to replace it with remoter, more alien ones.

Another aspect of the Pareto effect in international tourism generation is what has been called 'the squeeze on the generating regions' (Seaton 1996b) which refers to the international trend for the same few countries to be prioritised as the main target markets by more and more destinations. A 1994 WTO study predicted that by the year 2010 over 40 per cent of all arrivals in Europe would come from five countries: France, Germany, Japan, UK and USA (WTO 1994: 70). Germany is an extreme instance of this trend; in arrivals Germany was the leading generating country in the world during the mid-1990s, only overtaken in receipts by the USA (WTO 1995a: 108). Ninety per cent of German trips were taken in Europe, of which 73 per cent were in Western and Southern Europe in 1994. Germany has historically been the leading market for Austria (around 60 per cent of arrivals between 1992 and 1994), Switzerland (around 30 per cent between 1992 and 1994), France (around 20 per cent of arrivals 1992–4), Norway (around 20 per cent of stays 1992–4) and it was the third major generating country for Ireland and UK. In addition, Germans were a priority market for many other European countries and international long-haul destinations. Never was so much owed by so many destinations to so few generating countries.

Given the importance of these few main generating countries there will be a need for destination planners to find more refined measures of targeting and promoting *within* a generating country, to establish a marketing competitive advantage. Market targeting based on a whole country will be inadequate to succeed in competitive markets where so many are targeting so few. This may involve the deployment of a whole range of additional segmentation measures – geodemographics, life style, etc. The British Tourist Authority in 1993 targeted their Japanese promotion specifically at Japanese women over the age of forty who constituted only 7 per cent of the total population (Seaton 1996b: 366).

In a marketplace in which many destinations are wooing a few main target markets it will increasingly be important to achieve competitive advantage in promotional materials. This may involve:

- testing materials (adverts, brochures, etc.) with the target consumers; and/or
- comparative testing of promotional materials against those produced by other destinations vying for the same market.

Tourist officers in both Ireland and Scotland have conducted research with target markets in Britain and mainland Europe to determine existing consumer perceptions of their destinations, and then used the findings to develop promotional materials (Dunlop 1997; Seaton and Hay 1998).

In conclusion then, globalisation for many destinations and attractions will continue to mean a primary promotional targeting of a few countries and regions which are likely to be not those most remote geographically but those closest. This law of proximity will operate with particular force in the global expansion of short-break travel.

Segmentation and cultural difference

The truth that even in the age of globalisation destination targeting largely revolves around a few important markets can be seen in the *de facto* market segmentation practices of destination agencies. If global consumers were evenly spread and becoming similar across the world (as one strand of the 'global consumer' thesis infers) then market targeting and segmentation would be getting less important. In fact the practices of national destination agencies suggest that the opposite is the case: the more advanced NTOs are putting more effort than ever before into identifying the differences, not the similarities, between their major markets. The real global trend in destination marketing is differentiation of efforts, not an assumption of one-world, cultural convergence.

The cultural differences factor has affected the tourism plans of several international destinations. In 1988 a 63 per cent increase in Japanese tourists to Australia over one year encouraged the Australian Tourism Industry Association and the Asian Studies Council to commission special studies of the Japanese tourist recognising that, 'misunderstandings and communication difficulties can often arise when those with different languages and cultural backgrounds meet' (Platt *et al.* 1988). The results were two guides on the Japanese tourist, the first on language, the second on culture and communication. In the mid-1990s Bord Failte, the Irish Tourist Board, produced and published detailed market profiles called, *Know Your Market* on their main generating countries – Britain, US, Germany and France (Bord Failte 1996a). Bord Failte also published a five-year study of trends in all their main generating countries (Bord Failte 1996b). The Scottish Tourist Board currently publishes a detailed international marketing plan which differentiates strategic activity by country (Scottish Tourist Board 1996). The Australian Tourist Commission produce regularly updated trend reports on their main markets.

Another instance of disaggregation rather than aggregation in market targeting is the efforts of several international destinations to identify and target specific behavioural segments within national tourist demand. In the US and Canada there has been a renaissance of interest in the cultural tourist as a differentiated behavioural category that began in the early 1990s (Tighe 1991; Ontario 1993) and was consolidated by a national study, carried out in the US by Travelscope, that produced a detailed profile of historic and cultural travellers revealing them to be older, better educated and more affluent than others (TIAA 1997). In 1995 Bord Failte in Ireland published an international study of the aging market (Bord Failte 1995). It is also awareness of the specific needs of particular prime markets, rather than global convergence, that makes France, the number one tourism nation in Europe, heavily promote battlefield attractions in northern France which appeal to its main markets there, the British and Belgians.

The global importance of the domestic market: the perils of global awareness

The major globalisation texts place emphasis upon the importance of exploiting global opportunity, invariably conceived as markets overseas. In tourism terms the most important global trend to emerge from research is, paradoxically, not the importance of overseas markets but the home one. In India domestic tourists outnumber foreign visitors by 45–1. In France and Spain around 75–80 per cent of the holidaying population does so at home. In Japan the importance of the domestic market may be suggested by the country's intensive monitoring of its own population movements at home and abroad (Mangiboyat 1996). Domestic tourists do not, of course, gladden the hearts of national destination agencies, since they do not create incoming revenue, or improve balance of trade figures, but they may create employment and contribute to local economies.

The attraction market illustrates the same lesson. An international study of tourism attractions (theme parks, museums, galleries, etc.) carried out in twelve countries in 1998 (Scottish Enterprise 1998) revealed that for all kinds of attraction the domestic market was the prime one, particularly out of season when domestic groups such as school parties, clubs, associations and senior citizen tours may constitute the main customers. Even for the paradigm example of the global hospitality organisation, McDonald's, its single largest source of profit is its home market (Ritzer 1996). Travel agents are almost wholly based on servicing domestic travellers, even if they sell them holidays abroad. Tour operators mainly service their own nationals in their own countries. This is why the overwhelming weight of tourism promotional activity takes place not in a global arena but in one's own back yard, even if hi-tech developments such as the Internet may increasingly be used to reach it.

An example of over-reacting to globalisation and misjudging the domestic customer is the British Airways corporate design catastrophe where the company relaunched itself with an expensive new livery which included individuated, multicultural tail fin designs, and the elimination of the national flag – all in the name of global adjustment. In just over a year BA had to reinstate the original livery due to the volume of protest (including a broadside from ex-prime minister Margaret Thatcher) from its main market, UK flyers who constitute 40 per cent of its business. The debacle happened because the company ignored the oldest marketing adage in the book – that the retention of one's existing customers must always be the first consideration when premeditating any major corporate change. The campaign was based on two other elementary marketing mistakes: the adoption of a new corporate positioning using the evidence that *employees*, rather than *customers*, liked it (as the late David Ogilvy observed forty years ago, employees always tire of company advertising, and favour change, before the public); that most comic of corporate mutations – the large, established enterprise that dreams of being young and trendy again, and dresses up as a young swinger, instead of understanding that, in an area like air travel, traditional design and the safe, 'square' image of a national carrier are forms of brand equity that should never be tossed aside. By entering into the world of tail-fin tarting up, and aircraft body-painting British Airways was effectively aligning itself with Lauda Airlines, Easy Jet and other recent small carriers (for whom visual gimmickry may be a necessary way of creating awareness).

In short, the most significant danger posed by globalisation in market terms, may be less one of under-response, than over-response. One of the least studied, and probably most lethal, factors in multinational enterprise (MNE) culture is the secret craving by executives, in the routinised and controlled world of the modern corporation, for novelty, which makes them highly susceptible to the latest fads offered by the managerial pundits and design gurus of the air-terminal book stall.

Globalisation and promotional planning?

The globalisation of best practice

In the last decade there has been considerable dissemination of know-how in the theory and practice of tourism promotion. There has been a growing sophistication of tourism practice within tourism organisations and within academia, particularly in the more developed world, that means that tourism planners now have access, if they are prepared to look for them, to published sources on best practice in many places. The activities of international tourism agencies such as the World Tourism Organisation, the OECD, the World Travel and Tourism Council, and the Pacific Asia Travel Association (PATA 1999) have not just promoted tourism as a world economic priority, but also published reports and statistics that have contributed to professional knowledge of tourism. The annual work of the WTO in gathering and codifying, for example, world tourist trends (WTO 1995a, 1995b) and, more recently, in publishing data on NTO budgets and promotional spending (1995b), has offered possibilities for detailed comparative analysis of destination agency performance which at least one destination, Scotland, has utilised (Seaton 1996b). Many individual national tourism organisations have also published their own studies, including: the guideline to tourism planning by the US Travel and Tourism Admini- stration (Missouri 1991); several studies by Bord Failte in Ireland which has been particularly well funded by EU grants (Bord Failte 1995, 1996a, 1996b), and by the UK Tourist Boards which have commissioned and published innovative promotional evaluation studies, e.g. into both the effects of advertising and the impact of film representations on Scotland's tourism (Seaton and Hay 1998). There has also been international expan- sion of tourism consultancies and the entry of general management consultancies into tourism. Above all, the last two decades have seen the rapid evolution of tourism as a university research field which has stimu- lated the publication of more than ten tourism journals, and the convening of many international conferences, through which both academics and practitioners have been able to exchange experiences and research.

The results of all this activity mean that it is now possible to distinguish some convergence in destination marketing and promotional practices in Europe and the developed world. Some of these may be briefly summarised:

1   There is a growing tendency in Europe, where many destination agen- cies have been under pressure to account for their impacts, to produce quantified marketing plans with measurable objectives and outcomes that would not have existed ten years ago.
2   Internationalisation of tourism has also meant that benchmarking, which will be discussed and illustrated later, is becoming increasingly important as a method of monitoring competition and learning from best international practices.
3   Another trend, visible in Spain, Ireland, Canada (Meis and Wilton 1998) and the UK, is the spread of *branding* as a destination concept. This broadly means promoting the image of a destination through advertis- ing that establishes a 'personality' for the destination, and under which other non-tourism products may be subsumed. It is still too early to judge

whether this *branding* of a country is a realistic aim, and whether, in net results, it is much more than a new name for destination imaging.

4   A major development focus in many destinations has been the promotion and development of IT programmes which have resulted, among other things, in: destination websites, CRS systems in TICs, and information kiosks. As a European Union conference on information technology in tourism noted, travel and tourism is now the leading E-commerce application on the Internet (EU 1999). A book-length study already exists of tourism organisations on the Internet in Europe (Marcussen 1999). The increasing speed and volume of this activity means that it is unlikely that any destination will be able to maintain competitive advantage in Internet usage and web site design for long, due to the ease of benchmarking Internet developments that will allow other destinations to catch up on the leaders, provided regular monitoring of destination sites in maintained.

5   Another spreading practice is the provision of multilingual promotional materials, targeted at prime markets, in hotels, destination guides, and in the signage of international destinations.

6   In many international cities tourism authorities now market city card schemes which offer tourists, for an all-inclusive fee, a combination of travel, free or reduced attraction admissions, and discounts at participating retail outlets. Versions of these city cards exist in places as different as Budapest, New York, Leeds, and Helsinki.

7   In several parts of the world systematic research has been conducted into the usage and impacts of Tourist Information Centres (Welcome Centres in the US). Reports of this work have been published, among others, in the USA, Scotland and Wales (Fesenmaier and Vogt 1991; Lennon and Mercer 1994; Wales Tourist Board 1995);

8   There is some evidence that destination agencies are coming to recognise the limitations of advertising as a medium for promoting destination images, given the fact that promotional budgets are usually inadequate to achieve a significant 'voice in the market place', and are recognising the importance of public relations campaigns, designed to maximise general media coverage in the hugely expanding output of travel and tourism journalism in the press and on TV throughout the West.

9   Finally, it is worth recognising the increasing importance of relationship marketing, efforts directed to retaining existing customers and get them to make return trips. This depends upon keeping updated databases of past customers and targeting special activities, deals and promotion at them to develop loyalty.

Are international campaigns feasible?

For more than thirty years a debate has intermittently raged in the advertising and marketing circles of MNEs about the feasibility and desirability of standardising advertising planning, so that common campaign strategies and programmes, rather than many, can be developed for most or all the countries in which the MNEs operated (Whitelock and Chung 1989). Discussed as long ago as the 1960s and early 1970s, before the word 'global' replaced 'international' (Elinder 1966; Fatt 1967; Buzzell 1968 and Britt 1974), the possibility was most forcefully presented in a classic article by Levitt more than a decade later. He argued that: '. . . global companies sell the same things the same way everywhere and different cultural preferences, national tastes and standards are vestiges of the past . . .' (Levitt 1983: 93).

The question of a common approach to international promotion by individual organisations is obviously one of great relevance to National Tourist Boards and tourist attractions with an international market. However, once again, caution must be a watchword. From what has been said earlier it should be apparent that it may not be necessary to promote too globally, but focus on the domestic market and a few prime overseas ones, rather than many.

Moreover, even when a destination agency wishes to target a number of generating countries homogenisation of promotional planning may be difficult or impossible. The task of destination promotion can be divided into two key functions:

- trip generation and destination choice; and
- trip influence.

The first takes place in the country of the tourist; the second in the tourist destination. The first is thus communication across frontiers; the second is communication at the host destination, the temporary home of the tourist. The latter may indeed lend itself to common promotional approaches since trip influence is mainly concerned with the distribution of promotional messages, once tourists have arrived at a destination, in locations where they are known to assemble, e.g. hotel rooms and lobbies, tourist information centres (TICs), attractions, etc. In such areas promotional materials may easily be standardised for the main generating countries, provided they are multilingual.

Promotional materials produced by destination agencies to generate trips are more difficult to homogenise because of cultural differences in target segments and also international media variations. Moreover, promotion is often adapted at the national, regional and local levels precisely to make the product – whether a brand, corporation or destination – seem to offer something specially attractive to the target market. Promotion is a key element that makes large, international organisations look sympathetic with the market. It is no accident that Coca-Cola, an American MNE with no tradition of support for the European national game, lived soccer, slept and breathed soccer promotionally in Europe during the French World Cup, or that McDonald's have employed the England soccer hero, Alan

Shearer, as the spearhead of their advertising in the UK for two years. Similarly many American multinationals operating in India suddenly developed a mania for cricket (a game with even less of a history in the US than soccer) at the time of the World Cricket Cup in June 1999.

In summary the opportunities for global campaigns are limited due to the need to reflect the interests and cultural preferences of different target markets. Though it may be just about possible for an MNE to sell oil or hi-tech electronics with the same corporate campaign in several countries, it is less possible for tourism organisations to promote London to the Japanese pleasure traveller in the same way they might do so to French businessmen. Though the strategic message of destination campaigns may be a common one (e.g. Spain's changing emphasis, in its main generating countries, to its cultural attractions, as well as it sun and beach products) there will normally be exceptional differences in the way in which the strategy is delivered to specific markets. Destination agencies will, in most cases, need to develop promotional 'horses for courses' that recognise cultural differences in different market segments.

The competitive impact of globalisation

The competitive implications of globalisation for every tourism organisation, destination agencies included, are enormous. Among major factors that have contributed to competitive pressure: the impact of a world shrinking beneath the withering, all pervasive reach of IT-driven communications; the falling price of long-haul travel; lowered political barriers to travel; and the now universal assumption among both large governments and small regional administrations, that tourism is a major development opportunity, so that just about every country and many of the regions within them want to attract tourism and, through the wonders of the old and new technologies, have the capability to promote themselves internationally – and thus attack the franchises of current players.

Evidence of this fundamentally changed competitive environment is so widespread that a few examples will serve to illustrate it. There is now virtually no town or region in the developed world that does not have a tourism strategy that would have not existed a generation ago. Lower air fares mean that New York and Florida are now competing for UK short-break tourists with traditional British cities and close European capitals. The fall of the Berlin Wall, and the opening up of Eastern Europe, has already produced bonanza growth for destinations like Prague and Budapest at the expense of traditionally strong destinations like Switzerland and Austria which declined between 1984 and 1994. By the mid-1990s the Eastern Mediterranean was growing faster than many established European destinations. On the Internet it is hardly possible to type in the name of any place, however small and distant, that will not produce a range of tourism promotional pages and products.

How do organisations cope in marketing and promotional terms with this expanded competition?

Benchmarking: a response to globalisation

One of the responses to the reality of greater competition is benchmarking. In the markets for physical goods, particularly fast moving consumer goods, where fierce competition has been a fact of life for half a century, benchmarking has been a routine marketing activity. The large corporations have constantly evaluated competitive market performance, researched their own products against rival brands, and monitored competitors' promotion. It has only been in the last decade that the possibilities and benefits of tourism benchmarking have been appreciated by tourism organisations. The primary aims of benchmarking are:

- to analyse an organisation's or destination's position in relation to others;
- to derive benefit from identifying best practice, state-of-the-art methods which may be used as models for future developments by the benchmark sponsor;
- to compete more effectively with the organisations benchmarked; and
- to derive benefit from negative instances. This is a much less recognised aspect of benchmarking, but learning theory suggests that best practice may be learned as much from avoiding the bad, as from pursuing the best models.

The Scottish Benchmarking Programme 1996–9: a state-of-the-art initiative

Between 1996 and 1999 the two leading national destination agencies in Scotland, the Scottish Tourist Board and Scottish Enterprise, commissioned a systematic programme of benchmarking from a research group of whom one of the present writers was director (Seaton 1996a; Scottish Enterprise 1998). The aims were to appraise: Scotland's overall destination performance; its attractions' performance; its human resource performance; and its activity holiday performance. Though the research programme used a variety of methodologies (literature searches, Internet scans, and primary data generated by direct observation and key informant interviews) it was based on a common philosophy of tourism benchmarking which had been formalised into a seven-point criterion of best practice. Some of the main criteria were:

- The organisations chosen for study must be *comparable ones to the sponsoring organisation and/or cutting-edge* ones likely to reveal, not just where the sponsoring organisation stands, but how it could be improved through 'creative borrowing' from developments elsewhere.

- The comparisons must be systematic. There is a danger that bench-marking can end up as impressionistic and anecdotal; it is easy to produce an almost infinite set of comparisons outside of systematic or coherent schema.
- Systematic comparisons can only be made through the choice and application of a *consistent number of evaluative dimensions*.
- The evaluative dimensions must be *relevant* ones which address the central *critical success factors* pertaining to an organisation's market.
- Benchmarking studies require both quantitative and qualitative appraisals.

Benchmarking Scotland's overall tourism performance 1984–94

The first of the benchmarking studies conducted within the Scottish programme was an analysis of Scotland's performance, over a ten-year period, compared to European tourism as a whole, and also compared to the performance of a number of chosen international destinations (Seaton 1996b). In order to conduct the analysis a number of performance dimensions were developed that included:

- trends in arrivals;
- trends in revenue;
- share of market for European tourism;
- positions in European arrivals/revenue league tables;
- numbers of tourists attracted per NTO dollar spent;
- number of tourists attracted per marketing/promotional dollar spent; and
- costs of attracting one tourist.

The study revealed quantitative evidence of Scotland's poor performance over a ten-year period, and suggested the reasons for it which enabled a programme of remedial action to be devised which included a five-year marketing plan, aimed to win back some of Scotland's lost franchise. One of the findings of the study was that a rough typology of destinations emerged which each was associated with different marketing and promotional developments (see Table 5.1).

Benchmarking international tourist attractions

Another study benchmarked international tourist attractions in order to develop ideas for improving the performance of attractions in Scotland, about 70 per cent of which required some kind of public subsidy.

The sample of attractions chosen for study were in ten destinations which comprised: those that were most nearly comparable to Scotland in

Table 5.1 Matrix of destination types and their market/marketing/promotional features

Size	Mature	Developing
Small	(Austria, Switzerland, Scotland) Medium tourist volume Decline or lower growth Declining market share Dependence on 1 to 3 markets	(New Zealand, Ireland, Norway) Low tourist volume Faster growth Small, but improving, market share High promotional/tourist ratios
Large	(France, UK, Spain) High tourist volume Lower growth Declining market share Market dependency more widely spread Low NTO/tourist ratios	(Eastern Europe, East Mediterranean) Medium tourist volume Fastest growth Low promotional cost per tourist

size and tourism products; and those which afford examples of 'cutting edge' attraction development and management. The ten international destinations were:

- **Similar destinations to Scotland.** Four destinations were chosen because they included some from among the countries defined by the World Tourism Organisation (WTO 1995a) as Northern Europe (Denmark, Ireland/Northern Ireland, Finland and Norway) in which Scotland is located. Two more were included since they comprised some of the countries grouped by WTO as Western Europe (Netherlands and Germany); and
- **Cutting-edge destinations.** The other four destinations were chosen as leading players in international tourism development. USA as the world's biggest tourism destination; France as Europe's leading destination; Singapore as a fast growing urban destination; and Canada because of the strength of its tourism administration over many years.

The analysis of developments in the international attractions focused on twenty main *evaluative dimensions* believed to embody the *critical success factors* in attraction management. They included the following:

- marketing and information system initiatives;
- promotional and publicity innovations;
- seasonality extension initiatives;
- consumer-directed technological innovations;
- repeat visit incentives;
- spend more on-site initiatives; and
- stay longer initiatives.

Each of the attractions was evaluated through a methodology that included: direct observation by researchers as 'mystery guests'; interviews with key informants from attractions' organisations, consultancies, academia, convention bureau, tourist boards and TICs; and content analysis of the promotional materials published by the attractions.

The results provided a wealth of data which formed the basis of a conference symposium, organised by the sponsors of the research, Scottish Enterprise (Scottish Enterprise 1998). Among the most interesting findings was in the technological innovations analysis where, against expectations, it was found that interactive innovations were less popular than the extensive use of video and film offered at many newer attraction; this offered a salutary warning against uncritical acceptance of technology-driven, rather than consumer-driven, innovations.

Benchmarking destination agency objectives

Benchmarking has also been used, not just to assess managerial tourism solutions, but also to compare strategic aims and objectives. A survey conducted in 1992 for the Department of Employment by the management consultancy Touche Ross, assessed the tourism priorities of twelve European Union tourism destinations. The national goals included :

- increase the spend of international visitors which gained twice as many maximum scores as any other objective;
- redistribute tourism geographically;
- economic objectives realised only middle level scores, in particular increasing the size of the industry and increasing employment; and
- redistribution of the tourism industry and reducing seasonality were important to France.

The study then showed how different objectives could be associated with different instruments, agencies and performance indicators. Below five of the objectives are linked to instruments, agencies and performance indicators.

Summary

Benchmarking is likely to grow internationally, where tourism organisations have the resources to fund it, as part of the way of managing in an environment of expanding competition. Organisations that identify, measure and compare are most likely to compete best in a global theatre.

However, one caveat should be made to this prospective expansion – access to data. For many years destination and attractions' management has been a very gentlemanly world, in which organisations have happily shared or made public their information (which is why the WTO, among

Table 5.2 Taxonomy of NTO tourism policy

Goal	Instrument	Agency	Performance indicator
Attract more foreign visitors	International destination marketing	National Tourism Offices overseas	1. Number of foreign tourists 2. Per cent growth in visitor numbers
Generate higher expenditure per visitor	Targeted campaigns aimed at higher income groups	NTO jointly with travel trade	1. Average expenditure per tourist 2. Total foreign exchange earnings from tourism
Reduce seasonality	Out of season marketing/promotion	NTO + private sector	Annual tourist/visitor monitors
Spread tourism geographically	Regional product development and marketing/promotion	Regional tourist agencies (ATBs/LECs)	Visitor analysis by area
Enhance the efficiency of NTO	NTO board and management policy	Ministry responsible for NTO	1. Enquiries handled per person 2. Press articles and other free publicity 3. Brochures distributed and other physical indicators 4. NTO cost-per-tourist/tourist £ 5. Arrivals per NTO $ spent

Adapted from Touche-Roche, 1992.

others, has so much data to collate in their reports). Such openness would be unknown in MNE operations, where all information is classified and guarded obsessively. In the past the reasons destination agencies and attractions' organisations, particularly public sector ones, have been so ready to share and publish information has been threefold: as a requirement to their members; for commercial reasons (research reports have been sold as commodities); and as good public relations evidence of their own performance. Research conferences have been staged to show what good research tourism organisations are doing. In short, dissemination has been part of the accountability process. In the context of global competition it may be argued that such openness is counter-productive. National and regional data on tourist arrivals and revenues, including the results of visitor studies, constitute valuable market information of great utility to competitors; it would be inconceivable, for example, for Mcdonald's or Disney to disclose detailed customer information in the way that destination agencies routinely do. Still less would they publish the results of promotional evaluation programmes into the impacts of advertising or TICs. It may well be that in the new millennium destination agencies will recognise the competitive value of the information they are currently making available, and be much less ready to 'hand it over on a plate' to any interested party.

Global symbiosis: the socio-cultural impact of tourism promotion

If, as we have argued earlier, tourism promotion by a single organisation is rarely exposed to a very wide international audience, except on Internet, there is another sense in which tourism as an aggregate network of representations, is becoming a major element of globalisation. Tourism with its the images of escape, fantasy, far-away exotica, dream-worlds and otherness, is now a ubiquitous presence internationally, not just in promotion specifically selling tourism, but as an element in many other kinds of promotion and publicity. It is thus part of a global consumption ethic:

> The new consumption ethic which was taken over by the advertising industry by the late 1920s celebrated living for the moment, hedonism, self-expression, the body beautiful, paganism, freedom from social obligations, the exotica of far-away places, the cultivation of style and the stylization of life.
>
> (Featherstone, 1991: 174)

Tourism is now synergistically associated, through visual imaging, with the selling of numerous other commodities such as cars and petrol (advertisements always reflect the leisure use, not the work use, of cars), pop music and fashion, and is thus an overt constituent of 'lifestyle', the 'aestheticisation of life' and the 'promotional culture' (Wernick 1991), that postmodern commentators have been diagnosing. Visual images of multinational promotion, particularly in metropolitan cities and urban conurbations, centralise the idea that travel is good, glamorous and high status.

> The central idea is that postmodern cities have become centres of consumption, play and entertainment, saturated with signs and images to the extent that anything can become represented, thematised and made an object of interest, an object of the tourist gaze.
>
> (Featherstone 1991: 185)

Dann has demonstrated how tourism is now explicitly linked, in life-style magazines like Conde Nast *Traveller*, to the sale of other luxury items which depict a placeless, global consumption ethic. These cross-product advertising presentations are anchored in the multi-referentiality of travel and tourism as a tie-concept that brings together an international hedonism of food, jewellery, clothes, cars, entertainment, etc. – Lafant's 'tourist neoculture' (Dann 1998).

However, this may not be completely new. Travel has always been a status badge, associated through the life styles of the classes that could afford it, with privileged access to other goods. Indeed, tourism choice may be seen as an important, hierarchically derived, form of taste discrimination to add to those other kinds of consumer decision and aesthetic choice that Bourdieu (1984) has so brilliantly shown to be related to social position, occupation and family status (Seaton 1999a). However, as we have argued earlier, this tendency towards a world consumer culture does not mean that a homogenised global market has emerged for tourism. As Warhurst, Nickson and Shaw have concluded:

> Even within a global culture of consumerism, consumer needs, wants and demands across the globe may continue to vary. In short, it is one thing to argue that the world's economic activity is becoming dominated by consumerism and market transactions and quite another to then insist that this market and its consumers are homogenized. The two phenomena should not be conflated: a domineering ideology of consumerism does not equate with a single world market.
>
> (Warhurst *et al.* 1998: 12)

Globalisation and the promotion of SMEs – the case of BookTownNet

The term 'globalisation' has, for a decade, mainly been associated with multinational organisations. However, rapid technological development means that, in the twenty-first century, global marketing strategies have become an option for small and medium-sized enterprises (SME). SMEs are the dominant force in the tourism industry. These small or very small enterprises are rarely able to afford big investments in marketing budgets or technological innovation necessary to keep pace with global competition and development. However, through strategic alliances and cross-industrial cooperation, based on the new global technologies, they may be able to achieve a critical promotional mass impossible as single organisations. An important rethinking of globalisation, particularly in large international organisations like the European Union, has been to focus on the role of IT in tourism SMEs. A case study of this is BookTownNet, an EU supported research project under evaluation at the University of Luton in the UK, which is designed to investigate the commercial and tourism impact of IT on an SME consortium network, based in peripheral regions of five countries.

Book town tourism and the EU BookTownNet research project

BookTownNet (BTN) is a consortium of small book towns in Europe developed to use information technology to achieve global reach. A book town is a small rural town or village, usually of between 500 and 1,500 inhabitants, in which second-hand and antiquarian bookshops are concentrated, thus creating a critical mass which allows the town to be branded as a unique destination concept, a book town. Bookshops are often complemented by artisan enterprises related to the production of books and writings, such as paper production, calligraphy, printing, book design and illustration and traditional bookbinding.

Book towns have been shown to create regeneration in peripheral, rural areas through the stimulation of new retail sectors and cultural tourism (Seaton 1996c, 1997, 1999b). They create economic purpose for older, unoccupied buildings, generally without extensive or expensive restoration, thus contributing to the conservation of architectural heritage. In tourism terms they provide a significant impetus to the local economy through the recovery of the local hotel and catering industry. Book towns have enormous potential as a form of branded destination attraction, appealing in general to upper-income, educated visitors, who constitute the market for second-hand and antiquarian books across Europe – and also speciality travel.

BookTownNet is being partly financed by the European Commission's information technology Directorate, DG XIII, and the project is aiming to develop a telematics-based organisational structure within and between the book towns in the form of Intranet and Internet services that will ensure that visitors have continually coordinated information about all the book towns.

The local partners in the BookTownNet include the following rural towns.

- Hay-on-Wye, UK;
- Bredevort, The Netherlands;
- Redu, Belgium;
- Montolieu, France; and
- Fjærland, Norway.

In each book town a user group of an average of 15 local SMEs (book dealers) is being established, thus involving about 75 local SMEs.

The application is being tested and demonstrated in normal operational activities in SMEs at each site and aims to meet actual needs for response time, level of interactivity, etc. Due to the fact that the SMEs involved in the BookTownNet project are very small, the cost associated with the establishment of each application site must be at a non-expensive level, typically within about 5,000–15,000 ECUs.

In addition to fostering commercial transactions and links between the towns through the Intranet, the external (Internet) arm of BTN aims to achieve the following objectives:

- Create a virtual book town organisation.
- Market the book town network as a pan-European tourist trail.
- Provide information about activities in the book towns to the global book-lovers community.

The project thus aims to target a specialist tourism niche market, book buyers and enthusiasts, supplied by SMEs, through a partnership in IT. In so doing BookTownNet may provide a blueprint for the development of other niche market destination products based on retail or cultural specialisation. A think-tank (EC DG XIII 1998) on future strategies recognised that Europe is facing competition from the new tourism destinations in Asia and America, profiting from lower labour costs and from the reduction of the cost of long-haul travel. The world market share of travel and tourism of the twelve European Union members has been declining since the mid-1990s. A shift in European tourism orientation is required if this trend is to change: the focus must extend beyond traditional mass-3s's (sex, sun and sand) tourism to exploit smaller, more focused products that capitalise on unique aspects of Europe's heritage, culture and environment. The problem is that such niche market products are likely to be delivered by SMEs with small resources for distribution and promotion, particularly in the rural areas and in the southern part of Europe. Information technology provides a mechanism for overcoming this. In a digital economy, it is possible for networks of small SME suppliers to use digital distribution channels, in addition to traditional ones, to access consumers and that implies investment in hardware, software and the skills necessary to exploit and manipulate new media. SMEs normally do not have the skills and economic resources required to exploit the opportunities of the digital economy. They need to disseminate their product information, knowing what distribution options are available to them. They need to be able to present their products to their target markets at minimal distribution cost through access to affordable electronic channels that are known to be used by their target markets without being forced to become computer and telecommunication experts. This requires access to reliable and accurate advice and expertise to make the right technological choice. They need support for innovation, for learning best practices, for improving the quality of services, in order to improve their competitiveness. Networking approaches are needed which increase their market relevance by maintaining the autonomy of the individual players. These approaches have to take into consideration the different types of suppliers (hotels, restaurants, but also organisers of cultural events) and their different, maybe conflicting, interests.

The working framework assumptions for BookTownNet are that:

* information is locally generated and stored, but is accessed and used on a global basis;
* content should be multilingual and have a multimedia format;
* commercial transactions require security and reliability;
* different platforms, architectures and data models which have to co-exist;
* data must be updated asynchronously and in real time; and
* different kinds of users must be catered for (user/interface issues) – access levels to Intranet.

The cluster approach

In many ways BTN is following the principles of the 'cluster approach'. The cluster concept was recommended by a European Tourism Think Tank (Nachira 1998) and supported in Directorate General XIII's subsequent papers relating to information society technologies (IST) for tourism. This approach focuses on using the knowledge and strengths of players with different but complementary areas of expertise or specialisation. Clusters could be:

* Geographical – several regions share the same knowledge, training, systems.
* Vertical – cooperation between different actors in the value chain.
* Technological – different players sharing the same base technological modules.
* Focal – concentration of action around a central actor, such as an education and training centre or a destination management organisation (DMO).
* Lateral – different business sectors sharing the same systems relating, for example, to transport, accommodation, sport, culture.

This approach could lead to the launch of a series of complementary projects, supporting the marketing of tourism services. The outcome would be a network of projects and applications, reusing the same components, methodologies and information. There is scope for extending the BTN project, incorporating the options outlined above. For example the project already involves five countries within the European Union and there are plans to develop an international Book Town Association and open up membership to other book towns in other countries. In addition to a geographic cluster this would also enable a focal cluster with the Association as the central actor. There is also scope to extend the project laterally with the development of interoperable systems involving local DMO organisations and private sector tourism interests. There is little point in BTN 'reinventing the wheel' and building in its own tourist information systems when there is the potential for mutual collaboration with tourism players.

The potential for technological and vertical cooperation and for transferring the BTN concept to other retail specialisations could provide a route to the future for the globalisation of tourism SMEs. The transfer and upgrading of the BTN knowledge base could result in improvements in the following areas:

* Technological modules – new or better functionality (i.e. multilingual, multimedia presentation, virtual reality, data mining, intelligent agents, improved user interface).
* Information content and services – new or more detailed types of data (e.g. concerning accommodation, multi-modal transport, cultural events, specific services).
* Regional networks – making existing systems more interoperable (i.e. with existing regional tourist information systems, with large transport information systems interconnecting air, road, bus information).

BookTownNet represents the European Union's commitment to encouraging SMEs to globalise through consortia utilising IT and, particularly, Internet to achieve a promotional critical mass, impossible for them individually. Other comparable projects, supported by the EU, have been running in Spain (Doncel 1999) and Italy (Falchero 1999) under the 'Enjoy Europe' programme which federates tourism organisations and destination agencies in comparatively little known parts of Europe in Internet marketing and promotional activities.

BookTownNet represents a number of unique innovations in IT projects supported by the European Union. First, the average size of bookshops in the book towns is smaller than that in any other SME project (the shops have on average just over one full time employee, and most have a turnover of less than 50,000 ECU per annum). Second, it is the first EU tourism research project focusing on the possibility of retail specialisation (through books) as a tourism focus in rural areas. Third, it represents the co-branding potential of place and private industry, by virtue of the fact that the existence of a critical mass of book retailers in the towns allows each to be marketed in terms of a *unique tourism destination concept*, a book town, that generates tourism as well as promoting retail sales of books. There are now more than a dozen book towns in Europe, and several rural towns concentrating their economic activity around retail specialisations that include: fashion; kitchen hardware and warehouse 'seconds' (goods that are slightly faulty and sold cheaply).

It is too early yet to evaluate the final impact of the BookTownNet research. At the time of writing the Intranet network had been inaugurated between book dealers in the five towns, while the promotional Internet site, designed to attract book buyers and promote the towns to visitors, had recorded over 1,200 external 'hits' in its first month. The case provides a timely reminder that through partnerships and consortia small destinations and SMEs may achieve international promotional synergy through IT, and that operational globalisation is not exclusively a matter for MNEs, as much of the literature suggests.

Conclusion

This chapter has offered a critique of globalisation in relation to tourism marketing and promotion. It has distinguished five different dimensions of globalisation (locational, consumer, promotional, competitive and cultural) and suggested that there are differences in the extent to which tourism is associated with each. The main effect of globalisation on tourism marketing has not been the visible homogenisation of promotional campaigns directed to world markets by tourism organisations, but in converging approaches to strategic planning, and the gradual emergence of state-of-the-art managerial techniques which include: refinements in market segmentation, use of consumer research, and the application of benchmarking to the tourism sector. One of the most significant effects of tourism promotion and publicity worldwide has been to contribute to the evolution of a global ethos of consumption, through the aggregate and cumulative dispersion of tourism and destination imagery, alongside other kinds of luxury commodities in media representations of placeless, postmodern life-styles. Finally, the chapter has identified the potential consequences of globalisation for SMEs as an antidote to more usual focus on globalisation and MNEs.

References

Bord Failte (1995) *'The overseas market for seniors'*, Dublin: Bord Failte.
Bord Failte (1996a) *Know Your Market: United States; Know Your Market: Great Britain; Know Your Market: Germany; Know Your Market: France*, Dublin: Bord Failte.
Bord Failte (1996b) *Markets 1991–1995*, Dublin: Bord Failte.
Bourdieu, P. (1984) *Distinction*, London: Kegan Paul.
Britt, S. H. (1974) 'Standardizing marketing for the international market', *Columbia Journal of World Business*, 9: 34–45.
Buzzell, R. D. (1968) 'Can you standardise multinational marketing?', *Harvard Business Review*, Nov-Dec: 102–13.
Dann, G. (1998) 'The pomo promo of tourism', *Tourism, Culture and Communication* 1: 1–16.
Doncel, A. (1999) 'Promoting cooperation amongst European tourism regions through interoperable systems', paper given to *Tourism in the Information Society Conference*, 12 November, Brussels.
Dunlop, A. (1997) *Consumer Journeys*, Dublin: Bord Failte.

Economic Research Associates (1998) *Report on International Best Practice Attractions*, Glasgow: Scottish Enterprise.

Elinder, E (1966): 'How international can European advertising be?', *Journal of Marketing* 29: 7–11.

European Commission Directorate-General XIII (1998) *Conference Brief, Information Society Related Research RTD & Tourism*, Brussels.

European Commission Directorate-General XXIII (1999) *Conference Brief, Tourism in the Information Society Conference*, 12 November, Brussels.

Falchero, C. (1999) 'Enjoy Europe – Teletourisme initiative – experience of the regions of Southern Italy: Thenet project', paper given, *Tourism in the Information Society Conference*, 12 November, Brussels.

Fatt, A. C. (1967) 'The danger of local "international advertising"', *Journal of Marketing* 31: 60–2.

Featherstone, M. (1991) *Consumer Culture and Postmodernism*, London: Sage.

Fesenmaier, D. R. and Vogt, C. A. (1991) 'Exploratory analysis of information use at Indiana State Welcome Centers', in *Tourism: Building Credibility for a Credible Industry*, Proceedings of 22nd Annual Conference of the Travel and Tourism Research Association, pp. 111–22, Bureau of Economic and Business Research, Graduate School of Business, Utah.

Go, F., Pine, R. and Hanlon, P. (1995) *Globalization Strategy in the Hotel Industry*, London: Routledge.

Hanlon, P. (1996) *Global airlines: Competition in a Transnational Industry*, Oxford: Butterworth-Heinemann.

Lennon, J. J. and Mercer, A. (1994) 'Service quality in practice: Customer service in Scotland's Tourist Information Centres', *International Journal of Hospitality Management* 13(2): 231–49.

Levitt, T. (1983) 'The globalisation of markets', *Harvard Business Review*, May-June: 92–102.

Mangiboyat, A. (ed.) (1996) *Japan Travel Blue Book 95/96: A Comprehensive Guide to Japan's Outbound Market*, Tokyo: Travel Journal Inc/J. T. Moritani.

Marcussen, C. H. (1999) *Internet Distribution of European Travel and Tourism Services – The Market, Transportation, Accommodation and Package Tours'*, Denmark: Research Centre, Bornholm.

Meis, S. and Wilton, D. (1998) 'Assessing the economic outcomes of branding Canada: Applications, results and implications of Canadian tourism satellite account', paper delivered, *29th Conference of the Travel and Tourism Research Association,* June 1998.

Missouri, University of (1991) third edn *Tourism USA: Guidelines for Tourism Development*, Washington: US Travel and Tourism Administration.

Nachira, F. (1998) 'EC main emphasis in research and development in the field for the next 5 years', *Proceedings of the ENTER conference IT and the Dynamic Tourism Marketplace: New Partnerships, New Competition through Virtual Enterprise* (http://www.ifitt.org/enter), Istanbul.

Ontario, Government of (1993) *The Cultural Tourism Handbook*, Ontario.

Pacific Asia Travel Association (1999) *Issues and Trends 1998*, PATA.

Platt, A., McGowan, V., Todhunter, M. and Chalmers, N. (1988) *Japanese for the Tourist Industry: Culture and Communication*, Melbourne: Hospitality Press.

Ritzer, G. (1996) 'The McDonaldization thesis: Is expansion inevitable?', *International Sociology* 11(3): 291–308.

Scottish Enterprise (1998) *New Horizons: International Benchmarking and Best Practice for Visitor Attractions*, Glasgow.

Scottish Tourist Board (1996) *International Marketing Plan 1995/1996*, Edinburgh: STB.

Seaton, A. V. (1996a) 'Destination marketing' in, A. V. Seaton, and M. M. Bennett, (eds) *The Marketing of Tourism Products*, pp. 35–76, London: International Thomson Business Press.

Seaton, A. V. (1996b) *The Comparative Evaluation of Tourism Destination Performance: Scotland and European Tourism 1985–1994*, Edinburgh: Scottish Tourist Board.

Seaton, A. V. (1996c) 'Hay on Wye, the mouse that roared: Book towns and rural tourism', *Tourism Management*, 17(4): 379–82.

Seaton, A. V. (1997) 'The book towns phenomenon: Retail specialisation and rural tourism', *Insights*, British Tourist Authority, July 1997: D1-D5.

Seaton, A. V. and Hay, B. (1998) 'The marketing of Scotland as a tourist destination, 1985–1996', in, R. MacLellan and R. Smith, *Tourism in Scotland*, pp. 209–40, London: International Thomson Business Press.

Seaton, A. V. (1999a) 'Why do people travel? Introductory perspectives on tourist behaviour', in L. Pender, '*Marketing Management for Travel and Tourism*', pp. 174–214, Cheltenham: Stanley Thornes.

Seaton, A. V. (1999b) 'Book town tourism and rural development in peripheral Europe', *International Journal of Tourism Research* 1(5): 389–99.

Theobald, W. (1994) *Global Tourism*, Oxford: Butterworth-Heinemann.

Tighe, A. (1991) '*Research on cultural tourism in the United States*', TTRA 22nd Annual Conference, 9–13 June, California.

Travel Industry Association of America (1997) '*Profile of travelers who participate in historic and cultural activities*', Washington: Travelscope Survey.

University of Missouri-Columbia (1991, 3rd edition): *Tourism USA: Guidelines for Tourism Development*, US Department of Commerce.

Victoria State (1996) *Tourism Victoria Research Update*, Melbourne, Australia.

Wales Tourist Board (1991) *1990 Marketing Campaign Tracking Study*, Cardiff.

Wales Tourist Board (1995) *Tourist Information Centre Users*, January.

Warhurst, C., Nickson, D. and Shaw, E. (1998) 'Globalization under question: Political, economic and cultural considerations', in T. Scanduri and M. Serapio, *Research in International Business and International Relations*, Stamford: Jai Press.

Wernick, A. (1991) *Promotional Culture*, London: Sage.

Whitelock, J. and Chung, D. (1989) 'Cross cultural advertising: An empirical study', *International Journal of Advertising* 8: 291–310.

World Tourism Organisation (1994) *Global Tourism Forecasts to the Year 2000*, Madrid: WTO.

World Tourism Organisation (1995a) *Market Trends in Europe1994*, Madrid: World Tourism Organisation.

World Tourism Organisation (1995b) *Budgets and Marketing Plans of National Tourism Organisations*, Madrid: WTO. (Landmark opportunity for benchmarking which no other industry would voluntarily provide; specifies target markets.)

6 Globalisation of tourism demand, global distribution systems and marketing

Norbert Vanhove

Introduction

Globalisation is a generic term comprising three basic elements:

- First, there is the geographical element. The term covers intraregional and interregional travel. In French we speak of the 'mondialisation' of tourism or its expansion to global scale.
- Others see globalisation in terms of convergence in world tastes, product preferences and lifestyles, which leads to growing standardisation and market homogenisation. In other words, there is a trend towards similar customer preferences worldwide.
- A third basic element is the existence of internationally similar practices, such as distribution systems, marketing practice, product development, etc. We find this interpretation in Go's contribution (1996) at the AIEST Congress in Rotorua: 'Briefly stated, globalisation (that is, the social act of reproduction of a global culture), involves the process of organising or establishing structures and commercial activity by institutions on a world-wide scale'. The first and second characteristics are more demand oriented; the third is supply oriented.

Within the framework of this contribution, globalisation covers all three aspects with an emphasis on long-haul demand. To analyse the globalisation of the market, we must examine four major aspects. In the first section we will attempt to demonstrate globalisation in tourism. A number of data are illustrative in this respect. Furthermore, we will examine what the future might hold for this phenomenon. Globalisation cannot be explained only in terms of traditional factors such as aviation and air fares, tour operating, exchange rates, etc. Can we find a further explanation of globalisation in a number of macro-economic or social trends and specific trends in tourism demand? Both sets of trends have an impact on globalisation. This is the subject of the second section. Globalisation cannot be dealt with without reference to the role of the rapid evolution of information technology – computer reservation systems, specific reservation systems, Internet, etc. – during the last two decades. This is the topic of the third section. Globalisation (and global distribution systems, the Internet included) is confronting most destinations with management and marketing questions. What is the impact of it on marketing strategy? What is the impact of it on the distribution channels? These very difficult and complex questions are the topic of the fourth and last section.

The globalisation of tourism demand

General evolution

In 1997, international tourist arrivals – according to WTO estimates – increased to 613 million, whilst international tourism receipts reached US$444 billion. International tourist arrivals have witnessed an uninter-rupted growth since 1950, although the growth rate is declining (see Table 6.1).

Over several decades, growth rates have proved to be resilient – at least on a global scale – to factors such as economic recession, variable exchange rates, terrorist activities and political unrest in many parts of the world. Of course, there has been a levelling off of movements due to specific circumstances and the steady growth has slowed down or even turned into a small decline on several occasions (1967–8, 1973–4, early-1980s and early-1990s).

It is also important to underline that international tourism receipts at a global level have increased faster than other important export sectors. According to WTO-GATT-UNCTAD sources, the average annual growth in current terms of tourism was 9.6 per cent for tourism, 7.5 per cent for commercial services and 5.5 per cent for merchandise exports.

Origin-destination matrix of international tourist arrivals

Based on information provided by the WTO in Table 6.2 we compare the origin-destination matrix of international tourist arrivals 1980 and 1990. A matrix based on more up-to-date data is not available.

Table 6.1 International tourist arrivals

Year	Growth rate (%)
1950–60	10.6
1960–70	9.1
1970–80	5.5
1980–90	4.8
1990–97	4.3

Table 6.2 Origin-destination matrix of international tourist arrivals, 1980 and 1990 (in %)

Origin/destination	Europe	Americas	East Asia and Pacific	Africa	Middle East	South Asia	World
1980							
Europe	86.6	10.8	14.4	61.1	22.7	49.3	63.4
Americas	8.8	84.2	11.7	4.9	5.6	10.0	25.0
East Asia and Pacific	2.3	4.6	70.4	1.6	2.4	12.2	7.7
Africa	1.5	0.4	0.4	30.9	2.6	1.4	1.8
Middle East	0.6	0.0	0.5	1.0	63.2	2.5	1.6
South Asia	0.3	0.1	2.5	0.4	3.6	24.6	0.6
Total	100.0	100.0	100.0	100.0	100.0	100.0	100.0
1990							
Europe	86.1	12.9	7.9	40.2	17.2	40.8	56.3
Americas	8.2	79.9	5.5	3.5	2.7	8.6	20.7
East Asia and Pacific	2.9	6.5	84.6	1.7	16.9	13.8	17.7
Africa	1.9	0.3	0.3	45.3	7.4	2.1	2.6
Middle East	0.6	0.2	0.3	9.2	49.2	4.9	1.9
South Asia	0.3	0.3	1.5	0.3	6.5	29.7	0.8
Total	100.0	100.0	100.0	100.0	100.0	100.0	100.0

Source: Information provided by the WTO.
WTO, Global tourism forecasts to the year 2000 and beyond, Madrid, 1995.

Table 6.3 Intraregional flows during the last decade

	1980	1990
Europe	86.6	86.1
Americas	84.2	79.9
East Asia and Pacific	70.4	84.6
Africa	30.9	45.2
Middle East	63.2	49.2
South Asia	24.6	29.7

One expects a decrease of intraregional flows during the last decade. To a certain extent this is the case. We must take into account that Europe and the Americas generate 77 per cent of all arrivals (see Table 6.3).

Taking into account the growing economic importance of East Asia, one can conclude that Table 6.2 indicates a growing globalisation process in tourism demand.

The growth of international tourism during the last decades of the twentieth century was dominated by the price factor. The tour operators' ability to deliver previously inaccessible warm-climate destinations at prices afford-able to a substantial part of the population of especially Western Europe and North America, played an important role in this growth. These low prices depend on low air fares, the negotiating power of the tour operators towards the hoteliers (made possible by low wage levels) and favourable exchange rates. An airfare Brussels–New York (excursion tariff) was much cheaper in 1998 than it was in the mid-1960s.

The importance of the price factor in the destination pattern of international travel flows can be illustrated by the recent evolution in international travel flows to Spain, a first generation destination for international tourism. Spain (mainly the Spanish seaside resorts) witnessed a noticeable increase during the 1980s, but the evolution and the prospects seemed gloomy at the turn of the decade. Most observers referred to the excesses of standardised, unplanned tourism development in most seaside resorts in the country. Contrary to the expectations at that time, Spain – due to the low air fares and especially the successive devaluations of the peseta – witnessed a 1990s boom in tourist arrivals, albeit at low prices. It is obvious that the globalisation of products and demand is moving forward, especially in the mass market for beach resorts.

The combination of low air fares (made possible by overcapacity, a low dollar rate up to the beginning of 1997 and the opening of long-haul destinations for charter flights) and low hotel rates in most long-haul destinations (made possible by low wage rates and/or attractive exchange rates) created the basis for the increasing popularity of long-haul travel. Offering an exotic flavour as an incentive in the competition with the first generation destinations, a product/price combination was created which could compete with a part of the traditional short- and medium-haul demand. This gave rise to booming destinations like Thailand, Indonesia, the Dominican Republic and others.

Prospects for international tourism

Two major sources are available on forecasting the future of international tourism: the Economist Intelligence Unit (EIU) and the World Tourism Organisation (WTO). The EIU produces forecasts on international tourism at irregular intervals. The latest publication (Edwards 1992) is based on quantitative assumptions and projection techniques. The global EIU forecasts for the period 1995–2005 are summarised in Table 6.4.

Table 6.4 Global EIU forecasts 1995–2005

	1995–2000	*2000–2005*
Short- or medium-haul trips	4.8	3.6
Long-haul trips	6.8	6.1
All trips	5.1	4.0
All nights	4.7	3.9

Global growth rates of the total number of trips are estimated at between 4 and 5 per cent for the decade, with a somewhat higher rate for the period 1995–2000 than for the period 2000–2005. The EIU expected a further decrease in the length of stay of the trips, especially in the period 1995–2000.

The forecast results for the period 1995–2005 are very similar to those achieved in the 1980s. A relatively new element in the EIU forecasts is that travel from a number of major origin countries, especially in Europe, is expected to be increasingly constrained by ceiling limitations to travel. A major feature, especially since about 1985, has been the significant increase of long-haul travel. The decline in air fares (in real terms), improved air access and the rising demand for long-haul holidays versus traditional short- and medium-haul destinations (provided the additional holiday cost involved remains reasonable) were major forces in this development. The EIU forecast a further increase of long haul from 13.3 per cent in 1989 to 15.4 per cent in 2000 at world level (market share of total international trips).

Table 6.5 summarises the latest WTO (1998) forecasts (1995–2020) of international tourist arrivals per destination region. A distinction has been made between interregional and intraregional travel. Although we have some problems with forecasts for such a long period, the results sustain our thesis of globalisation.

On a world scale, long-haul travel is predicted to increase its market share from 18 per cent in 1995 up to 24 per cent in 2020, which is globally in line with the EIU forecast share of long-haul travel.

The evolution of the intraregional/long-haul split of international tourist arrivals is very unequal per receiving region. Table 6.6 gives the share (in per cent) of long-haul arrivals per receiving region in 1995 and 2020.

Table 6.5 Forecasts on inbound tourism by region, 1995–2020

Receiving regions	International tourist arrivals (millions)				Average annual growth rates (%)			
	1995	2000	2010	2020	1995-2000	2000-2010	2010-2020	1995-2020
Europe	334	386	526	717	3.0	3.2	3.1	3.1
East Asia/Pacific	81	105	231	438	5.2	8.2	6.8	7.0
Americas	110	131	195	284	3.6	4.0	3.8	3.8
Africa	20	26	46	75	5.4	5.7	5.1	5.5
Middle East	14	19	37	69	6.4	7.1	6.5	6.7
South Asia	4	6	11	19	5.6	6.8	5.8	6.2
Total	563	673	1,046	1,602	3.6	4.5	4.4	4.3

Source: WTO (1998); *Tourism: 2020 Vision.*

Table 6.6 Long-haul arrivals per region

	1995 (%)	2020 (%)
Africa	42	36
Americas	23	38
East Asia and Pacific	21	17
Europe	12	15
Middle East	58	63
South Asia	76	86

By and large more originating and receiving countries will be involved in this process of globalisation.

Globalisation within the framework of trends in tourism supply and demand

A number of fundamental factors were given in the preceding section which explain not only the high growth rate of international tourism but also the globalisation of tourism demand. In this section we pay more attention to a number of trends in tourism demand and tourism supply as exponents of, or with a possible impact on, globalisation.

In the 1990s, many authors and organisations dealt with changes in tourism demand: Opaschowski (1985); Krippendorf (1987); WTO (1990); Poon (1993); Vanhove (1993); Middleton (1994). Trends in tourism demand can be divided into three main groups. These are:

- changing macro socio-economic factors;
- trends in tourism demand; and
- trends on the supply side.

Macro socio-economic trends

Higher incomes and income elasticity of tourism demand

Income and mainly discretionary income is after all the dominant factor in explaining tourism demand. In most parts of the world national income has increased during the last two decades. There are no indications that this trend will change in the medium term, although the growth rate is likely to be very unequal between countries. The factor that is important for us is the high income elasticity of tourism demand (Martin and Witt 1989). It is impossible to put forward an average income elasticity, but 1.5 to 2 is a very acceptable figure for long-haul tourism demand. This means that, with an annual GNP growth rate of 3 per cent, tourism demand is increasing by 4.5 to 6 per cent a year.

Demographic changes

It is a well-known fact that population in developed countries is stagnating and ageing. This is not only changing demographic structures but also the composition of the family. What will be the consequences for tourism? There are many. We limit ourselves here to one aspect. Senior citizens and/or empty nest families become more important. They have above-average financial resources and travel frequently. Many of them are confronted with earlier retirement. They seek opportunities for personal development.

Older tourists have a specific motivation. As these elderly people have considerable travel experience already and their physical condition is good, ageing will not provide a lowering of holiday participation in the medium term. Poon (1993) describes the implications of demographic trends as follows:

> The ageing of the population in the developed countries goes hand in hand with the change in values described above. In general, consumers will be more mature in planning their holidays, and they will not be so easily satisfied as were their former counterparts. They will listen more carefully to their inner selves and to their pocketbooks to determine what they really need and can afford. To the most important groups, travel will no longer offer the excitement of the new. Instead they will be looking for the different and for the enjoyable. They will want to be surer and surer that their holidays will be both pleasurable and worthwhile.

More attention to the environment and the growing importance of ecotourism

The environment is becoming one of the main concerns of our society. Everyone is aware of this. Tourist resorts, however, are sometimes much less worried about it. The holidaymaker, though, wants to be released from the negative aspects of his environment. The landscape in all its diversity is the basic element, the main ingredient, the raw material of tourism. It is the very essence of tourism and constitutes its driving force. From the tourist's point of view, the attraction of the landscape consists in its diversity and the contrast it offers with his daily environment. It is the degree of this contrast that will determine the attraction for tourists. There is therefore a search for the real and authentic. European travellers want nature to be more prominent in their vacations (Poon 1993). A direct consequence of this movement is a boosting of ecotourism in Europe and the USA. We agree, ecotourism can mean a number of things and is often old wine in a new barrel. It is generally used to describe tourism activities which are conducted in harmony with nature, as opposed to more traditional mass tourism activities. According to Hawkins (1994), ecotourism offers opportunities for many developing countries (Moore and Cater 1993).

Certain resorts or regions are running the risk of or have already fallen victim to overdevelopment and subsequent rejection by tourists. This is a second consequence of this movement.

Maslow's need theory and new travel motivations

To what extent is there a relationship between the globalisation of tourism demand and Maslow's need theory? This is a very difficult question to answer. The travel literature indicates that travel motivations can fit into Maslow's hierarchy of needs model. Mill (1992) adds two other words 'to know and understand' and 'aesthetics' to the original list. This might explain the success of culture tourism and thematic tourism (e.g. Orlando in Florida). They probably make sense and are related to the preceding section and to our subject in general.

Changed values

In some of the preceding sections, there are indications of changed values such as a growing consciousness of nature and a search for the real and authentic. There is much more, however. Individualisation is gaining in importance in our society. For the tourism sector this implies that 'a' consumer no longer exists. We speak of 'this' consumer. The consequence is evident. The tourist product has to be adapted to 'this' consumer. We can refer to a set of building blocks where the parts are assembled differently according to the personality of the consumer. Poon (1993) indicates two other changed values. She is convinced there are growing signs today that the fashion for sun is beginning to fade. The sun is no longer sufficient to build a viable and sustainable tourism industry. Destinations have to begin to offer 'sun-plus' holidays such as *sun plus spas, plus nature, plus fishing*. The second additional changed value is the search for the different. 'The new traveller wants to experience the inexperienced, see the unexpected, gain impressions of new cultures and a new horizon.'

Changing lifestyles

Krippendorf (1987) argues that society has moved through three phases between the industrial era and the present. In the industrial era, tourists were drawn from a population that 'lived to work'. Since about two decades people 'work to live'. Travel motivations changed from 'to recover, to rest, to have no problems' to 'to experience something different, to have fun, to have a change, to be active'. But today there is a third phase which has been described as the desire to experience 'the new unity of everyday life'. In this phase, the polarity of work and leisure has been reduced.

The vacation motivations of this group include:

- to broaden their horizon;
- to learn something new;
- to encourage introspection and communication with other people;
- to discover the simpler things in life and nature;
- to foster creativity, open-mindedness; and
- to experiment, take personal risks.

Some speak of a global lifestyle. The latter is a consequence of improved educational levels and the revolution in communication technology. The world is becoming increasingly cosmopolitan with all its people influencing each other. This globalisation process has many impacts on and implications for tourism. The most fundamental of these is the fact that increased travel is both a reason for, and the result of, the global lifestyle (WTO 1990).

Tourism demand modifies quickly and is no longer always coherent. The fast modifications reveal themselves in the Mediterranean region. The market share of the Mediterranean countries in the European market varies from year to year. But demand is not always coherent either. We may expect that a chairman of a large company takes his holiday in a five-star hotel, in an exotic destination. But is that really so? Not any more. That same chairman may also ask for a bicycle holiday, which will take him from one place to another. And this proves that we no longer deal with 'the' consumer but 'this' consumer. This is a clear illustration of the so-called hybrid consumer. 'Healthy lifestyle is another trend . . . The practice of healthy living reflects itself in holiday and tourism lifestyles and is responsible for the proliferation of health spas, saunas, fitness centres, "fat farms", gyms, massage . . . and other such additions to many hotels and resorts' (Poon 1993: 130–3).

The European Union

Europe represents a very high percentage of international arrivals. Europe has many tourist attractions to offer. However, not all of these arrivals are long-haul trips. Further, European integration has two consequences of great importance for European tourism: frontier formality withdrawal and the Euro. The Schengen agreement aims at abolishing frontier controls between European countries allowing people to travel without passports or border checks. It came into force in 1995, and in the late 1990s nine countries were included in the Schengen agreement. Other countries are negotiating for membership. Not only Europeans but also non-Europeans can move much easier between the member states. One visa gives access to all Schengen countries.

There is a widespread conviction that the Euro will be good for tourism. The customer will have three benefits: (a) simplicity, (b) transparency, and (c) economy. As such, the Euro will stimulate the intra-European tourism flows.

Trends in tourist demand

In this section we deal with a number of trends within tourism demand. There is no doubt that these trends are partly related to the macro socio-economic factors which we put forward in the preceding section. Here we distinguish six such trends.

Fragmentation of annual holidays

In many, if not all, European countries the main holiday has become shorter, but at the same time many holidaymakers take two or more holidays a year. Furthermore, many holidaymakers take short holidays (one to three nights) in between. This fragmentation is the main reason for shorter holidays. But we are also confronted with time-efficient product development. In a period of more leisure time, an increasing proportion of the population finds less time to travel and is in a 'time poor–money rich' situation. The outcome of this trend is an increasing number of products that offer the tourist the maximum excitement in a minimum of time (WTO 1998).

More independent tourists as opposed to mass tourism

In the early 1990s Middleton (1991) posed the question 'Whither the package tour'. In his contribution the author demonstrated that the UK outbound market for traditional forms of air-inclusive package holidays reached maturity in the mid-1980s and has been in decline since. The evidence also shows that profitable future growth for both tour operators and resorts lies in developing new forms of IT products. Mature tourists look for the core advantages of packages (price, reliability, etc.) without the traditional requirement and stigma of travelling and staying together in highly visible groups in chartered flights and hotels. The following statement by Middleton is very important: 'So far as possible, customers should not be aware of being labelled and identified as tour groups. They will, of course, continue to be bound by the specific times and product options which are the basis on which bulk purchase prices can be obtained from suppliers.'

Poon (1993) holds similar ideas. For her, new tourism exists if and where the following six conditions hold:

1   The holiday is flexible and can be purchased at prices that are competitive with mass-produced holidays (cruises v. land-based holidays).
2   Production of travel and tourism-related services is not dominated by

scale economies alone. Tailor-made services will be produced while still taking advantages of scale economies where they apply (yield management).

3 Production is increasingly driven by the requirements of consumers.

4 The holiday is marketed to individuals with different needs, incomes, time constraints and travel interests. Mass marketing is no longer the dominant paradigm.

5 The holiday is consumed on a large scale by tourists who are more experienced travellers, more educated, more destination oriented, more independent, more flexible and more green.

6 Consumers look at the environment and culture of the destinations they visit as a key part of the holiday experience.

New types of holidays and special interest

It is not surprising to learn that, in the light of changing values and lifestyles, new types of holidays and recreation arise, under the slogan: 'to experience something during the holiday'. The holidaymaker wants to enjoy his holidays thoroughly, thus resulting in an increasing interest in holidays devoted to sports or other hobbies, urban tourism, natural, health, cultural, adventure and language holidays, etc. We speak of 'targeted product market development' (especially theme-based) oriented to one or a combination of the three Es: entertainment, excitement and education.

There is a net polarisation of tourist tastes: the comfort-based and the adventure-oriented. With respect to the latter there is a trend to travel to high places (mountain), under water (tourist submarines) and to the ends of the earth (e.g. Antarctic Peninsula).

Next to the desire to experience something, an increasing demand for animation also arises. Many tourists need to be incited to discover their own capacities and to develop them within the framework of holidays with a real content. That is why tourist 'animation' is very important. This can take different forms: movement, social life, creative activities, education and discovery, self-discovery, quietness and adventure.

More quality conscious

Another trend can be summarised as the search for 'more quality'. This is in agreement with the trend towards new forms of holidays. More quality does not mean more luxury, but what the Germans call 'Erlebnistiefe', holidays with a meaning. This is what Martin and Mason (1987: 112) clearly emphasise:

> Different types of people will make new and varied demands on the tourism products. For example: older people will look for better quality

and more secure surroundings while single people seek more social contact through tourism. In addition, there will be accompanying shifts in what people want out of their lives, which will affect their choices as tourists. Likely changes in attitude are:

- the development of greater awareness of the range of tourism choices available and demands for a higher standard of service and value for money from tourism operators; and
- growing concern about the quality of the tourism experience in all senses, including the nature of the facilities used, the state of the environment visited, and the health-enhancing (or detrimental) features of the activities undertaken.

The attraction of the country(side), quiet holidays with a content, with attention to the environment, health and art cities are predominant trends. In the United Kingdom, the term 'green tourism' is a perfectly integrated concept, which goes beyond 'rural tourism' (Green, 1990).

More experienced and educated holidaymakers

Holiday participation rates speak for themselves. A holiday a year has for a long time been a must for most people from developed countries. More experienced does not only mean 'more quality conscious' but also:

- more need of variety;
- desire for communication and personal attention on holiday;
- need for greater variety and choice; and
- more active, more adventure.

As well as having more experience, tourists are very often better educated than they were in the past.

More flexible tourists

New consumers are more flexible. What does this mean for Poon? She underlines two characteristics:

1   New consumers are hybrid in nature and consume along unpredictable lines.
2   New consumers are more spontaneous (e.g. shorter lead times before booking a holiday).

Trends on the supply side

The tourism sector is confronted with four main supply trends:

- more receiving countries and regions;
- concentration;
- revolution in information technology; and
- transport technology.

Every year new regions are developed as destinations. The world is becoming increasingly explored and adventure tourism is in the picture (mountains, tourist submarines, ends of the earth, etc.). The process of concentration of power within the subsectors which make up travel and tourism goes on. We refer to the airline alliances (with four dominant groups (WTO 1998)), the accommodation sector (consolidation of hotel groups) and tour operating.

In addition to horizontal and vertical integration, many travel and tourism companies are utilising a diagonal integration strategy, whereby they establish operations to offer products or services which tourists commonly purchase but which are not directly part of the tourism product (Poon 1993; WTO 1998). Here we are confronted with diversification. Airline companies are offering insurance, etc.

Transport technology will lead – *ceteris paribus* – to either a further decrease in transport costs or shorter travel times. Larger and faster aircraft are being developed; new and faster high speed trains forecast (the 500 kilometres per hour levitation train between Hamburg and Berlin) and even larger cruise ships are under construction; the American *World City* is designed to carry 6,200 passengers (WTO 1998).

Air transport and air fares are and will be influenced by several other trends:

- air deregulation and an influx of low cost airlines;
- further privatisation of the world's national airlines. Major carriers take share holdings or acquire privatising airlines. This phenomenon is very well phrased in the WTO report (1998) 'Privatisation can be seen to be fuelling the spread of globalisation'; and
- aircraft yield management by having the right type of aircraft for given routes.

The conclusion of this subsection is clear. The concentration of airlines, hotel groups, tour operators and car rental firms will lead to economies of scales and to economies of scope. This should lead to lower prices. However, it also contains a danger. Large groups might gain too powerful a hold on the markets.

Impact of tourism demand on globalisation

What is the impact of all the above-mentioned trends on globalisation of tourism demand? We hereby refer the reader to an opinion poll conducted among the members of the Tourist Research Centre on the occasion of the 1996 meeting in Bergen as well as recent discussions with colleagues.

In the medium term seven of the above-mentioned trends will influence the tourism globalisation process. They can be divided into two subgroups. The first, which will have the greatest impact on globalisation, is composed of:

- higher income levels in developed and newly developed countries;
- trends on the supply side (new destinations, concentration, information and transport technology); and
- more experienced holidaymakers.

In the second subgroup we retain four trends :

- new travel motivations;
- more flexible and spontaneous tourists;
- new types of holidays and special interests; and
- demographic changes and improved health among the retired.

Globalisation and information technology

Factors responsible for globalisation of tourism demand

Globalisation of tourism over the last two or three decades of the twentieth century cannot be explained in terms of demand factors only. However, they are important when one considers Baum's (1995) key factors in explaining the growth of international tourism. Nevertheless it would be unwise to overlook the great influence of technology, as we have noted in the preceding sections. Technology has two different aspects: transportation and communication. Without changes in aviation (air fares included) the globalisation of tourism would have been impossible. Probably the same applies to the changes in information technology.

Nonetheless, demand and technology together are still an insufficient explanation. Supply factors, such as international hotel chains and the increasing number of cruise carriers, are important and the role of the destination countries is very often underestimated. Many developing countries consider tourism as a main source of wealth. The benefits are sometimes great (Vanhove 1997). The incentive is often related to demand as well. The broadening of tourist interest not only leads to a greater diversification of tourism development in established destinations but more destinations are entering the tourism market.

Another factor that we should mention is 'mass-media'. Its impact on the consumer is tremendous. People can be informed about worldwide events on a daily basis.

Information technology

One of the essential factors that has stimulated the globalisation of tourism is the revolution in information technology and more generally in communication. 'Information is the lifeblood of the travel industry' (Sheldon 1995). It connects tourists (travellers), tour operators, travel agents and tourism industry suppliers. For Poon (1993) information is the cement that holds together the different producers within the tourism sector – namely cruise lines, airlines, hotels, tour operators, travel agents, car rentals and many other suppliers. It is essential to keep in mind that the links between and among tourism producers are provided not by goods, but by a flow of information. These information flows consist of data, services and payments. In the case of consumers, information is received in the form of advertising, promotion, counselling, bookings (e.g. airline tickets) and a matching information transaction flows from consumers to suppliers (e.g. payment). Information technology is essential for the efficient and timely processing and distribution of all necessary information.

Computer reservation systems and global distribution systems

Information technology is a generic term. In fact there is a system of information technologies. The largest and most important information systems in the tourism sector are the computer reservation systems. The US Department of Transportation defines a CRS as 'a periodically updated central data base that is accessed by subscribers through computer terminals'. They have emerged as the dominant technology.

Computer reservation systems (CRSs) were developed by the airlines. They are primarily tools used by airlines to maintain inventory control of their seat offerings. They have played a significant role in facilitating increasing volumes of travel over the last two decades. A significant shift is now occurring towards global distribution systems (GDS) and increased competition amongst airline groups seeking to broaden and strengthen their product distribution through developing regional global CRSs. A GDS takes the inventory from a CRS (or from many of them) and distributes it via travel agents and other distribution outlets. A GDS has no specific airline inventory control functions other than to 'report back' (French 1998). Since the early 1990s their function has expanded to include many other travel products (e.g. accommodation, rent a car, etc.) and to embrace alternative means of distribution to travel agents, such as Internet (see Figure 6.1).

The leading GDSs are Sabre, Galileo, Amadeus and Worldspan. Table 6.7 outlines number of characteristics of the leading GDSs.

Sabre leads in North America in terms of travel agency locations. Amadeus/System one has the lead in Europe. Galileo can claim to have

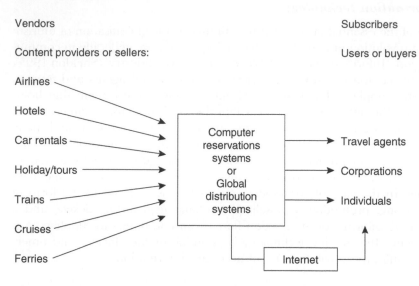

Figure 6.1 Transaction flow of global CRS/GDS industry.
Source: Merrill Lynch, and borrowed from French (1998).

Table 6.7 The major GDSs, 1997

GDS	Agency locations	Terminals (1996)	Participating airlines (1995)	Flight segments booked (in millions)[b]
Sabre	33,453	124,800	377	329
Galileo	36,614	125,000	525[a]	308
Amadeus/System one	42,328	111,400	n.a.	245
Worldspan	17,325	49,500	420[a]	121

a 1997; b Year to September 1997.
Source: Travel distribution report and French (1998).

the best balanced world representation (French 1998). All four are oper-
ating worldwide. Besides the leading GDSs there are a number of regional
CRS vendors (see Table 6.8). In 1996 the five regional CRS vendors in
Table 6.8 represented 18 per cent of the locations.

Are CRSs the cause or consequence of increased travel? According to
the WTO (1990), the development of global CRS capabilities is due to
anticipated changes in the pattern and requirements of the global travel
market. Business is moving away from international towards global operations and hence the requirement for travel global booking and reservation
systems. But the change in the tourism sector itself, including a growth in
long-haul travel, a market trend away from package tour products and a
more demanding and affluent public has also activated the development
of global CRS.

Table 6.8 Regional CRS vendors

	Locations (1996)	Areas(s) of operation
Abacus	6,600	Asia/Pacific
Axess	6,700	Japan
Jets	3,000	Latin America, Europe, Africa, Asia, Pacific
Infini	6,200	Japan
Topas	2,400	Korea

The consequence of these demand trends is an increasing requirement for CRSs to offer more travel information, faster processing time and more comprehensive booking and reservation functions as well as providing enhanced management accounting information (see yield management) WTO (1990). Poon expresses the same idea:

> The applications of technology to the travel and tourism industry allow producers to supply new and flexible services that are cost-competitive with mass, standardised and rigidly packaged options. Technology gives suppliers the flexibility to move with the market and the capacity to diagonally integrate with other suppliers to provide new combinations of services and improve cost effectiveness.
>
> (1993: 153)

For a number of years, most GDSs have had a strategy to become more independent. No GDS has more than about 12 per cent of its bookings from one carrier (French 1998) and not one airline has a majority of its bookings processed by one GDS. The looser link with the airlines also follows from fund raising on the capital market (e.g. Sabre in 1996 and Galileo in 1997).

Rapid diffusion of CRS in Europe

We are now confronted with a very rapid diffusion of technology and CRSs in particular. In 1988, CRSs were used in more than 95 per cent of US travel agencies. Today they are found in close to 100 per cent of all US travel agencies. At the beginning of the 1990s, the diffusion of CRSs in travel agencies was lower in Europe. It varied between 25 per cent in the UK to 88 per cent in Scandinavia. The slow diffusion in the UK was due to the dominant position of package tours that are distributed through tour operator systems based on view-data technology. In the UK market there is also more polarisation of business and leisure travel agencies with the business segment being more automated (more dependent on bookings on scheduled flights).

But in the 1990s, in Europe, travel agency CRS locations have grown rapidly and the diffusion is above 80 to 90 per cent in many countries.

The four leading GDSs increased their locations in Europe from 31,300 in 1993 to 44,800 in 1996.

Travel agencies rely on GDSs for their information retrieval and bookings. They are used most extensively for air travel but more and more for accommodation and car rental and to a lesser extent for tours and other tourist products (e.g. events). Some GDSs are also adding imaging technology to their terminals. This makes it possible for tourists and travel agents to view pictures of tourism products (e.g. hotels, maps, etc.).

Other information technologies

CRSs are not the only information technology. Furthermore, CRSs are more popular with business-oriented travel agencies than those specialised in leisure products. Nevertheless, there is a high penetration rate in all travel agencies in the largest European markets. Other reservation systems are:

* hotel reservation systems;
* reservation systems owned by independent companies (e.g. UTELL);
* videotext (e.g. Prestel in the UK and Minitel in France);
* tour operators very often have their own product reservation system (e.g. Thomson Open-line Programme (TOP));
* car rental reservation systems; and
* national distribution systems (several of them are integrated into a GDS).

But not only the suppliers and those in the trade are using information technology. Computer technology is also used by public or private tourism agencies to create databases of their resort facilities. These databases can be consulted in the destination and in the market regions.

Benefits of CRSs and other information technology

To obtain a general idea about the impact of technology in tourism in general and on the globalisation of tourism demand in particular, we have to define its possible benefits and/or opportunities. We refer here to two interesting documents.

According to Sheldon (1995) information technology has the potential both to increase the efficiency of travel firms, and to make products accessible to the travelling public. It can improve the service quality and contribute to higher tourist satisfactions. Sheldon believes the key benefits are:

Travel suppliers

* higher efficiency and productivity of travel industry firms;
* good channel to promote products;
* CRSs facilitate trip planning;

- management of last seat (room) availability;
- advantages for internal operations (efficient control of inventory, yield management); and
- ability to store customer profile.

Travel agents

- able to access up-to-date on-line information on most facilities; and
- ability to store customer profile.

Consumers

- products more accessible;
- higher quality;
- higher tourist satisfaction;
- individuals can research trips for themselves; and
- knowing in advance the facilities and events at the destination.

The second source is Poon (1993). She makes a distinction between four key impacts:

- efficiency improvement of production;
- quality improvement of services provided to consumers;
- the generation of new services; and
- it will engender the spread of a whole new industry 'best practice'.

The Internet

A special information technology and computer reservation system which has evolved spectacularly is the Internet. The Internet is in fact a network of networks. Two of its most used functions are electronic mail and the World Wide Web. In the late 1990s, the number of Internet users was rather limited in relative terms. The USA, Canada and Finland are exceptions. But the number of Internet users is growing fast. According to Network Wizards, in July 1997 there were more than 19.5 million Internet hosts (313,000 in 1990). It is more difficult to assess the number of Internet users. In 1997 the total adult online population in the USA was estimated at 56 million and an additional 16 million adults intended to connect to the Internet in 1998 (Smith and Jenner, 1998). The same source refers to the Nua survey for the world total of 90.9 million online in November 1997 of which 54 million are in the USA and Canada, 20.2 million in Europe and 8.6 million in Japan.

We are not too far from the point where every home in the developed world will have a personal computer and possibly a modem. Each person will be able to access many information systems all over the world.

Of course there is a difference between owning a computer connected to the Internet and making use of the Internet. However, referring to International Data Corporation, Smith and Jenner put growth at 71,000 new users per day logging onto the Internet for the first time during 1997.

Tourism is a very popular subject on the Web. The well-known Yahoo directory offered about 5,000 travel sites in 1997 (each of which would have a number of pages). What is now important for our subject? Many of these sites act as travel agents. The development of online booking services either through the Internet or through company intranets specialising in corporate business travel concern not only the traditional travel agent but also the GDSs (French 1998). The Internet is a very real threat. According to French, there are estimates that in 1997, travel products worth US$300 million were booked in the USA over the Internet from PCs. In relative terms this is a tiny fraction of the value of travel products booked, but Internet use is growing exponentially. According to Jupiter Communications, consumer spending over the Internet was expected to reach in 1998 a total of US$5.8 billion; of which $1.8 billion would be for travel (Smith and Jenner 1998).

In the face of this evolution GDSs are adopting a new strategy. They are attempting to supply the products Internet needs, rather than be sidelined entirely by parallel distribution networks. French gives a good summary of the Internet activities of the four major GDS systems at the end of 1997 (see Table 6.9).

The joint venture of Worldspan/Abacus and the software giant Microsoft with its Expedia Internet travel booking system is quite remarkable. However, the Internet still has a number of problems. First, there is an

Table 6.9 Internet activities of four leading GDSs, end 1997

	Galileo	Sabre	Amadeus/ System one	Worldspan/ Abacus[a]
Consumers (direct)	Airline branded products: Swissair Skysurfer United Connection US Airways Priority TravelWorks BA Executive Travel	Travelocity	None	Expedia E-Travel
Corporate (direct)	Unnamed product Travel Solutions	Business	None	None
Travel agents	Agency Connection Travelpoint.com	Planet Sabre Turbo Sabre	Corporate Tripsolution	Odyssey

Source: Merrill Lynch and borrowed from French (1998).
a Abacus is hosted on Worldspan's data centre. Due to these two systems' equity, operating and marketing alliance, they are listed as one system.

internal Internet capacity problem which leads to long communication times. There is a second capacity problem. Not all PCs have a sufficient capacity to handle a transaction in due time. Third, the problem of reliability of financial transactions (e.g credit cards) is not yet guaranteed (Smith and Jenner 1998). Maybe, this last problem will be solved in the near future. Even with 100 per cent safety, the consumer – especially the European – must be convinced of the credibility of the system. Furthermore, access to CRSs requires special capabilities and is time consuming for an incxperienced traveller (even with an Easy Sabre or an Easy Worldspan). But very soon (not longer than in the medium term) the Internet may well become a threat to the travel trade.

We agree with the following statement by Sheldon (1995: 138), an expert in tourism information technologies: 'This will also make a wider variety of products available to the travellers, until ultimately any individual can access information on any travel product from any location in the world.' This is certainly one important aspect of globalisation of tourism demand.

Globalisation and marketing strategy

The globalisation of tourism is a reality. In acknowledging this fact, the issue for all organisations involved in the worldwide market is, therefore, how to respond to this globalisation process. How must an organisation adapt its strategy in general and its marketing and communication strategy in particular. Though these questions are obvious, the answers are not, as will be made clear in this section. A second issue of this section is the impact of global distribution systems on the trade.

In the global market, two completely distinct strategies can be implemented: a global strategy or a multi-domestic strategy. According to Yip and Coundouriotis (1991) a global strategy 'is a process of worldwide integration of strategy formulation and implementation'. The emphasis lies on what is common rather than on what is different. This contrasts with a multi-domestic approach which allows an independent development of strategy by country or regional unit. In that case the emphasis is placed on what is different rather than on what is common.

While a global strategy seeks to maximise worldwide performance through sharing and integration, a multi-domestic strategy seeks to maximise worldwide performance by maximising local competitive advantage, revenues or profits by adapting the strategy completely to the needs and the characteristics of the local markets.

According to Yip, in developing a worldwide global strategy three steps are necessary.

1   Develop the core strategy. It is the basis of sustainable competitive advantage. It is usually developed for the home country first.

2   Internationalise the core strategy through international expansion of activities and through adaptation.
3   Globalise the international strategy by integrating the strategy across countries.

The difference between a multi-domestic and a global strategy lies logically in the third step. Let's take a simplified hypothetical example. A sun and beach destination, for instance, has first established a strategy for selling its tourism product to the local market. The second step is developing appropriate strategies to sell its tourist product in other countries. In this step, the approach used in each international market is different and not coordinated. Each country manager can adapt his strategy and marketing tools to the needs of the market and do so independently of the other country managers. The third step involves integrating the different strategies into one coherent global strategy. Many multinational companies know the third step less well since globalisation runs counter to the accepted wisdom of tailoring for national markets (Douglas and Wind 1987). That is why in practice, one can see that global marketing strategies in their extreme form can only be found in limited cases. Most of the time there is some adaptation to the local market.

A striking non-tourism example, for instance, is the fact that an important producer of canned products offers more than sixteen different varieties of plain tomato soup throughout Europe. British people like this soup to be more orange than red; in other countries salt must be added or extracted, etc. Indeed, cultural differences, with their resultant differing demand patterns, constitute the key difficulty for global marketing.

According to Yip (1991), organisations that use global strategy levers can achieve one or more of the following benefits:

*   cost reductions;
*   enhanced customer preference;
*   improved quality of products and programmes; and
*   increased competitive leverage.

However, globalisation has also its drawbacks:

*   increased management costs: through increased coordination, reporting requirements and even added staff;
*   reduction of an organisation's effectiveness in individual countries if overcentralisation hurts local motivation and morale;
*   product standardisation can result in a product that does not entirely satisfy *any* customer;
*   uniform marketing can reduce adaptation to local customer behaviour; and
*   integrated competitive moves can mean scarifying revenues, profits, or competitive position in individual countries.

The question whether a global strategy must be implemented or not, or if so, to what degree, becomes still more complex when analysing some recent trends. Here we examine only two examples.

More homogeneous consumer versus remaining and sometimes reinforced customer differences

In the last few decades the world has become 'the' marketplace. This strong globalisation of business has stimulated a globalisation of culture: international trade, world travel and mass electronic communications have created tremendous pressures for a global homogenisation of products, lifestyles, architecture, food and eating habits, entertainment and many forms of everyday behaviour (Ritchie 1991).

Though there is a serious pressure for homogenisation, major differences still exist between different geographical markets in terms of culture, living habits, religion, etc. Moreover, one can witness a counter-reaction to this global homogenisation process.

> There has been an increasing effort on the part of many societies and cultural groups to consciously undertake efforts to strengthen and develop clear cultures and their supporting values. What seems to be emerging is a rather paradoxical situation in which cultural diversity is thriving in a sea of homogenisation.
>
> (Ritchie 1991: 152)

Global marketing versus micro-segmentation

The reaction of population groups to this globalisation of the economy is also witnessed at the level of the individual. Naisbitt (1993) stated that the 1990s would become the years of triumph of the individual, while Popcorn (1992) speaks about ergonomics. People strive more and more for personal development and self-realisation. Poon (1993) denominates this trend as new tourism, hybrid consumers, etc.

The crucial element within these trends is that consumers want more and more individualised products and services which are tailored to their needs, which is exactly the opposite of the ideal standardised product in the context of global marketing.

Implementing global strategies in the tourism sector

By its nature, the tourism sector is confronted with a global marketplace. Moreover, the number of generating and receiving markets has dramatically increased in the last few decades, making the tourism sector more than ever a player in the global market. Whether a global strategy can be applied in the tourist sector or not and to which degree is again not an easy question and depends on several factors such as: the nature of the organisation, the nature of the supplied products, the homogeneity of the markets, and so on.

Nature of the organisation

It is easier for a hotel chain, which can control each element of its marketing-mix, to apply a global strategy, than for 'an NTO', which cannot always control elements such as the price level, third party communication actions, etc.

Nature of the supplied products

Some tourism products are more compatible with global strategies than others. An island targeting the international divers' markets can more easily apply a global strategy than a country with different tourist attractions, appealing to different market segments and geographical markets. In the latter case, a global strategy is not impossible, but surely more difficult. The more homogeneous the perception of the product and the clearer the concept in a global market, the more easy it is to apply a global strategy.

Let's take the example of the Maldives. As it is used to focusing on the international divers' market, this archipelago could easily approach international markets with a global strategy, as the international segment is clearly determined, the travel motives and travel patterns quite homogeneous, and the selling concept universally applicable. This means that in the marketing strategy used, the Maldives could use the same positioning, communication messages, product proposals, etc.

At the other extreme, it is clear that for a country such as Belgium, for instance, applying the same global market strategy (including market segmentation, positioning, communication messages, product proposals, etc.) for the neighbouring, medium and long-haul markets would be suicidal. Travel motives and travel patterns of American or Japanese tourists, for instance, are completely different from those of French, Dutch or German tourists. Even within these latter neighbouring markets some differences can be found, which necessitate a specific approach.

Some interesting examples can be observed in the hotel sector. Take, for instance, the example of the Jamaican owned SuperClubs chain. This chain offers a global concept that can be easily promoted worldwide. The SuperClub hotel chain caters to different segments of the worldwide market in its five locations in Jamaica, offering all-inclusive holidays for:

- couples only at its couples resort;
- families only at Boscobel;
- persons who want to be 'wicked for a week' or hedonism, at Negril;

- persons with a focus on health nature, sport and 'a touch of Jamaica' at Runaway Bay; and
- guests who want grand luxury at Grand Lido.

Due to the clarity of the concept, which can easily be understood world-wide, this chain could target worldwide segments characterised by the same travel motives and travel patterns.

Another aspect which is of an influence is the uniqueness of the product. When the tourism core product is unique (e.g. the Norwegian city of Bergen and its fjords) the application of a global marketing strategy may be easier (distribution, promotion, etc.).

Homogeneity of the markets

The homogeneity of the markets is of course to a large extent determined by the product concepts offered. The more homogeneous the market pref-erences and travel patterns, the easier it is to apply a global market strategy, as we have illustrated with the examples of the Maldives and Belgium.

However, the existence of a clear and internationally valid product concept is not a guarantee of success, as witnessed by the example of Eurodisney. The Disney Company tried to copy their successful Disneyland concept in Europe, with the known results. After only a few months, the original concept had to be adapted to the specific needs and wants of European tourists and even the name was changed to Disneyland Paris. It is clear that the issue of whether or not and if so to what degree global strategies must be implemented is a very complex one and that there is no such thing as an easy answer. Globalisation offers many benefits, but at the same time serious drawbacks. That is why this debate has been going on for many years already. In the early 1980s the paradigm 'Think global, act local' found many supporters within Europe. The idea was to establish a strategy which was as global as possible and as local as necessary, in order to realise the highest possible economies of scale.

In recent years this paradigm has changed somewhat to 'Think local within global positioning boundaries' (Naert 1992), meaning adapt your strategy to the local needs without losing the advantages of a global approach. In this view the local field has gained somewhat in importance, though in fact the difference from the previous paradigm is in our view rather limited. We can conclude that an organisation should analyse its own situation and strike a balance between the benefits and drawbacks of implementing global strategies. As a general rule we can state that the more homogeneous the market preferences and the more universal the product concept, the more likely it is that a global market strategy can be successful.

The above-mentioned ideas can be summarised in a general schema of communication strategy (see Figure 6.2). In the matrix, we distinguish two

		MARKET (consumers, countries)	
		HOMOGENEOUS	HETEROGENEOUS
M E S S A G E	GENERAL	Global communication (corporate identity)	Segment communication (several products to one segment)
		1	2
		3	4
	SPECIFIC	Product communication (one product to all markets)	Specific communication (one product to one segment)

Figure 6.2 Globalisation and communication strategy.

types of consumers or markets (homogeneous and heterogeneous) and two distinctive messages (general or specific).

In practice more and more destinations apply either a segment communication strategy (e.g. 'Spain suitable for families'. In this case we refer to all products in which families are interested such as beaches, hotels, theme parks, etc.) or a product communication strategy (e.g. art cities to all markets). This approach is more efficient. The link with homogeneity of markets or homogeneity of tourist products is quite clear. A specific communication strategy can be ideal but is very expensive to apply. Segment, product or specific communication strategies are very often complemented with a global communication (e.g. the famous corporate identity of Spain during the 1980s and the beginning of the 1990s): 'Spain: everything under the sun' as the base line and Miro's sun as the isotype.

Communication strategy 2 or 3 or 4 (see Figure 6.2) combined with a global communication is called an 'integrated communication'. This is very close to the paradigm 'Think local within global positioning boundaries'. However, success is still not guaranteed. In most cases, global strategies will always have to be adapted in some degree to the specific requirements of the markets. The future answer to the issue of global strategies will probably be given by production and information technologies. The first enables mass customised production, the latter world-wide communication while adapting the message to the individual consumer.

A final basic remark may be relevant. We have seen that most new destinations depend on the same twenty or more generating markets. Due to a lack of financial resources the choice of markets is a part of the marketing strategy. This choice depends on several factors. Market choice implies a *de facto* reduction of globalisation.

Global distribution systems and marketing

Information technology in general and global distribution systems in particular have had a great impact on the globalisation of tourism. The Internet can only enhance this impact. But for their part distribution systems are not without consequences for marketing in tourism. This is a very vast and complex subject. Here we will limit ourselves mainly to a number of issues related to the Internet.

First, the consumer becomes an important player in the tourism system. He/she defines who will be retained on the shortlist, who is allowed to provide information and who will receive the booking in due time. The consumer can compose his products and buy directly all the components of that product (flight, hotel room, rent a car, etc.) from his PC at home. Besides the traditional value chains of purchase, production, marketing and sales, a virtual value chain has been created: the collecting, organising, selecting and distribution of information.

Second, because of the new media, travel agents are losing their monopoly over access to the reservation systems of all important players in the market (Walle 1996; Centre of Tourism Management 1996; NBT 1998). See Figure 6.3.

Competition comes from new online travel agents such as Microsoft, Internet Travel Network, CNN Interactive Travel, and big companies such as air carriers (American Airlines) and hotel chains (Mariott International), which see an opportunity in direct selling. In late 1995, British Midland was the first airline in Europe to introduce a reservation booking system with online payment via credit card on the WWW, called Cyberseat (Marcussen 1997). The new players are successful because of their competence in their core business (e.g. Microsoft: knowledge of new media and information technology, and CNN: brand name).

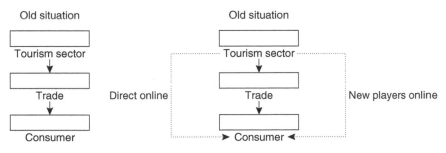

Figure 6.3 Digital revolution in the tourism sector.

The consumer can get information and make reservations online anywhere, anytime. Several hotel chains in the US account for thousands of reservations a month (Kotler, *et al.* 1999). Restaurant companies (e.g. Pizza Hut) are also using the Internet as a distribution channel. The capability of transmitting colour photographs to millions of people across the globe makes the Internet an interesting channel. Hotels, restaurants and others can tangibilise their products through the use of colours and videos. At present, it is popular for simple products (e.g. tickets); tomorrow for more complex products. Traditional suppliers are also exploring the new opportunities of information technology as an instrument for cost reduction and cost control (see the policies of KLM, Lufthansa and British Airways). In fact the role of travel agents – especially business travel agents – is not regarded as essential in the distribution of their products and the added value of these intermediaries tend from time to time to be questioned (Centre of Tourism Management 1996).

Third, the GDSs – and indeed the airlines themselves – face a serious challenge from electronic online developments such as the Internet. Here we refer to a conclusion by French (1998).

> These new systems have the potential to cut out the middlemen – principally the travel agents, but also possibly the GDSs themselves – by selling airline inventory direct to consumers via an extremely cheap global network linked to home- or office-based PCs. The GDSs, often in partnership with their airline owners, are seeking to meet this challenge by aligning themselves with the Internet and offering Internet products, rather than ceding these products to non-airline (perhaps even non-travel industry) new entrants. The competitive battle within airline distribution, which has focused on GDSs for most of the 1990s, is thus swinging away from GDSs towards online direct-to-consumer distribution systems.

Here we would like to make a remark. What are the strategic objectives of the CRSs? Do they want an online direct-to-consumer distribution channel or do they indirectly encourage the travel agents to make use of their GDSs? Anyhow, one channel is less costly.

Fourth, the position of the trade will change. The old-style travel agent is indeed in danger. This is more the case for business travel agents than for consumer travel agents. The majority of shopping transactions on the Web are business-to-business. Many business travel agencies are afraid of electronic ticketing (not all problems are solved). But this is already widely used in the USA.

According to Smith and Jenner (1998) selling to consumers is slightly more difficult. Surveys of Internet users reveal a significant pattern of discontentment. People cannot find the sites they want. If they find the sites, they cannot find the information on the sites, and when it comes to placing an order they worry about giving information over the Internet (e.g. credit card details). Nevertheless, there are good reasons for supposing that sales over the Internet, including travel sales, are going to become important in the near future. However, it must be stressed that not all bookings via the Internet are reducing the activities of travel agents; many of those bookings are replacing reservations by phone or fax.

The most crucial issue in the survival of travel agents in the distribution chain is the change of their core business as ticket providers to become full service providers. The capacity to take full advantage of information technology and stay at least one step ahead of their clients through expertise and knowledge of the travel business and/or holidays is the major challenge. The revenues expected must come from the sale of complex products and possibly consumer fees (Centre of Tourism Management 1996). Some travel agents supply clients with software such as Galileo View which allows corporate clients to make their own itineraries or requests which are subsequently passed on and processed by the travel agent.

A real advantage of a travel agent is the personal contact with the customer which new players cannot provide (the travel agent as a counsellor, provider of good service, reliable information and a certain insurance when something goes wrong). However, this is not sufficient. They must be able to offer the best travel solution. On the other hand, CRSs give an opportunity to travel agents to compose holidays in a flexible manner answering the aspiration of the tourist 'new style' (see second section of this chapter). Poon (1993) has a rather optimistic view with respect to the role of travel agents:

> The increased importance of the travel agent is related both to the need for 'high touch' to complement an increasingly technology oriented industry and the need for human intelligence to sift the incredible volume of information and services that technology, deregulation and competition make now possible.

Fifth, in the online economy we can expect a battle of brands. Brands will have an even greater importance than in a traditional economy (NBT 1998). A brand is an identifying mark for consumers who cannot see the wood for the trees. A brand may also give confidence.

Sixth, small firms have new opportunities. More and more smaller firms have access to the reservation systems. Two years ago, in a town such as Bruges 12 per cent of the hotels (mainly family hotels) were affiliated to the Galileo system. Family hotels can work together via a network and compete with the hotel chains. Travelocity has already installed such a network on her reservation system by which very small family hotels without computer-link but via fax service (printers) take part in a worldwide network. Many local English bed and breakfast hotels have also become global in this way.

Seventh, tour operators are also under pressure though less than travel agents. Also tour operators go online to make their products accessible to the individual consumer. Tour operators have one big advantage. They can provide economies of scale for the two most important items of a tour operator's holiday: transport (air) and accommodation. In theory it is possible for a consumer to collect 200 potential buyers for a destination and to shop for the best price. The reality is not that easy. But in the future more sophisticated software will make it possible to combine components and supply the offers worldwide. The only adequate answer to this threat is to supply more than a combination of components: experience, services, knowledge of destinations, and so on.

Eighth, the Internet becomes not only an additional distribution channel but also an additional communication vehicle. This new communication instrument takes two forms. The first and the most important is the information provided by travel: the websites of countries, regions, resorts, tourism organisations and enterprises.

The expanding use of the Internet/WWW will create a trend towards more direct sales between supplier and consumer within tourism (Marcussen 1997). WWW homepages have the potential to become powerful advertising and marketing tools. Web users are a market of well educated and affluent groups of users (Cano and Prentice 1998). If a region or firm chooses to make a WWW presentation it is very important to get it registered in the relevant places so it can be found easily by those who might be interested. The lay-out and make-up of a beautiful and efficient WWW presentation is a professional's job (see also Cano and Prentice 1998).

The second is advertisements, in the form of banners. At the end of 1997, around 15 per cent of commercial web sites carried advertising banners (Smith and Jenner 1998). As technology progresses, these banners become more and more sophisticated, making use of movement and music. As an advertising medium, the web is still in its infancy. But also here we see an enormous growth. According to the Internet Advertising Bureau, the expenditure on the Internet in 1997 was over three times more than in the previous year.

Ninth, the Internet creates a new form of word-of-mouth publicity, which is much more dangerous than the traditional one. Good or bad holiday experiences can be distributed on the Internet in a minimum of time. This phenomenon, together with the growing quality consciousness of tourists, places greater emphasis on the slogan 'perform or disappear'.

Tenth, the reservation systems permit the building of an interesting databank. This can be an excellent source for segmentation and direct marketing. Some new players can obtain the data from different sources. So Microsoft gets its information not only from travel activities but from software sales as well. A buyer of computer golf games can be informed, via an e-mail, of a golf weekend in one or more places.

Conclusions

Tourism demand and leisure holidays in particular are still dominated by intraregional destination flows. However, the market share of interregional flows is becoming significant and is growing worldwide. The forecasts for tourism demand for the next decade indicate an increased market share of long-haul destinations which is after all a good indication of the globalisation of tourism demand.

Globalisation of demand is by definition linked to the existence of similar customer preferences worldwide. The underlying factors for the globalisation of tourism demand can be divided into five main groups:

1 Cheap air fares and/or package tours from tour operators. This is the basic determinant in/for the increase in long-haul trips.
2 The impact of a number of socio-economic trends, especially higher income levels, demographic changes and new travel motivations.
3 But globalisation is also stimulated by changes such as new types of holidays and more experienced holidaymakers and above all trends on the supply side.
4 The emergence of new destination countries and regions across all continents.
5 The rapid development of information technology. The influence of the diffusion of CRSs was great and the role of the Internet will become important.

The answer to the question of how an organisation should adapt its strategy in general and marketing and communication strategy in particular to the process of globalisation is not easy. The application of a global strategy depends on the nature of an organisation, the nature of products supplied and the degree of homogeneity of the markets. The paradigm should probably be 'think local within global positioning boundaries'.

Information technology has created benefits for suppliers, travel agents and consumers and generated new services. In the near future the Internet as a new distribution system will have an important role as a distribution channel and will be an additional communication instrument. There are several indicators which support this evolution:

• the consumer becomes an important player in the tourism system;
• the trade is losing its monopoly;
• GDSs face a serious challenge from online electronic developments;
• the travel agent old style is in danger; however, the travel agent new style has still a role to play;
• in the online economy 'brands' will be of greater importance;
• the Internet provides new chances to small firms;
• the tour operator business is less under pressure due to the economies of scale they can provide;

- the Internet becomes an additional distribution channel;
- the Internet creates a new form of word-of-mouth publicity; and
- reservation systems provide the opportunity to build up interesting databanks which will be useful for an efficient segmentation policy and direct marketing.

The globalisation of tourism demand and the global reservation systems are creating a new environment for the tourism sector in general and the trade in particular. For all players in the tourism sector creative marketing is the right answer.

Bibliography

Archdale, G. (1993) 'Computer reservation systems and public tourist offices', *Tourism Management*, February.

Baum, T. (1995) 'Trends in International Tourism', *Insights*, March.

Bruce, M. (1991) 'New technology and the future of tourism', in S. Medlik (ed.), *Managing Tourism*, Oxford.

Cano, V. and Prentice, P. (1998) 'Opportunities for endearment to place through electronic "Visiting": WWW Homepages and the tourism promotion of Scotland', *Tourism Management* 19(1).

Centre of Tourism Management (1996) *'The spawn of information technology in travel business: challenges and opportunities*, Rotterdam (unpublished).

Douglas, S. P. and Wind, Y. (1987) 'The myth of globalisation', *Columbia Journal of World Business*, Winter.

Edwards, A. (1992) 'International tourism forecasts to 2005', *The Economist Intelligence Unit*, London.

Emmer, R., Tauck, C., Wilkinson, S. and Moore, R. (1993) 'Marketing hotels. Using global distribution systems', *Cornell Hotel and Restaurant Administration Quarterly* 6.

Forrest, J., Wotring, C. and Brymer, R. (1996) 'Hotel management and marketing in the Internet', *Cornell Hotel and Restaurant Administration Quarterly* 3.

French, T. (1998) 'The future of global distribution systems', *Travel and Tourism Analyst* 3.

Go, F. M. (1996), 'Globalisation and corporate organisation', *Globalisation and Tourism*, 46th Aiest Congress, Rotorua.

Green, S. (1990) 'The future for green tourism', *Insights*, September.

Hawkins, R. E. (1994) 'Ecotourism: opportunities for developing countries', in W. Theobald (ed.) *Global Tourism. The next decade*, London.

Knowles, T. and Garland, M. (1994) 'The strategic importance of crisis in the airline industry', *Travel and Tourism Analyst* 4.

Kotler, Ph., Bowen, J. and Makens, J. (1999) *Marketing for Hospitality and Tourism*, New York, 2nd edn.

Krippendorf, J. (1987) *The Holidaymakers: Understanding the Impact of Leisure and Travel*, London.

Marcussen, C. H. (1997) 'Marketing European tourism products via Internet/WWW', *Journal of Travel and Tourism Marketing* 6(3/4).

Martin, C. A. and Witt, S. F. (1989) 'Tourism demand elasticities', in S. F. Witt and L. Moutinho, *Tourism Marketing and Management Handbook*.

Mcguffie, J. (1994) 'Accommodation: Crs development in the hotel sector', *Travel and Tourism Analyst* 2.

Middleton, V. T. C. (1991) 'Whither the package tour?, *Tourism Management*, September.

Middleton, V. T. C. (1994) *Marketing in Travel and Tourism*, Oxford.

Mill, R. (1992) *The Tourism System*, London.

Moore, S. and Cater, B. (1993) 'Ecotourism in the 21st century', *Tourism Management* 2.

Naert, Ph. (1992) *European Market Strategies Post 1992*, Ipo-Management School, Antwerpen.

Naisbitt, J. (1993) *Megatrends, Het Spektrum*, Wijnegem.

Nederlands Bureau Voor Toerisme (1998) *Digitale Revolutie in de Toeristenbranche*, 'S Gravenhage.

Opaschowski, H. W. (1985) 'Neue Urlaubsformen und Tourismustrends', in *Trends of Tourist Demand*, Aiest 26.

Poon, A. (1993) *Tourism Technology and Competitive Strategies*, Wallingford.

Popcorn, F. (1992) *Trends Van Overmorgen. Consumentengedrag in de Jaren Negentig*, Amsterdam.

Ritchie, J. R. B. (1991) 'Global tourism policy issues: an agenda for the 1990s', *World Travel and Tourism Review* 1.

Sheldon, P. J. (1995) 'Information technology and computer reservation systems', in S. F. Witt and L. Mouthinho, *Tourism Marketing and Management Handbook*, Student edition.

Smith, C. and Jenner, P. (1998) 'Tourism and the Internet', *Travel and Tourism Analyst* 1.

Truitt, J., Teye, V. and Farris, T. (1991) 'The role of computer reservation systems', *Tourism Management*, March.

Vanhove, N. (1993) 'Les Tendances Micro- Et Macro Économiques Dans Le Tourisme Européen', *Teoros International* 1.

Vanhove, N. (1996) *Globalisation of Tourism Demand: The Underlying Factors and the Impact on Marketing Strategy*, 46th Aiest Congress, Rotorua.

Vanhove, N. (1997) 'Mass tourism – benefits and costs', in J. Pigram and S. Wahab (eds), *Tourism Sustainability and Growth*.

Walle, R. H. (1996) 'Tourism and the Internet: opportunities for direct marketing', *Journal of Travel Research* 1.

WTO (1990) *Tourism to the Year 2000. Qualitative Aspects Affecting Global Growth*, Discussion Paper.

WTO (1995) *Global Tourism Forecasts to the Year 2000 and Beyond*, Madrid 1.

WTO (1998) *Tourism: 2020 Vision*, Madrid.

Yip, G. S. and Coundouriotis, G. A. (1991) 'Diagnosing global strategy potential: the world chocolate confectionery industry, *Planning Review*, January/February.

Xxx (1996) 'Faut-Il "Croire" Au Réseau Internet?', *Espaces* 137.

Part III

Globalisation and competitiveness in tourism

7 Organisation of tourism at the destination

John Swarbrooke

Introduction

Destinations are the core of the tourism product. The desire to visit them is the main motivation for most tourist trips. They provide all the services that the tourist needs to sustain them during their vacation. Furthermore, it is in the destination that the physical, social, and economic impacts of tourism – both positive and negative – are experienced.

In short, destinations are at the very heart of tourism, whether they be coastal resorts, historic cities, spa towns or mountain villages. However, they are at the same time, an amazingly complex phenomenon, as we shall see shortly. This complexity makes their management and marketing an extremely difficult task. Destinations are currently being affected by many aspects of globalisation, in terms of both the supply and demand sides of tourism. In this chapter, we will explore the impact of globalisation on destinations today, and in the future. Before we can do that, however, we need to begin by exploring the concept of destinations in detail.

The nature of destinations

Destinations possess a number of characteristics that differentiate them from most other elements in the tourism system, which ensure that their management and their marketing are particularly difficult tasks. These include the following:

1 *Destinations are not single, homogeneous products.* They are, instead, the physical packaging for a wide range of individual tourism products such as visitor attractions, hotels, restaurants, bars and transport services. Destinations are very much like 'Lego' sets. They contain large numbers of different 'building bricks' from which each tourist can build their own destination product or experience. As with Lego sets very different products or experiences can be created from identical sets of 'building bricks'.

2 *Destinations are multi-purpose places,* very few of which are totally devoted to tourism. Most destinations also have to function as residential

centres, industrial production areas, zones of agricultural production and retailing centres. Simultaneously, they have to serve the needs of both residents and visitors. Thus a major challenge is the requirement to seek to reconcile these often conflicting roles and needs.

3 *In some destinations, tourism is a long-established tradition, while in others it is a relatively new phenomenon.* This difference influences both local attitudes towards tourists, as well as the capacity and quality of the tourism infrastructure.

4 Destinations exist at a number of different geographical levels that are interrelated, as can be seen from the example of the small historic French town of Sarlat, illustrated in Figure 7.1.

5 At the same time, tourists often view places as a single destination that are not seen as such by local government. *Bureaucratic boundaries are immensely important to governments but are largely irrelevant to the tourist.* This often leads to the public sector bodies that market destinations doing so in ways which do not match tourist perceptions of the geographical extent of the destination, which limits the effectiveness of such marketing.

6 *Marketers often invent powerful brand names for destinations that have very imprecise geographical boundaries and may not fully represent the nature of the area.* For example, we have the 'French Riviera' (France), 'Shakespeare's

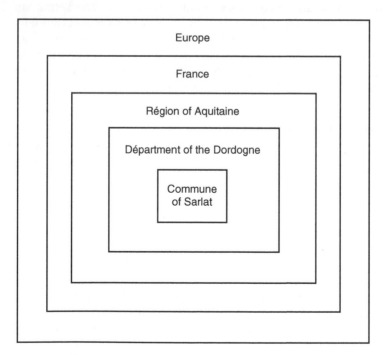

Figure 7.1 The geographical hierarchy of destinations in France.

Country' (UK), 'Fjord Country' (Norway) and the 'Holyland' (Israel and Palestine).

7 *Tourists travel greatly varying distances to visit destinations, from a few kilometres to thousands of kilometres.* Some are people who will understand the culture of the destination while others will be from countries with very different cultures. This can cause problems between hosts and guests, and the growth of long-haul travel is making this an ever-greater issue.

8 *The boundaries between individual visitor attractions and destinations are blurring* as theme parks, for example, grow in size and develop their own on-site accommodation and restaurants, together with secondary attractions. Already, Disneyland Paris, for example, is as large as many small seaside resorts or rural tourism destinations. This blurring is also being seen with the growth of resort complexes such as Center Parcs that also have all the components of a traditional destination.

9 *Most destinations are managed and marketed by public sector bodies while most of the individual products within the destination are owned and controlled by private sector enterprises.* Thus, these public sector bodies have relatively little control over the quality, pricing, and distribution of these products. This makes their task particularly difficult and explains their apparent emphasis on the promotional element of the Marketing Mix or four 'P's, where they can have an influence through brochures and advertising campaigns.

10 *Because of the role of the public sector, with its political and social agendas, tourism management in destinations often has a wide range of complex objectives rather than simple profit maximisation or market share optimisation.* Tourism can be used to:
    • diversify rural economies and regenerate cities where traditional industries are in decline;
    • improve an area's image in the hope that this will encourage inward investment in other industries;
    • maintain the viability of local services such as shops and theatres that are largely used by residents;
    • increase local pride; and
    • help justify and/or fund infrastructure improvements that will also benefit local people.

There can be tension between some of these objectives and the aims of entrepreneurs who generally have simpler, more personal objectives such as increasing the turnover of their businesses. Furthermore, conflict can also arise between the public sector objectives that are longer term and the short-term world in which most entrepreneurs have to spend their lives.

11 *Tourists have perceptions of destinations, even those they have not visited, which may not be based on reality.* These perceptions will cover issues such as

the destination's attractiveness for different types of holiday, its accessibility, and its safety for the tourist. It is these perceptions, not the reality, which determine the purchase decisions made by tourists.

12 *The appeal in some destinations is based on their physical attributes, such as beaches and mountains, while others are based on their cultural resources,* such as the traditional way of life, food, local heritage, and theatres. This has implications for everything from seasonality to the demographic profile of visitors.

13 *Destinations offer their visitors only shared use rights* and conflict may perhaps arise because different segments with different needs and tastes are in a destination at the same time, such as young pleasure-seekers, families and elderly tourists.

14 *Those visiting a destination often have very different motivations.* For example, a visitor to Crete may wish to sunbathe all day and go out at night *or* may wish to visit archaeological sites most of the time. It is difficult to anticipate and satisfy these very different markets.

15 *The satisfaction, or otherwise, of a tourist with their stay in a destination can be heavily influenced by factors outside the destination manager's control* such as weather or transport delays on the way to the destination.

16 *Most tourists visit destinations for relatively short periods of time* ranging from one day to a couple of weeks. This limited duration of stay can make it very difficult to put problems right before the tourist leaves.

17 *Externally-based organisations often play an enormously important role in marketing destinations to tourists.* Often these organisations – airlines, tour operators, and hotel chains – have little or no commitment to the destination. For them it is just a commodity. They also may not have a high level of detailed knowledge of the destination so they may give inaccurate or misleading messages about it to potential visitors.

While not exhaustive, this list of characteristics of destinations does hopefully illustrate the diversity and complexity of the phenomenon of destinations. It is now time for us to move on to look at the issue of globalisation and its relevance to destinations.

The scope of globalisation

The concept of globalisation has a number of dimensions in respect of tourist destinations, as follows:

- the globalisation of the tourism industry, the supply side of the tourism equation;
- the globalisation of the market, the demand side of tourism;
- the globalisation of communication technologies;
- the globalisation of the media; and
- the globalisation of political power.

We will now look at each of these briefly, in turn. The globalisation of the supply side in tourism, itself, has a number of elements. First, it is fair to say that a number of major airlines and hotel chains are already global players operating and selling their products across the world. However, this phenomenon is growing as more and more tourism enterprises seek to sell their products to consumers in other countries. At the same time, the growth of long-haul travel is means that tour operators who a decade ago operated largely exclusively in their own continents are now operating in all or most of the continents of the world.

The process of supply side globalisation is slowly weakening the ties that have to date bound most tourism enterprises to their national, regional or local roots. As they become more footloose there is a danger that they will become even less committed to, and concerned about, individual destinations. They will simply be able to play one destination off against another for their own benefit, regardless of the implications of this for the destination in question.

The globalisation of tourism operations also brings particular management problems for tourism organisations as they seek to come to terms with unfamiliar employment practices, business cultures and government systems.

The globalisation of the demand side, the tourism market, takes two main forms as follows:

- the growing desire of tourists to travel to new areas of the world in search of novel experiences. This trend has been reinforced by the growth of the concept of 'ecotourism' so that even the Arctic and Antarctic are growing in popularity as destinations; and
- the rise of new generating countries which have not previously been noted for their outbound tourism, but which are now generating growing numbers of leisure tourist trips to other countries. In recent years this phenomenon has been seen dramatically in countries such as Taiwan, South Korea and Russia. The current economic climate in these countries has now slowed this growth but this is likely to be only temporary. At the same time there has been a steady growth in outbound tourism from countries not traditionally associated with outbound tourism in the rest of Asia, Africa, the Middle East, and South America.

For destinations, this globalisation of demand implies a need to become skilled at welcoming tourists from a variety of different cultures.

Third, we are seeing the globalisation of communication technologies. Increasingly, the development of communication technologies such as the Internet, and computer conferencing, for example, are making national boundaries less meaningful. They allow both tourism suppliers and tourists to communicate with each other on a truly global basis. The IT corporations which dominate the provision of Internet services are seeing tourism as a particularly lucrative field for their services.

Next, we have the globalisation of the media with the rise of multi-national newspapers, and television empires, and the growth of satellite television which allows the same programmes to be broadcast worldwide at the same time.

Finally, there is the globalisation of political power; in other words, the growth of supra-governmental organisations such as the European Union and ASEAN. These organisations are taking an increasing interest in tourism and are keen to promote their regions as tourist destinations to the global market. They also encourage cooperation between their member states in terms of tourism policy as in other areas of activity. Global institutions such as the World Bank are also playing a growing role in funding tourism-related projects.

Key topical issues in destination management

While we are mainly concerned with the impact of globalisation, we need to recognise that destinations are also currently facing a range of other key issues. These include:

1   The continual growth of competition from newly developed destinations or from destinations that formerly had only a regional market but which are now attracting tourists from around the world.
2   The fact that many established destinations are losing out to the new competitors and are in decline, with a need for rejuvenation.
3   The increasing homogenisation of destinations due partly to the globalisation of supply and demand as well as to the tendency for destination management agencies try to copy what has worked for other destinations. For example, most Mediterranean destinations are now seeking to establish themselves as destinations for golfers, because this strategy is perceived to have been successful in Spain and Portugal.
4   The continuing problem of seasonality which leads to the under-utilisation of expensive capital resources, means that many destinations are looking for ways of extending their season.
5   Many destinations which grew up by serving the needs of the mass market are now seeking to re-position themselves and increase the benefits, as well as reducing the costs, of tourism by attracting 'better quality' tourists. This usually translates as higher-spending tourists. At the same time there is a desire in some destinations to actively discourage what are seen as poor quality tourists from visiting. For instance, some destinations would like to 'de-market' themselves as far as the so-called 'lager-louts' – young, heavy drinking, poorly behaved tourists – are concerned.
6   The growing recognition that no single public sector agency has the skills, resources, or even the mandate, to manage and market desti-

nations on its own. The 'buzz word' in destination marketing today, therefore, is partnership, involving partnerships between:

- local destination marketing agencies and local tourism industries;
- local destination planning and marketing agencies and local communities; and
- local government and regional and central government agencies.

7   The growth of 'newer' markets such as eco-tourism, activity holidays, and health tourism, for example, and the desire of many destinations to attract these markets which are seen to be both lucrative and good for a destination's image.

8   The apparent increasing interest of consumers in the quality of the destination environment. This is a real problem for those destinations that grew up quickly in a largely unplanned fashion, and where the infrastructure is inadequate to support peak tourist numbers, resulting in visually unattractive buildings and pollution problems.

For destination managers, the challenges posed by globalisation specifically, will be in addition to these other issues, several of which are related to globalisation.

The scope of destination management

Destination management is a broad area of activity that has four main elements. These are illustrated in Figure 7.2. To varying degrees the globalisation of supply, demand, communication technologies, the media, and political power will influence all four elements of destination management over the coming years.

However, before we discuss this influence in more detail, we need to recognise that:

1   These four elements are heavily interrelated. For instance, accommodation quality is a key element of the tourism product. Likewise planning constraints or the lack of them help determine the environmental impact of tourism which in turn affects the quality of the tourism product.

2   These four elements are rarely managed, as a whole, by a single agency. Usually responsibility for them is split between different organisations, as we can see from Figure 7.3 that shows the situation of a hypothetical seaside resort in the UK.

Even within Golden Sands Borough Council itself these issues will be dealt with by different departments, including Economic Development, Planning, Environmental Health and Education. This fragmentation of responsibilities for the different aspects of destination management is a potentially serious problem. It can prevent the development of leadership,

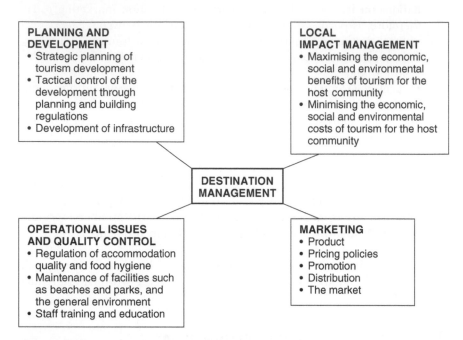

Figure 7.2 The scope of destination management.

vision, and a sense of shared purpose, all of which are vital if destinations are to be successful in the era of globalisation.

Those partnership bodies that have grown up, such as the Visitor and Convention Bureaux in the USA and UK, while a step in the right direction, do not generally go far enough. Their concern is largely with marketing rather than with destination management as a whole. This can worsen the negative impact of tourism in a destination because it leads to more effective marketing of the destination but brings no equivalent improvement in the planning and management of the destination. Thus, the volume of tourists may rise without any commensurate increase in the capacity or quality of infrastructure to accommodate these extra tourists.

The implications of globalisation for destinations

The first point to make is that the five types of globalisation noted earlier in this chapter already affect, and will affect all of the four elements of destination management to some extent. These relationships are reflected in highly stylised form in Figure 7.4. The width of the lines depicts the relative strengths of the links.

Having established the complexities of the links, let us now go on to look at each element of destination management in turn.

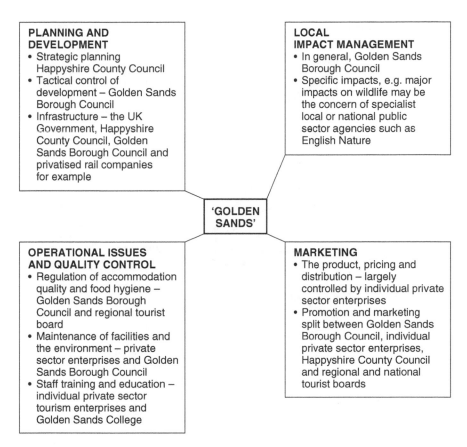

Figure 7.3 Destination management responsibility in a hypothetical UK seaside resort.

Planning and development

There is a danger that the globalisation of the supply side in tourism could lead to the growing influence of large, powerful, externally based, multi-national corporations which could use their power to subvert local, destination level planning controls. Such organisations can afford to have good planning lawyers and run effective public relations campaigns to convince local opinion about the merits of their proposals. They can also afford 'gifts' for the local community to improve their image such as a new swimming pool, equipment for schools, or road improvements. Such organisations will also usually have good contacts with government ministers and may even be able and willing to pay bribes where this is standard local practice to get their own way.

Types of globalisation   **Elements of destination management**

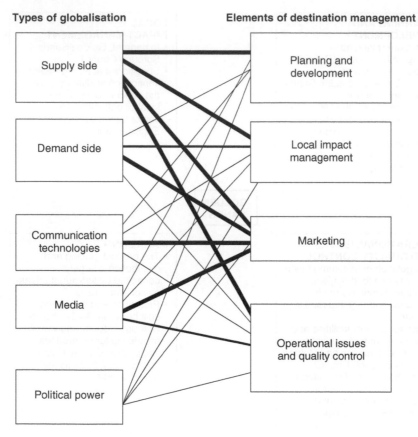

Figure 7.4 Relationships between types of globalisation and the elements of destination management.

On the other hand, many locally based enterprises may lack the financial resources to allow them to match the influence of the large global corporations; but, they may still have enough local political power to influence decisions in their favour.

At the same time, however, the globalisation of the demand side of tourism may help strengthen planning in destinations. Where the tourists come from countries with well developed planning systems and high levels of environmental awareness they may, over time, influence first the industry, and then the public sector in destinations, to enforce stricter planning controls to protect resources which are enjoyed by tourists. Foreign tourists may also help the host community to appreciate the value of traditional landscapes which local people may have taken for granted. This could lead to local pressure for conservation and more proactive planning.

The globalisation of the tourism market has other impacts on destinations. For example, the arrival in a destination of experienced travellers from other countries may lead to demand for both the development and the 'internationalisation' of the tourism infrastructure. This latter phenomenon, whereby tourists expect to have access to familiar international-style airports, roads and hotels, can lead to a loss of local character, a dilution of destination individuality, and a scale of development which is too large for the location.

The rise of global communication technologies has a limited impact on destination planning and development. However, if the destination is heavily dependent upon business tourism, the rise of computer, video, and satellite conferencing may well reduce the overall demand for business travel The growth of a truly global media means that more and more people around the world will be made aware of the strengths and weaknesses of planning and developments in tourism destinations, which may put pressure on destinations to make improvements.

Finally, the globalisation of political power is affecting destination planning and development, in several ways, including the following:

1   There has been a growth in global bodies that bring together governments on a cooperative basis, such as the World Tourism Organisation (WTO). The WTO has published several guides on national and regional destination planning which feature examples of good practice, in the hope perhaps, that other countries and regions may seek to emulate them.

2   Supra-governmental entities, notably the European Union, have been taking an increasingly proactive stance on destination planning and tourism development. For instance, the European Union:
     • provides funding for tourism infrastructure projects such as roads and airports;
     • encourages strategic planning by often only giving funding for individual tourism projects where they are part of a broader integrated area development strategy.

Overall, therefore, we can see that globalisation is likely to make planning and development a more complex, problematic activity for destinations, but its impact is likely to be a mixture of positive and negative effects.

Local impact management

We noted in Figure 7.2 that the main dimensions to local impact management in destinations are the twin desires to:

•   maximise the environmental, economic, and social benefits of tourism for the host community; and

- minimise the environmental, economic, and social costs of tourism for the host community.

Globalisation has implications for these two sides of the same coin in a number of ways. Increasing the benefits of tourism for the destination population while reducing its costs is usually thought to imply local control and ownership, so that:

- the interests of the host population will be given priority; and
- the vast majority of tourist expenditure will stay within the area.

However, the globalisation of the supply side in tourism is threatening this as follows:

1   The growing power of transnational corporations can make it more difficult for locally based, smaller-scale enterprises to become established and survive.
2   Such corporations are footloose, and can, at relatively short notice and low cost to themselves, withdraw from destinations, causing great problems. This is particularly true of tour operators and airlines that usually make little direct capital investment in destinations. The power of these corporations and the fear of what will happen if they pull out of a destination gives them great strength when negotiating with local suppliers. Thus, they can demand prices and terms which are to their benefit at the expense of the local industry.
3   As these enterprises are externally based a significant proportion of the money paid to these corporations by tourists for their holidays will not reach the destination.

Economically, therefore, we can see that the process of globalisation of the supply side of tourism can have very negative impacts on destinations. However, in terms of the environmental and social impacts of tourism, the position is more balanced between costs and benefits. The arrival of global operators in a destination can have social and environmental benefits as follows:

1   Foreign airlines and tour operators may bring new ideas and technologies to the destination in terms of environmental management, for example.
2   International hotel chains may offer local staff the chance to travel to units around the world to work and gain experience.

However, this latter example illustrates the fact that some of the apparent benefits of globalisation can have a negative dimension. In other words, in this case, some of the ablest and most dynamic young people from a destination may emigrate, to work abroad, for ever, thus depriving the local community of their talents.

The globalisation of the demand side in tourism is leading to interesting issues in destinations, particularly relating to cultural differences between host and guest. The rise of long-haul travel is taking people to destinations with very different and traditional cultures, while often the motivation to visit such places may not be a wish to experience the culture, but rather the desire for a new and better sea, sun and sand type holiday. The growth in outbound Northern European tourism to Goa, Thailand and the Dominican Republic is an example of this phenomenon.

The arrival of large numbers of tourists from different cultures in destinations with strong cultural traditions can cause social problems over time. Until recently, Cuba attracted mainly tourists who were generally sympathetic to the political regime. Now, the economic crisis has forced the country to attract all kinds of tourists. Many now have no interest in the politics of Cuba and see it simply as a cheap place to get a suntan and have fun. The lack of tourist sensitivity to the local situation means that tourism is having a major impact on Cuban society, as well as the economy.

Long-haul tourists also often seem to expect the same standards of comfort and hygiene as they have at home, and do not recognise and appreciate the problems faced by developing countries. Tourists may reject destinations unless they take the ethically questionable approach of operating a two-class system of hygiene and food supply for example, where tourists enjoy the first-class system.

The new global tourist is a footloose creature, keen for new experiences, and eagerly targeted by tourism organisations and destinations alike. They show little long-term commitment to most destinations, and do not really seem to care about the welfare of the local population. This means they will have no compunction about moving on to the next inexpensive beach holiday destination.

However, there is also a positive side to the globalisation of tourism demand. The rise of long-haul travel means that destinations which are keen to enjoy the economic benefits of tourism but which are geographically isolated from the main generating countries can now realistically hope to attract tourists provided that they market themselves effectively. This is good news for countries as diverse as New Zealand and South Africa, India and Argentina.

Likewise, the globalisation of communication technologies is also a potential benefit for destinations in that it allows tourists all over the world to find out about lesser-known, newer destinations. On the other hand, this also creates more competition for well-known, established destinations. Thus, this trend could increase the benefits of tourism for the former type of destinations while reducing them for the latter category.

The globalisation of the media can help destinations increase the benefits they receive from tourism, in two ways, namely:

- by raising awareness of the problems which tourism currently causes in these areas and increasing interest in sustainable tourism; and
- by publicising the merits of destinations through holiday programmes and wildlife programmes, on television, for example.

At the same time, the globalisation of political power could help destinations, if supra-governmental organisations such as the European Union were to act to regulate the tourism industry in the countries concerned to improve the situation for destinations. This regulation could include everything from limits on ownership by transnational corporations to environmental standards to wage rates. However, there seems little evidence that this is happening currently. Instead, the Single Market is encouraging transnational ownership while most tourism-related European Union legislation seems more concerned with consumer protection such as the Package Travel Directives.

On balance, it appears that globalisation is a threat to destinations in terms of the local economy, as well as the environmental and social impacts of tourism.

Marketing

We will consider the impact of globalisation in destination marketing in terms of the Marketing Mix – product, price, place and promotion – as well as the market itself.

Globalisation has major implications for the destination product, particularly in relation to the globalisation of the supply and demand sides of tourism. As tourism organisations seek to expand internationally, they tend to build on expertise gained from their own country, and endeavour to offer similar products to consumers in other countries. This has already occurred in the hotel sector with those international chains, which offer a standardised product whether the hotel is in Brussels or Bangkok, Brisbane or Barcelona, Bogota or Baltimore.

However, we are also seeing the same now in other sectors of tourism. In the visitor attractions sector we are experiencing the internationalisation of the theme park market, for example. The major theme park operators are currently looking to develop new theme park projects in Europe and Asia. At the same time, consultants who specialise in theme park projects gain experience on projects all over the world, and then pass this globalised experience on to their clients wherever they may be. This can lead to a rather homogeneous approach to theme park development and management worldwide.

This globalisation of the attraction product is also being seen in the museum sector with the use of inter-active computer-based interpretation; for instance, to interpret everything from the lifestyle of prehistoric man to pop music in the 1960s.

The growth of global communications technologies is also having an effect on other sectors of the tourism industry in destinations. For example, for the benefit of business travellers, many hotels are now installing computer access points in hotels. At the same time, airlines now offer 'smart cards' selling check-in facilities at airports for the convenience of travellers.

There is also a less obvious way in which globalisation of the supply side is affecting the destination product worldwide, namely where new destinations seek to achieve competitive advantage by copying product developments that have proved successful in more established destinations, particularly those they see as their main competitors. This phenomenon is clearly seen in a number of ways, including:

1   The adoption of the all-inclusive hotel concept in some established Mediterranean destinations, following the success of this format in the Caribbean.
2   The introduction of new special events and festivals because they have been seen to be an effective way of putting other destinations on the world tourism map, such as the Edinburgh Festival, the Munich Beer Festival, and the plethora of arts festivals and sporting events that have been created in recent years.
3   The development of facilities in destinations for activity-based tourism, ranging from diving to bungee-jumping, ski-ing to whale-watching.

This globalisation of the supply side in tourism is a real threat to the future of many destinations. It is leading to the homogenisation and standardisation of the destination product around the world. This means that the main way of differentiating many destinations increasingly is on the basis of price alone, which reduces the potential economic benefits of tourism and means that any destination will lose out if a lower cost destination enters the market. This emphasis on competition based on lower price will also make it difficult to find the money to invest in improving the quality of the destination product.

The destination product is also being affected by the globalisation of political power. For example, the European Union is influencing product development in destinations through its funding programmes such as the LEADER scheme in rural areas and its grants for new visitor activities in areas where traditional manufacturing industries are in decline. At the same time, the European Union is developing new brand identities for some destinations, through its 'European Cities of Culture' initiative, whereby each year a different city is given the title or 'brand name' of 'European City of Culture' for one year. For many past 'European Cities of Culture', this brand identity has been used after the year in question to attract new market segments. The cases of Glasgow, Dublin and Lisbon are just three examples of this phenomenon.

Globalisation of supply and demand in tourism is also influencing pricing policies, by making competition global, so that beach resorts in the Mediterranean are now competing with resorts in the Caribbean and Asia. This is putting real pressure on those beach destinations in countries where the general standard of living is higher than that in competitor countries.

At the same time, the globalisation of political power is also affecting the pricing of holidays in general, and the destination product specifically. For example, in Europe the Single Market and airline liberalisation is leading to a steady increase in competition and a reduction in the price of air tickets across all fifteen countries of the European Union.

The globalisation of communication technologies is transforming the role of place or distribution, in tourism. For destinations that have little local control over how their product is distributed, the rise of the Internet offers the potential for them to gain more control over its distribution. If they can develop their own products they can now distribute them directly to consumers at a relatively low cost via the Internet. This could reduce their dependence on externally based tourism organisations, especially tour operators.

The globalisation of communication technologies is also changing the face of promotion in tourism. Indeed the Internet, for instance, is blurring the distinction between place and promotion, for consumers can use it to gain information and make bookings directly, using the same technology. Thus, destinations can now, at relatively low cost, promote themselves inexpensively via the Internet. They can communicate the messages they want to potential tourists rather than simply relying on those being offered about their destination by foreign tour operators.

Globalisation is also affecting the tourism market itself. The globalisation of the supply side has intensified competition between tourism organisations and between destinations. This has led to imaginative product innovation designed to differentiate an organisation's products or even a whole destination from its competitors. Thus, tourists are now being offered products they would not even have contemplated a few years ago such as snow-boarding, and heli-ski-ing, for instance.

The globalisation of the demand side is obviously also bringing major changes to the nature of the world tourism market. People from one continent and culture are increasingly travelling to other continents and cultures for their vacations. At the same time, we are seeing growing numbers of tourists travelling from countries that have not traditionally generated international tourist trips such as Russia and South-East Asia.

Both trends, while being a great opportunity for destinations, also represent a major challenge for them. It means they increasingly have to meet the needs of tourists who are from very different cultural backgrounds. This situation is worsened by the fact that many of these tourists may be the type who want the same things – food, levels of comfort, and language – that they are used to at home.

This is a problem for some destinations, because, for example:

1   Popular brands of food and drink from the tourist's country may be unavailable in, or at least very expensive to export to, remote destinations.
2   The desire of Western tourists for alcohol may be at odds with the prohibition of alcohol, for religious reasons, in some destinations. Religious considerations may also mean that some foods which tourists enjoy eating may not be available.
3   In destinations with limited educational facilities it may be difficult to train staff to be able to speak the language of the foreign tourist.

The problems outlined above, if not resolved, may give tourists an unfavourable impression of a destination, so they will not visit again. It may also lead to them talking negatively about the destination to other people, when they get home, which may discourage those people from visiting the destination in the first place.

It is clear from this section that globalisation has massive implications for destination marketing and that, overall, it is perhaps more of a threat than an opportunity for most destinations.

Operational issues and quality control

Globalisation clearly has implications for a range of operational issues in destinations and for the broad area of quality control. We will illustrate this point by looking at a selection of such issues, including:

•   the regulation of accommodation standards;
•   food hygiene;
•   maintenance of facilities and the environment; and
•   staff training and education.

The globalisation of political power is having an impact on the regulation of accommodation standards in particular regions of the world. For example, within the European Union, the European Commission is seeking to enforce higher standards of fire safety in hotels in member countries. These standards may be far in excess of what national or local laws require currently in many countries.

At the same time, with the increasing globalisation of both the demand and supply sides of tourism, it may be necessary to develop a global system of accommodation classification. It is difficult for tour operators and consumers to know what to expect from a four-crown hotel in the UK, an 'A' class hotel in Greece, or a four-star hotel in Hong Kong. However, it will be very difficult to implement a single global system, and if there were to be such a system, it would probably disadvantage those destinations where standards in general are lower than average.

The globalisation of the demand and supply sides in tourism is also focusing attention on food hygiene, as tourists are increasingly visiting destinations with different – perhaps lower – standards of food hygiene. This can lead to tourists becoming ill, partly because of their own ignorance of food hygiene, as well as poor food handling and storage in hotels and restaurants. The problem is often exacerbated by the tourists' desire for foods they eat at home, and by the tourism industry's desire to reduce costs. The latter accounts, perhaps, for the growth of buffets which pose a hygiene problem because of the risk of cross-contamination, and the long periods for which the food may be displayed. Yet tourists like buffets, including those foods such as mayonnaise and seafood, which are risky from the point of view of the harmful bacteria which cause food poisoning. The food hygiene problem can adversely affect the reputation and popularity of a destination, as we saw in the case of the Dominican Republic in the mid-1990s. The existence of an increasingly globalised media also ensures that any food hygiene problems in destinations are communicated to potential tourists, worldwide, very quickly.

Globalisation of supply and demand is influencing the maintenance of facilities and the environment in destinations. Tourism organisations and tourists increasingly expect destinations to live up to the standards set by other leading destinations. For example, tourists and the industry are placing greater emphasis on the cleanliness of beaches.

At the same time, the globalisation of political power is also beginning to have an impact in this area. For example, in Europe, the European Union Blue Flag Scheme is setting Europe-wide cleanliness and quality standards for beaches.

Finally, everyone seems to agree that staff training and education is vital if destinations are to deliver a quality service to their visitors, and in this field globalisation is having a generally positive impact. Many global hotel chains, for instance, have well-developed training programmes which they use in all locations so that staff can become well trained and gain professional qualifications, even in places where this would otherwise be impossible. One the other hand, this training could be viewed negatively as the imposition of a standardised approach that takes little or no account of different local situations.

However, the globalisation of communication technologies is also allowing universities and colleges to offer training and education to people employed in tourism, anywhere in the world, through distance-learning courses, using the Internet, for instance. This is giving training and education opportunities to some people who would otherwise have no such opportunities.

Having looked at the impacts of globalisation in general, perhaps we should now briefly look at whether these impacts vary between different types of destination.

Globalisation and different types of destination

While it is still largely in the realms of speculation, it does seem likely that globalisation may affect different types of destinations in different ways, to different degrees. The author has attempted to illustrate these potential differences in Figure 7.5.

It is clear, therefore, that the impact of globalisation is likely to vary between different types of destination in response to a range of factors including, for example:

- the location of the destination;
- the level of economic development in the destination;
- political systems and policy in the destination;
- the level of development and power of the local tourism industry in the destination;
- the existing market for the destination; and
- the nature of the destination product.

However, it is also apparent that the ideas put forward in Figure 7.5 are largely hypothetical and generalised. The actual degree and type of impact globalisation will have in any one destination will depend upon the unique local situation and the actions of the key players. The suggestions made in Figure 7.5 are only potential impacts, they could be modified by action on the ground in a particular destination. At the same time, it is important to recognise that the impact of any single type of globalisation, such as the supply side for instance, in a particular destination, will probably have both negative and positive aspects. Finally, we also need to acknowledge that a destination could be affected positively by the globalisation of the media and communication technologies while at the same time being adversely affected by the globalisation of the supply side. Nevertheless, in spite of these complexities, as tourism becomes more globalised in different ways, we need to develop a conceptual framework that will allow us to predict the likely impacts of this globalisation in different types of destinations.

Conclusions

There is, as yet, little real empirical evidence concerning the impact of globalisation on destinations for it is still a relatively new phenomenon. However, as this chapter has hopefully made clear, it does seem likely that it will have a multi-dimensional, and significant impact, on destinations in the years to come. In short, it appears that globalisation is a fact of life that will not go away. Therefore, destinations will have to consider its likely impact on them and plan accordingly.

While it seems that globalisation may offer both threats and opportunities for destinations, as with much else in management, it is likely to be a

	← Greater impact	– Medium impact –	Lesser impact →
1. New vs. established destinations	New destinations seeking to establish themselves on the world map, who will be able to exploit the globalisation of supply, demand, media and communication technologies	Established destinations which are stagnant or in decline and are threatened by competition stimulated by the globalisation of the supply and demand sides	Established destinations which are still attracting new visitors and are already well integrated into the global tourism system
2. The existing market	Destinations which traditionally have attracted a regional or national market but that have the potential to attract tourists from all over the world, due to the effects of the globalisation of supply and demand	Destinations which already attract a largely international market	
3. Developed vs. developing countries	Developing country destinations where the tourism supply and demand is already largely controlled by externally-based organisations	Developed country destinations where tourism supply and demand is already largely controlled by externally-based organisations	Developed country destinations where tourism supply and demand is largely controlled by the local tourism industry
4. The level of resourcing of the destination marketing agency	Destinations with less well-resourced destination marketing agencies given that the globalisation of the media and communication technologies will reduce the cost of promoting destinations	Destinations with well-resourced destination marketing agencies which are already marketing their destinations effectively but will use the globalisation of the media and communication technologies to operate even more effectively	
5. Destination product quality	Destinations with lower than average product quality where weakness will be identified and exposed by the globalisation of supply, demand, media and communication technologies	Destinations with average product quality	
	Destinations with above average product quality who will gain an advantage through the globalisation of supply, demand, media and communication technologies, because all of these will lead to their product quality strengths being more widely publicised		
6. Mass market vs. niche market destinations	Mass market destinations where the existing markets are stagnant or in decline and there is a need to attract tourists from new markets	Niche market destinations that will seek to use globalisation of the media and communication technologies to attract tourists in a more cost-effective manner	
7. Political policy	Destinations where national and/or local government encourage the process of globalisation and encourages the involvement of externally-based organisations in the destination	Destinations where national and/or local government neither stimulate nor resist the impact of globalisation in the destination	Destinations where national and/or local government resist the process of globalisation and seek to insulate the destination from its effects

major threat only to those who do not seek to anticipate and manage its effects. A major danger for all destinations from globalisation is that the process of globalisation in general often involves trying to achieve economies of scale through the standardisation of products and service systems. This is a real threat to destinations that want to differentiate themselves from other destinations in order to:

- compete on the basis of what they offer rather than just being the cheapest producer of a standardised experience;
- develop brand loyalty and repeat business; and
- give tourists a motivation to wish to visit their particular location.

This potential standardisation is also a threat to the distinctive traditional cultures and lifestyles of destinations. This diminution of differences in culture and lifestyles often harms the social fabric of the host community. Equally importantly, this homogenisation appears to be at odds with the apparent desire of more and more tourists to:

- feel that they can tailor-make a vacation which matches their individual tastes and will be different from everyone else's; and
- continuously find and enjoy new and different tourist experiences.

Taken to extremes, therefore, globalisation could be a threat to the future of tourism as a whole, as well as just to the destinations themselves, by making the vacation experience less attractive and satisfying for the tourist.

Strategies for destinations in the age of globalisation

The author believes that if a destination wishes to, it can exploit globalisation for its own ends rather than being its victim. While it would be presumptuous at this stage to suggest a strategy that all destinations should follow in response to globalisation, it is perhaps legitimate to suggest some elements that might underpin any such strategy. These might be as follows:

1 Local tourism interests in individual destinations should develop mechanisms that allow them to cooperate effectively with each other so that they can present a united front against externally-based organisations and the globalisation of the supply side.
2 Destinations should learn all they can about the needs and tastes of the new markets they may wish to attract thanks to the globalisation of demand, and should then modify the product to meet these needs.

Figure 7.5 Different types of destination and the level of impact of globalisation.

3   Destination marketing organisations should endeavour to use the globalisation of the media and communication technologies, such as the Internet, for their own benefit. These two phenomena give all destinations the chance to sell their message to a wider audience at a lower cost than ever before.

4   The globalisation of political power and the growth of inter-government cooperation and supra-governmental organisations, should lead to regulation of the worst effects of globalisation on destinations. However, this should not be an excuse for protectionism that allows destinations to offer a poor quality product or charge too high a price for their product.

5   Destinations could perhaps also learn the lessons of the oil-producing countries and the OPEC organisation specifically. In other words, in an era of globalisation, destinations should perhaps work together to give themselves bargaining power in their relationship with major tourism corporations. Similar forms of destination could form marketing consortia. These consortia of eco-tourism destinations, spa towns or beach resorts for instance, could work together to:
   • prevent under-pricing and unfair price competition;
   • share resources to mount more effective promotional campaigns;
   • force tour operators to enter longer-term fairer agreements with destination suppliers;
   • prevent tour operators playing one destination off against another for short-term gain; and
   • share market intelligence and examples of good practice in destination marketing and management.

Consortia have already been seen to operate successfully in the visitor attraction field and even in the airline business, with the strategic alliances between different airlines. However, it could be difficult to persuade destinations that see themselves as competitors to cooperate for their mutual benefit. Nevertheless, in the era of globalisation, where the rules we lived by yesterday no longer seem valid, we may have to radically change our attitudes and adopt new approaches, such as cooperation between destinations.

The future

Globalisation is currently a feature of all industries, and indeed everyday life as a whole; it is not restricted to tourism and holiday-taking. It is increasingly making the traditional boundaries between different industries seem as irrelevant as national boundaries. In the longer term, it appears likely that today's major tourism industry players may well be replaced by corporations that are largely non-tourism related. The globalisation of the media and communication technologies brings the prospect of mega media and computer corporations taking tourism under their wings, seeing tourism as a sub-set of their wider global operations. Then, decisions affecting destinations may not even be taken by tourism specialists at all, but rather by media empire bosses or leaders of information technology corporations.

At the same time, the globalisation of supply, demand, the media, and communication technologies in tourism may force destinations to reverse a classic strategic management cliché if they are to survive in the era of globalisation. While corporations which seek to become global players talk of 'thinking globally, acting locally', destinations may instead need to:

'THINK LOCAL, ACT GLOBAL'

In other words, they may need to prevent the product standardisation and loss of local uniqueness which globalisation can cause, while taking advantage of the opportunities which globalisation offers for them to exploit new global markets.

Ironically, as the marketplace becomes ever more global, the uniqueness of individual local places may be the key to their survival and success as tourist destinations.

Bibliography

Archdale, G. (1994) 'Destination databases: issues and practices', in A. V. Seaton, C. L. Jenkins, R. C. Wood, P. U. C. Dieke, M. M. Bennett, L. R. MacLellan and R. Smith (eds) *Tourism: The State of the Art*, Chichester: John Wiley.

Ashworth, G. J. and Goodall, B. (eds) (1990) *Marketing Tourism Places*, London: Routledge.

Burns, P. M. and Holden, A. (1995) *Tourism: A New Perspective*, London: Prentice-Hall.

Douglas, S. P. and Craig, C. S. (1995) *Global Marketing Strategy*, New York: McGraw-Hill.

Ghoshal, S. (1987) 'Global strategy: an organising framework', *Strategic Management Journal* 8: 425–40.

Go, F. G. and Pine, R. (1995) *Globalisation Strategies in the Hotel Industry*, London: Routledge.

Heath, E. and Wall, G. (1992) *Marketing Tourism Destinations: A Strategic Planning Approach*, New York: John Wiley.

Horner, S. and Swarbrooke, J. (1996) *Marketing Tourism, Hospitality and Leisure in Europe*. London: International Thomson Business Press.

Inkpen, G. (1998) *Information Technologies for Travel and Tourism*. 2nd edn, Harlow: Longman.

Ioannides, D. and Debbage, K. G. (eds) (1998) *The Economic Geography of the Tourist Industry: A Supply-Side Analysis*, London: Routledge.

Kaynak E. (ed.) (1985) *Global Perspectives in Marketing*, New York: Praeger.

Laws, E. (1995) *Tourist Destination Management: Issues, Analysis, and Policies*, London: Routledge.

Laws, E., Faulkner, B. and Moscardo, G. (eds) (1998) *Embracing and Managing Change in Tourism: International Case Studies*, London: Routledge.

Poon, A. (1993) *Tourism, Technologies, and Competitive Strategies*, Wallingford: CAB International.

Ritzer, G. (1996) *The McDonaldisation of Society*, Revised edn, Thousand Oaks, CA: Pine Forge Press.

Smeral, E. (1998) 'The impact of globalisation on small and medium enterprises: new challenges for tourism policies in European countries', *Tourism Management* 19(4): 371–80.

Sharpley, R. (1994) *Tourism, Tourists and Society*, Huntingdon: Elm.

Swarbrooke, J. (1999) *Sustainable Tourism Management*, Wallingford: CAB International.

Swarbrooke, J. and Horner, S. (1999) *Consumer Behaviour in Tourism*, Oxford: Butterworth-Heinemann.

Urry, J. (1995) *Consuming Places*, London: Routledge.

Urry, J. (1990) *The Tourist Gaze: Leisure and Travel in Contemporary Society*, London: Sage.

Uysal, M. (ed.) (1994) *Global Tourist Behaviour*, Binghampton, NY: International Business Press.

Walters, M. (1995) *Globalisation*, London: Routledge.

8 Achieving global competitiveness in SMEs by building trust in interfirm alliances[1]

F. M. Go and J. Appelman

Introduction

This chapter explores cooperation and interfirm alliances between small and medium sized enterprises (SMEs) as a potential strategy in achieving competitive advantage within a global environment. It takes an integrative perspective approaching the question why small and medium sized enterprises in the rapidly growing tourism sector should adopt a collaborative as opposed to an internalization strategy. The former approach builds on the unique position of tourism SMEs to connect different organizations and public and private partners sectors when fulfilling customer demands. Due to its 'connective' nature, tourism has the potential to act as a change agent. In this respect a network comprised of agricultural and touristic interests could result in social and economic synergies in regionally oriented development schemes.

Networks are operationalized as consisting of value chains. The number and kind of value chains in such a network depend on customer demands and the vision that has been developed with all participating stakeholders. The aspiration of our research focus is to analyse the dynamism of tourism networks in the context of the national (The Netherlands) and supranational (European Union) competitive space. It concludes by reflecting on a number of issues that actors can operationalize to handle the problems stemming from this dynamism.

By the end of the 1980s the advent of the Single European Market had extended the geographical scope of many European companies, including SMEs, which previously competed in relatively sheltered local markets. It is expected that this shelter will continue to gradually whither away as long as the EU moves towards further integration and the supranational management of competitive space (Tulder 1996). Furthermore, Day (1990: 91) observed that consumer needs and lifestyles have become simultaneously more universal (global) and more indigenous (local). As a result the term 'mass-customization' has emerged, which implies that individual consumers can specify their preferences for a good or service delivery. In turn the simultaneous homogeneous and individualized market results in a more complex and diverse environment and means that firms must become more flexible to cater to shifts in demand. Accordingly, they will find it increasingly difficult to succeed by following industry conventions, producing well-defined commodities through the application of marginally differentiated technologies (Tremblay 1998).

SMEs strategies and tourism

Tourism is a highly complex compounded service brought about through the 'assembly' of different services that are being delivered by a network of companies that is often global in scope. The actual delivery of services is dependent on the way in which the consumer assembles a holiday and recreates himself once he arrives at a destination. A holiday is defined as a chain of services delivered in a certain order. The less structured a tourist travels the less it is possible to pre-structure the delivery of services and therewith to warrant a certain quality when delivering the service. The inability of entrepreneurs to control the quality of the service-delivery process and therewith the experience of the customer is further exacerbated in part by influence the external environment has on the valorization of a tourist experience beyond the control of a firm. From a business perspective a customer-based evaluation of a tourism experience is therefore dependent on many uncontrollable variables, in large part because it is the very interaction of the tourist with the environment that determines the degree of satisfaction.

Tourism firms have two generic strategies at their disposal in order to mitigate the problem of controlling the environment:

* First, a competitive strategy of 'internalization'. A company perceives itself as a provider of a bundle of specific services that can best be delivered by acquiring the various required components and placing them within a coherent company framework. Walt Disney's theme parks serve as an example of internalization strategy.
* The second strategy is aimed at collaboration. Learning, coordination and value-adding activities are far more central than the reduction of costs and control. In this way a network of collaborating small-scale firms can develop substantial competitive power. In the context of the Dutch tourism sector such an approach could prove fruitful, if only because 90 per cent of all tourism companies are SMEs. Due to the intrinsic peculiarities of delivering tourism services, the liberalization and deregulation of the market and the internationalization of the economy, the second strategy is an appropriate one for SMEs. However, to cooperate an organization needs to possess certain properties.

SME: the organization

In order to respond to changing market conditions SMEs need to adapt to the environment, develop new ideas and innovate. Innovation is a rather elusive concept. It is highly debatable what an optimal rate of innovation is. Enough to attain and maintain a competitive advantage but not so much that the firm innovates itself into oblivion. Which is to say running ahead of the market or focusing solely on development of new products and ignoring the market and clients. Innovation occurs as the alternatives to compete at the lowest possible cost by duplicating other entrepreneurs have been exhausted 'bringing about heightened competitiveness and fewer obvious opportunities for differentiation in the industry. In either case, an individual firm seeking further advantage must eventually make a major, risky, and expensive change' (Hickman and Silva 1987: 7). Cooperation, through interfirm alliances, connects different business systems and can itself stimulate creativity and innovation. Against such a backdrop, it is desirable to view the SME as an organization that has the physical resources (capital, equipment) core competencies, value-creating disciplines (tacit and operationalized), that allow a product to be delivered to customers at the best possible price/performance trade-off (Hamel 1991). Put differently, SMEs are essentially concerned with the acquisition of value-creating disciplines. The most sustainable value-creating discipline is one comprised of tacit knowledge. Such disciplines as the quality of face-to-face service delivery encounters are hard to imitate or buy especially when they are ingrained in the culture of an organization. Collaboration with other firms is a way to obtain such value-creating disciplines. It is this conclusion that drives the authors to make an argument for researching firms as networks, because such a view mirrors the day-to-day reality in regions containing SMEs. Collaboration calls into being the need for coordination, information and communication technology (ICT) is likely to play an increasingly significant role in this respect.

ICT

The increasing importance of ICT and the corresponding organizational transformations as drivers of change thus becomes a point of attention. We need to keep in mind that it is not technologies themselves that drive change but the way in which such technologies are organized to fit successfully into the daily life of the consumer and worker that determines the success of a new technology (Lash and Urry 1994; Pudney 1953). One of the most immediate technology driven impacts will come about through a further development of applications (forms of social organization) of communication and information technology. It is, for instance, likely that in the near future travel agents will need to redefine their business due to pressures from ICT. If they do not respond to these pressures in an innovative way it could result in bankruptcies and loss of jobs. If they do respond they could become expert information brokers and value chain coordinators (Sarkar *et al.* 1996).

Furthermore, competitive advantages to be derived from the Information Revolution, as opposed to the Industrial Revolution, follow a different logic. ICT does influence production and transportation processes but its distinctive character resides in enabling, enhancing and facilitating the organizational interfacing of firms, nations, business units, and local governments. Control and coordination of information is the key to competitive advantage for the modern-day firm. To this we would add that it is not only the coordination of information that counts: it is equally important to have access to information and know-how. Competitive advantages for firms can thus be created through:

- the access, control and coordination of information;
- the stimulation of creativity in the organization; and
- the strategic collaboration with other firms (Dyer and Singh 1998).

The EU and regions

The monetary and economic unification of Europe is progressing rapidly. Nation-states will become less important as points of reference for business and sources of identity for civilians and consumers. This process also affects the nature of competition (Piore and Sabel 1984). The effect of both processes is that regions such as Catalunya, Prato, Baden-Wurtemberg, Brabant, etc. will become more central as economic centres of growth and of tourism attractions as opposed to nations. Synergies between firms and clusters of SMEs in such regions could lead to a variant of industrial districts. We will label such a district, for the moment, a leisure service district. The Lake District in the United Kingdom is a 'pure' example in the sense that almost all activities are geared towards enhancing the value of the tourism experience. An example of a more mixed district is the city of Barcelona and the 'costa's' in the vicinity of Barcelona. It is mixed because this district is also an industrial centre within Spain. A new trend is emerging which is labelled for the moment eco-oriented regional development. It is in part driven by European Commission policy (EC 1991, 1995, 1997) to revitalize rural areas by looking for complementary competencies between at least three economic sectors: agriculture, tourism and transport. It has been supported by organizations that are responsible for nature development and conservation. The actual collaborative efforts and especially the monitoring are supported by the use of ICT. The drive toward regional development is exemplified by the 'regional culinary heritage' concept, a European initiative in which selected restaurants, food processing companies and farms connect within a region. They do so in order to deliver high-quality dishes that are part and parcel of the culinary tradition of a specific region. The raw produce and the major part of value-adding activities should happen within the region. Through a shared European-wide web-site participants can learn from each other and customers can inform themselves. For producers, expansion of the services delivered from such a site represents only a small step. Other services such as booking a hotel room, a fishing trip or visiting a local farm or museum can easily be added. Furthermore, international tourists with an interest in gastronomy could book, for instance, a European 'gourmand' holiday: tasting and enjoying, in other words experiencing a rich and diversified European culinary history.

Competitive advantages through collaboration: the central role of trust

The number and variety of interfirm relations and networks such as strategic alliances have accelerated during the last two decades. This process is also reflected in the proliferation of publications on aspects of inter-organizational relations (Gulati 1998; Nohria and Eccles 1992; Powell and Smith-Doerr 1994; Uzzi 1996, 1997). Partnerships deliver specific forms of rent. These forms of rent are called relational rent and defined as: '... a supernormal profit jointly generated in an exchange relationship that cannot be generated by either firm in isolation and can only be created through the joint idiosyncratic contributions of the specific alliance partners' (Dyer and Singh 1998: 662). According to a relational perspective, rents are jointly created and owned by partnering firms. Thus, relational rents are a property of the dyad or the network. A firm in isolation, irrespective of its capabilities or resources, cannot enjoy these rents. A relational capability only creates value when a firm is actually working together with another firm with the aim of creating added value that could not have been produced by either firm alone. To put it more sharply, the relation produces its own added value. Alliances generate competitive advantages *only as they move away from the attributes of market relationships*. Such a proposition implies a move toward cooperation between firms (or other stakeholders).

The increasing use of ICT is one of the drivers behind the creation of a 'global village'. Firms in such a village are far more open to public scrutiny and feel therefore the need to legitimize their actions. They increasingly feel the need to formulate a code of conduct and pay serious attention to it. Customers and (potential) partners have to trust a firm otherwise financial performance will be seriously affected. Who wants to buy from or work with an unreliable partner? In other words, firms become more and more part of their environment both ecologically and socially. They are increasingly related to other firms, other stakeholders and the natural environment. Financial profit as a proxy to measure a relational competitive advantage as proposed by Dyer and Singh is therefore, by itself, insufficient to assess improved performance as performance is multifaceted in nature. This is especially true for firms that are vulnerable to scrutiny from governments, NGOs and the public at large. For example, non-organic farming, especially the firms that produce genetically modified foods, tourism enterprises due to the customer involvement in the production process on site, and last but not least the intensive use of public space (monuments, squares, landscapes) are all under increasing public scrutiny. Performance measures that take ethical, social and reputational considerations into account will become more prominent. For instance, a lot of investment banks when assessing firms take the trustworthiness of firms in collaborations and the structure of the network of a company into account (personal communication, 14.10.98). Additionally, customers within the EU tend to focus more and more on health, security, novelty and quality when considering the purchase of food and leisure items. The industrial way of agricultural production has led to scandals such as BSE, public and scientific doubts about gen-tech foods, and the Dutch 'pig-crises'. The outbreak of Legionnaires Disease recently at a Dutch horticulture exhibition resulted in the death of twenty-one visitors and cast yet another dark shadow on the agricultural sector. Consumers respond to such events by looking for quality products that will not expose them to potential health-risks even if they have to pay a little extra. Firms that understand the importance of being perceived as an attractive partner will invest in interfacing with relevant organizations and corporate communication to boost their public reputation.

What other advantages can be gained from collaborating? What kinds of relational rents can be pursued? Dyer and Singh (1998) describe four categories of relational rent:

1   Investments in relational specific assets (e.g. joint investment in an ERP-system for a tourism destination, cooperative cold vegetable storage).
2   Substantial knowledge exchange, including the exchange of knowledge that results in joint learning (regional TQM systems, learning alliances).
3   The combining of complementary, but scarce, resources or capabilities (typically through multiple functional interfaces), resulting in the joint creation of unique new products, services or technologies (combining tourism and agriculture by emphasizing sustainability).
4   Lower transaction costs than those of competitor alliances, owing to more effective (= less costly) governance mechanisms.

Uzzi (1997: 35) finds the following positive effects of cooperation:

*   economies of time;
*   integrative agreements;
*   Pareto improvements in allocative efficiency; and
*   complex adaptation.

A considerable number of authors (Doz 1996; Uzzi 1997; Zaheer *et al.* 1998; Gulati 1998; Dyer and Singh 1998) stress the fundamental role that trust plays in establishing and maintaining affective, effective and efficient relationships in interfirm collaboration.

Trust

For the balance of this chapter, we focus on the role of trust as both the foundation and governance mechanism of sound interfirm alliances. Such alliances are the building blocks of a value chain. We will conclude with a short overview of the most important aspects of value chain development. Trust can be defined as: '. . . the extent to which a firm believes that its exchange partner is benevolent and honest' (Geyskens and Steenkamp 1995). Trust in a partner's benevolence is a channel member's belief that its partner is genuinely interested in one's interests of welfare and is motivated cooperatively to seek joint gain now and in the future. A benevolent partner subordinates immediate self-interest for joint long-term gain (Geyskens and Steenkamp 1995; Morgan and Hunt 1994).

The evolution of cooperation is intimately wound up with the evolution of trust. The evolution and amount of trust in a given relationship could be considered a proxy for the relational rents gained. The lower the amount of trusts in a relationship the less likely the chance that relational rents will be apparent. Conversely, the more embedded a relation between partners is the higher the likelihood of relational rents. Trust therefore becomes of increasing concern to firms; not only in interfirm relationships but also with stakeholders in the environment and towards customers.

There will be no customer loyalty without an investment of trust from both sides. The development of trust is accompanied by an increase in contacts between individuals from different organizations, who share not only precise, rational and goal-related information but also interact socially. Successful cooperating partners usually have a business-relationship embedded in more durable forms of social links such as friendship and family ties. Both kinds of relationship usually carry with them a large amount of trust. Uzzi (1997) found besides trust two other main functions of embedded relationships: fine-grained information transfer and joint problem-solving arrangements. Both functions can contribute to the creation and sustenance of competitive advantages.

Trust acts as the governance mechanism of embedded relationships. The primary outcome of governance by trust is that it promotes access to privileged and difficult-to-price resources, that enhance competitiveness and the overcoming of problems, but are difficult to exchange in arm's length ties. Unlike governance structures in atomistic markets, which are characterized by calculativeness, monitoring devices, and impersonal contractual ties, trust is a governance structure that resides in the (social) relationship between and among individuals. Because of the existence of trust there is lower uncertainty speeding up decision-making (economies of time) thus conserving cognitive resources that can be put to other uses, like sharing information or learning. Trust is both a governance mechanism as well as a way to promote economies of time. It is therefore relevant to know when actors will act trustworthy towards each other.

Granovetter (1992) sees three reasons why actors might act trustworthy in economic transactions:

- One is because it is in their social and economic interest to do so.
- The second reason why actors act trustworthy is because they believe it is morally right. Values and norms are supposed to steer behaviour.
- A third reason why actors might act trustworthy hovers between those two extremes; actors involved in business deals could see acting trustworthy as a part of the mutual regularized expectations.

Trust then is a characteristic of the personal relation between two actors. But it can become more than a personal characteristic when exchange is frequent and successful according to the actors involved.

Fine-grained information transfer is a second component of embedded relationships. The main features are that the information shared is more all-inclusive (holistic) and tacit compared to the price and quantity data in arm's length ties. Repetitive communication leads to an understanding of each other's product, knowledge and strategic options. An understanding of each other's frame of reference develops, increasing effective interfirm coordination. Because people know each other and trust each other the way is opened to start cooperating on new ventures and innovations could ensue in this way.

This is exemplified by joint problem-solving arrangements, the third component. When a problem arises, firms work together on solving this problem. Thus, obtaining feedback from each other that enhances learning and stimulates creativity in the form of the discovery of new combinations of products or ways of delivering a service.

It has become clear that the corporate structure or network represents the *framework* of conducting business in the information economy. Within such framework of organizational networks the issue arises how the *features and functions* of embedded ties and the role of trust can become linked and operationalized in the emerging networks. Embedded ties develop through third-party referral networks and previous personal relations. Therefore, trust plays a central role in almost any network. One actor connected to two other actors who do not have a direct relationship functions as a go-between or broker. The go-between not only brings two previously unconnected actors together but also, and this is important, transfers the qualities or characteristics of the relationship the go-between had with both other actors onto the new relation. Because the unconnected actors trust the go-between they take this trust, initially, with them in the new tie (Uzzi 1997: 48). As exchange is reciprocated, trust forms independently of the earlier tie and a basis for fine-grained information transfer and joint problem-solving is set in place. That this process occurs has also been proven in a quantitative way by Zaheer *et al.* (1998). The authors separated interorganizational and inter-personal trust when investigating the role of trust in interfirm exchange and assessed its effect on performance. Their conclusions point out that once con-tact between organizations has been established, and this part always starts with individuals, the personal relation becomes institutionalized. The interorganizational context becomes more prominent when exchange is carried out between organizations with an institutionalized pattern of deal-ings. They emphasize this point by using the following quote from an inter-view: 'Even though I may leave, the relationship (between our firms) will continue since what we've been doing goes beyond one or two people. Over the years there are a host of people who have worked together.'

These findings coincide with the model Doz (1996) puts forward with regard to the evolution of cooperation through learning in alliances. What is important is that trust lies at the roots of any kind of successful coop-eration. Although it need not be a starting condition, it is highly unlikely that the evolution of cooperation can be successful if trust does not surface rapidly.

In summary, the features of embedded ties, with trust as a governance mechanism, are: the mutual overcoming of problems, speeding up of decision-making processes, interfirm coordination and the promotion of learning. The three main functions of embedded relationships were trust, fine-grained information transfer and joint problem-solving arrangements. Through improved communication, trust and mutual feelings of obligation managers collaborate in different ways. By maintaining a balance between competition (arm's length contracts) and cooperation (embedded ties) a firm is more resilient in bad times and can become more innovative through learning thereby attempting to improve its performance in the process. In short, SMEs that manage their networks of contacts can be more compet-itive than similar firms that fail to do so. These findings are based on a study examining twenty-three firms in-depth. The premise of the same study was that: 'Economic theory makes strong predictions that social ties should play a minimal role in economic performance' (Uzzi 1997: 35). Traditionally, the Dutch tourism sector resembles such setting due to the lack of collaboration. At present, at different policy levels collaboration in the value chain and interactive policy-making are studied and applied so as to bring about integrated tourism development. A value chain is but one chain of actions executed at a certain time and in a certain order to satisfy the need(s) of a particular customer. In practice, most firms need to draw on a number of value chains depending on the demands a customer makes. In the remaining part of this chapter we therefore examine the potential application of the value network and interfirm alliances in the tourism sector.

Networks and interfirm alliances

The recognition that added value is often created through an interfirm alliance leads to the recognition of the importance of the network concept (Håkansson and Snehota 1997). The value network as proposed by Stabell and Fjeldstad (1998) relies on a mediating technology to link clients that wish to be interdependent (Thompson 1967). The value network is relevant for a tourism sector context for two reasons:

- First, information flows in this sector consist of a complex network of businesses engaged in providing entertainment, accommodation, food, transportation and communication to the traveller. In other words, the primary function of the tourism system is to deliver a pleasurable and satisfying experience. The ICT-applications should be viewed as the available mediating technology to coordinate and control information between the network players. In order to sustain competitiveness, SMEs must understand their roles, responsibilities and relationships within a given network and identify where value-adding activities are created in the network. The emergence of information technology and the intrinsic complexity of tourism require SMEs, therefore, to extend their horizons beyond traditional trade channels of distribution to the value network.

- Second, due to the unabated pressures of international competition it is very likely that the phenomenon of cooperation between firms, especially SMEs will spread (Go and Williams 1993). The rise of transnational corporations that base their competitive advantage amongst others on scale economies causes SMEs to turn into niche marketers and complement, where appropriate, transnationals that are engaged in the leisure sector, transportation and telecommunications. SMEs require an organizational structure to function effectively within the value network and a fiercely competitive market. The 'consumer-driven value chain' is introduced as organizational structure for SMEs to bundle and build the organizational capability that creates added value for stakeholders.

The consumer-driven chain is an excellent vehicle to transform constraints into opportunities. The following arguments explicate why the application of the consumer-driven chain model within a tourism context is appropriate:

1   The consumer-driven chain has been successfully applied in other sectors, including agriculture.
2   The consumer-driven chain creates more flexible and responsive attitudes amongst suppliers, so as to achieve a 'mass-customized' approach in service delivery.

3   The consumer-driven chain implies a higher degree of efficiency. The challenge is to achieve the highest possible yields against the lowest possible costs and losses.

4   The consumer-driven chain realizes the type of configuration enabling SMEs to achieve scale economies in production and marketing. In addition, cooperation and interfirm alliances may be an important precondition to be observed and listened to when public issues arise and to contribute to the resolution of public issues, such as environmental preservation and mobility, whilst stimulating economic growth.

5   The consumer-driven chain provides a platform to collectively generate, collect and diffuse knowledge. For example, through the operation of a collective customer database, which an individual SME could not afford (Anon 1998).

What does cooperation and/or interfirm alliances within the consumer-driven chain imply? It implies the horizontal and vertical integration and cooperation of firms. The operational aim then is to deliver the desired quality and variety to consumers in an efficient manner. In order to meet this aim SMEs must abandon the traditional orientation towards mass production and cost reduction. In other words, leave behind an 'industrial mentality' and move toward an 'information process mentality'. The adoption of an information mentality is required to reduce the fragmentation of tasks, alter the conventional perceptions of workers and job advancement and training as costs as opposed to investments.

SMEs need to transform toward a performance model that focuses on adding value for all stakeholders. Flexible production, teamwork, emphasis on skills and abilities and the notion that job advancement should be encouraged through certified skill training for everyone with the appropriate aptitude and attitude characterize such a model. Due to the introduction of information technology, the market and therewith the workplace is rapidly changing toward such a performance model. The increase in the information handling capacity of a value network produces new market structures and opens up opportunities for alternative service-delivery processes (Go *et al.* 1998). The application of the consumer-driven chain enables SMEs to capitalize on the challenges and opportunities in newly emerging markets. The central idea of the value chain approach is that horizontal and vertical harmonization and cooperation are essential when one seeks to provide the consumer with the demanded quality and desirable assortment of products in an efficient way, and to continue to do so. This train of thoughts leads to the formulation of a strategy that can be implemented cooperatively with relevant stakeholders.

1   Identifying consumer groups, including their needs and wants.

2   Reflecting upon the products and services that should be delivered, in which consumer preferences on the one hand and production opportunities on the other, are taken into account.

3   Verifying how to market the products and services with the best possible value added.
4   Identifying what types of cooperation would be most profitable. This would include cooperation with suppliers and buyers (vertical cooperation), as well as cooperation with producers of similar or related products or services (horizontal cooperation).

As a result of this process, newly formed value networks could position themselves according to three generic objectives:

1   Chain differentiation: the uniqueness of the product or service delivered. Chain differentiation is characterized by three keywords: innovation-strength, flexibility, and responsiveness, all essential qualities that are needed to be able to respond to the increasingly diversified consumer demands.
2   Integrated quality: in today's competitive arena, reliability is an ever increasingly important issue. In order to satisfy consumer demands for high quality, healthy, safe, durable and environmentally friendly products and services, it is essential to cooperate within the value chain.
3   Chain optimization: strategic chain objectives could also be formulated in relation to the efficiency and effectiveness of the production process. This involves the optimization of the mutual coherence of the nodes within the network.

Concluding reflections and remarks

We end by reflecting on a number of questions. *Is trust more than the latest business 'buzzword'?* Trust is a concept that is more relevant to our information-age than it was in the industrial age and therefore more attention is paid to the concept. Important effects of the information revolution are an increasing recognition of the interdependence of businesses and the increasing possibility of public scrutiny of businesses. Firms become more and more conceptualized as cooperating interdependent partners. Firms increasingly work in and through networks and chains but stay, in the very last instance, autonomous players. In such a setting trust between partners is imperative.

If trust becomes more important a logical question from a business point of view is: *'Is it even possible to reduce the idea of trust to managerial principles, strategies and tactics?'* We have shown that trust can and does fulfil different functions. Such functions can be managed but not controlled. Coordination and collaboration between firms through the development of a shared strategic vision and operationalized in value chains and networks summarizes the process. Trust evolves within this process from personal relations to an institutionalized interorganizational system. Trust starts as a prerequisite and evolves towards a governance mechanism. When this process is successful relational rents will surface, further strengthening the links between firms. As such, a strategy based on a vision that a firm can improve its performance by using the principles of collaboration and coordination can be feasible and desirable.

If market coordination fails to ensure a high degree of trust, will this lead to a swing toward more reliance on regulatory coordination? Market coordination never ensures a high degree of trust, because the former is characterized by calculativeness, monitoring devices, cost-based reasoning, and impersonal contractual ties. However, trust is a governance structure that resides in the (social) relationship between and among individuals or firms. Because of the existence of trust there is lower uncertainty speeding up decision-making (economies of time) thus conserving cognitive resources that can be put to other uses, like sharing information or learning. We have put forward the ideas of relational rent as proposed by Dyer and Singh (1998). They state unequivocally that alliances generate competitive advantages *only as they move away from the attributes of market relationships*. This does not mean a greater reliance on *regulatory coordination that imposed*, through law or trade-agreements, but opens up the way to *self-regulation within a network*. Ideally firms choose to be regulated on their own terms with the partners they wish to cooperate with. In practice dual forms of regulation will continue to be there, just as firms need to choose between competing and collaborating depending on the issue at hand. Nevertheless, we hope to have shown that SMEs can create and sustain competitive advantage through interfirm collaboration and the building of trust.

Note

1 This chapter was presented at the International Food and Agribusiness Management Association (IAMA) World Food and Agribusiness Congress in Florence, Italy, 14 June 1999.

References

Anon. (1998) *Een bredere weg naar een kansrijke toekomst*, Rosmalen: TREK-Brabant Haalbaarheidsstudie (oktober).

Day, G. S. (1990) *Market Driven Strategy Processes for Creating Value*, New York: Free Press.

Doz, Y. (1996) 'The evolution of cooperation in strategic alliances: initial conditions or learning processes?' *Strategic Management Journal* 17: 55–83.

Dyer, J. and Singh, H. (1998) 'The relational view: cooperative strategy and sources of interorganizational competitive advantage', *Academy of Management Review* 23(4): 660–79.

EC (1991) *Community Action Plan to Assist Tourism*, Brussels: Commission of the European Communities (24 April).

EC (1995) *The Role of the Union in the Field of Tourism*, Commission of the European Communities (4 April).

EC (1997) Report from the Commission to the Council, the European Parliament, the Economic and Social Committee and the Committee of the Regions on Community, *Measures Affecting Tourism*, Brussels: Commission of the European Communities (1 February).

Geyskens, I. and Steenkamp, J. B. E. M. (1995) 'Generalizations about trust in marketing channel relationships using meta-analysis', working paper: Catholic University of Leuven, Department of Applied Economics.

Go, F. M. and Williams, A. P. (1993) 'Competing and cooperating in the changing tourism channel system', *Journal of Travel and Tourism Marketing* 2(2/3): 229–48.

Go, F. M., Govers, R. and Heuvel, M. van den (1998) 'Towards interactive tourism: capitalizing on virtual and physical value chains', in *Information and Communication Technologies in Tourism*, Vienna: Springer Computer Science.

Granovetter, M. (1992) 'Problems of explanation in economic sociology', in *Networks and Organizations: Structure, Form and Action*, N. Nohria and R. Eccles (eds)

Gulati, R. (1998) 'Alliances and Networks', *Strategic Management Journal* 19: 293–317.

Hamel, G. (1991) 'Competition for competence and interpartner learning with strategic alliances', *Strategic Management Journal* 12: 83–103.

Håkansson, H. and Snehota, I (eds) (1997) *Developing Relationships in Business Networks*, London/Boston: International Thomson Business Press.

Hickman, C. R. and Silva, M. A. (1987) *The Future 500: Creating Tomorrow's Organizations Today*, New York: Penguin.

Lash, S. and Urry, J. (1994) *Economics of Signs and Spaces*, London: TCS.

Morgan, R.M. and Hunt, S.D. (1994) 'The commitment trust theory of relationship marketing', *Journal of Marketing* 58 (July): 20–38.

Nohria, N., and Eccles, G. E. (1992) *Networks and Organizations: Structure, Form and Action*, Boston: Harvard Business School Press.

Piore, M. and Sabel, C. (1984) *The Second Industrial Divide: Possibilities for Prosperity*, New York: Basic Books.

Powell, W. and Smith-Doerr, L. (1994) 'Networks and economic life', in: N. Smelser and Swedberg, *The Handbook of Economic Sociology*, Princeton University Press.

Pudney, J. (1953) *The Thomas Cook Story*, London.

Stabell, C. B. and Fjeldstad, O. D. (1998) 'Configuring value for competitive advantage: on chains, shops, and networks', *Strategic Management Journal* 19: 413–37.

Sarkar, M. B., Butler, B. and Steinfeld, C. (1996) 'Intermediaries and cyber-mediaries: a continuing role for mediating players in the electronic marketplace', *Journal of Computer Mediated Communications* 1(3) URL: HTTP:/shum.huji.ac.il/jcmc/vol1/issue 3/sarkar.html.

Thompson, J. D. (1967) *Organizations in Action*, New York: McGraw-Hill.

Tremblay, P. (1998) 'The economic organization of tourism', *Annals of Tourism Research* 25(4): 837–59.

Tulder, R. van (1996) *Internationalization and Competitive Space*, ERASM Research Project, Rotterdam: Erasmus University. Management Report Series (No. 289).

Uzzi, B. (1996) 'The sources and consequences of embeddedness for the economic performance of organizations: the network effect', *American Sociological Review* 61: 674–98.

Uzzi, B. (1997) 'Social structure and competition in interfirm networks: The paradox of embeddedness', *Administrative Science Quarterly* 42(1): 35–67.

Zaheer, A., McEvily, B. and Perrone, V., (1998) 'Does trust matter? Exploring the effects of interorganizational and interpersonal trust on performance', *Organization Science* 9(2): 141–59.

9 Education for tourism in a global economy

Tom Baum

Introduction

The process of globalisation is one that has impacted on all elements of socio-cultural, economic and political life. As other chapters of this book demonstrate, globalisation is evident both in terms of its most widely used contexts, relating to international trade agreements and the spread of global companies, products and services within the world's economy, and more subtly in the manifestation of global culture (music, theatre, communications) and leisure (sport, vacation tastes, fashion wear). Of course, these overt and subtler manifestations are closely allied and cannot be seen as independent entities.

In this chapter, we address globalisation across both these dimensions. Tourism, in the context of its impact, is both an economic component of the global economy and one that contributes to processes of homogenisation and increasingly common values and behaviour among both hosts and guests. Education, likewise, is influenced by demands of a global economy and yet also represents one of the key bastions of national identity (culture, language) and, in that sense, can work against forces of globalisation at a local level. In considering the relationship between economic and socio-cultural globalisation and the institutions and organisation of both tourism and education, there are no clear-cut answers and any assessment must be both tentative and qualified. In a sense, the jury must still be out on issues relating to the theme of this chapter because globalisation remains an ongoing process and both tourism and education are evolving phenomena, subject to diverse influences both internal and external.

The formal education process and globalisation

Education and educational systems have evolved in most countries in response to and as a reflection of complex combinations of historical, cultural, political, economic, geographical and technological factors. The interplay of these influences has contributed to the creation of institutional arrangements, systems and organisations that are distinctive at a national and regional level. Education, therefore, can never be judged in isolation of a full understanding and recognition of these influences.

1   Education, at a national level, is a product of a country's history because this, largely, has determined the role of various agencies and organisations (religious, political) in the provision of education and the role which successive legislative initiatives have played. Countries subject to the processes of colonial rule frequently retain key attributes of the educational system of the colonial power long after independence – this is evident in countries such as Barbados, India and Singapore among many others. A history that includes the joining of a number of independent political entities to form one federal state, as in Germany, sees separate educational systems operating within each of the *Länder* despite unity some 130 years ago. Irish education today has been significantly shaped by the role of the Catholic Church in providing schooling opportunities for the disadvantaged majority during the early years of the nineteenth century.

2   Culture, often allied closely to religious belief, is also a major shaper of education, both in a historic and contemporary sense. Attitudes to education and the value that a community places on schooling are, frequently, a product of the prevailing culture. Traditional Marxist analysis sees education as one route open to the working classes out of the class situation to which they are consigned, and this perspective contributed to the foundation of organisations such as the Mechanics Institutes and the Workers Educational Association in Britain in the nineteenth century. Likewise, minority communities have frequently embodied a culture which embraces education as a means of competing with majority domination – this was the response of cultures such as the Catholic community in Northern Ireland prior to the advent of the Civil Rights movement and of a number of Asian communities in Britain today. Cultural values relating to the education of women also greatly influence the level and structure of provision – the Taliban government in Afghanistan, for example, have determined that for cultural and religious reasons, women will be denied equal educational opportunity. Similarly, a very striking example of the impact of cultural and political change combining to influence education can be found in post-revolutionary Iran where Islamic values and educational systems replaced previously secular provision.

3   In most countries, education is a highly charged political issue and one that engenders considerable debate within the democratic political process. In most developed countries, public education demands one of the highest levels of exchequer expenditure and is also one of the major targets of donor aid to developing countries, 'generosity' which generally has clear strategic political objectives on the part of donor

countries. Changes in government frequently result in varying levels of change within the educational system and differing priorities given to aspects of provision – pre-school, primary, vocational, etc. Such change may be ideologically motivated, as was the case in China during the Cultural Revolution, or may have rather more pragmatic or fiscal objectives. President Clinton's commitment to boost education and training as means of enhancing the competitiveness of the American economy (Ashton and Green 1996) is one example of political change impacting directly on the focus and level of educational provision. Financial necessity influenced the introduction of fees for university students in the UK with the election of a new government in 1997, again demonstrating the role which the political process can have on the structure of education but also on levels of participation from within specific groups of society. In this case, levels of participation by mature students appear to have declined as a result of the new fee regime.

4   Economic influences on education are clear when issues of access to all levels of education and the quality of facilities and other aspects of provision in some of the poorest countries of the world are considered. There is a clear correlation between the economic prosperity of a country and the level of literacy to be found among its general population. Other indicators, such as levels of participation in secondary and higher education, are likewise correlated. Growing prosperity in countries such as Singapore has been directly reflected in investment and participation in all levels of education.

5   A country's geography is also of considerable influence on the nature of its educational provision. Widely dispersed populations, such as those in Australia, necessitate innovative forms of education such as pioneering broadcast schooling, while climate (for example monsoon) can influence the scheduling of educational provision. Geography also influences educational opportunity with rural populations frequently at a disadvantage when set alongside their urban peers. Peripheral location and perceived disadvantage that this engenders also stimulate new structures to combat such disadvantage. The new University of the Highlands and Islands in Scotland, like other open university models elsewhere, is designed to take provision to remote areas and to overcome some of the disadvantage of isolation through use of innovative models of provision and the use of technology.

6   Concern relating to educational access and participation is not only a factor of geography and, therefore, educational systems are influenced by the impact of changing social and lifestyle structures combined with ideological responses to such change. Disparity in participation levels in education on the basis of gender, ethnic origin and social class has encouraged initiatives to combat such inequality. Changes in the mode and method of educational delivery (part-time,

community-based, flexible access programmes) as well as measures to give recognition to informal learning in the home and workplace (APEL – the accreditation of prior experience and learning, for example – see Davies (1998a)) represent initiatives which reflect social and economic change in developed countries in such areas as increased female participation in the workforce and the requirements of the labour market for continuous updating and the acquisition of new skills. The concept of lifelong learning as a social but also an industrial issue is reflective of changing demands within both the workplace and in the expectations of stakeholders in the workplace, notably employees themselves. Davies identifies four key characteristics of lifelong learning which in themselves help to define the role which education in all areas, not least tourism, will need to play in the future. Lifelong learning:

- develops a person's competence throughout his/her lifetime. Competence includes knowledge, skills, capabilities, experience, contacts and networking, attitudes and values;
- includes all learning that takes place in different areas of life. It includes formal and informal education and training. Lifelong learning is not only about formal educational achievement;
- is a continuous development process which can be said to belong to an individual. Lifelong learning is not a one-off event, nor a short course; and
- provides an individual with the ability to live in a continuously changing world and to cope with a changing society and working life (Davies 1998b: 63–4).

7   Affirmative action programmes in education in the United States are also examples which have shaped educational provision in a significant way. Change is, furthermore, driven by lifestyle considerations as colleges and universities grapple with the reality of a student population working while studying and expecting curriculum and organisational compromise by institutions in order to permit this dual form of existence.

8   The technology which is available to and within education exerts increasing influence on the role of institutions and how they deliver learning opportunities to students. Technology can change the parameters of access by students to institutions of education in the form of various open access universities and colleges. It can facilitate the 'virtual university' concept (Teare *et al.* 1998) as well as changing the learning mode within conventional study programmes (Internet class notes, CD-ROM assessment packages). Technology opens up the boundaries of information access to create a global learning laboratory or library, permitting students access to worldwide information resources on an unprecedented scale. Technology also removes some of the international boundaries to learning, providing the potential for

global access to classes, programmes and qualifications in a way hereto-fore impossible.

The outcome of these varied influences is a diversity of educational systems at a national level and, within federal systems, at the level of the state or province as well. Each system exhibits its own features in terms of:

- age of entry and progression;
- pedagogic features;
- qualifications available;
- the status of various components;
- facility for transfer and profession within the system;
- links to work-based training providers;
- quality management systems;
- philosophical underpinnings of the system;
- the influence and acceptability of foreign providers;
- the role of private education within the system;
- nomenclature employed; and
- grading systems used.

Attempts, therefore, to create harmony between systems have been fraught with difficulty. At the simplest level, moves towards educational harmonisation have sought to grant recognition in one jurisdiction to schooling and qualifications gained in another. This is, however, by no means always a problem-free process as mutually acceptable criteria have to be established by which to evaluate different systems. Professional bodies in law, medicine, accounting and engineering have been at the forefront of such recognition and in some cases have had recourse to re-examination of candidates who have qualified from another jurisdiction. Accreditation through professional or regional bodies is based on rigorous examination of curricula to ensure smooth recognition and transfer between institutions and their students.

Rather more general qualifications equivalencies provide guidance to employers and educational institutions regarding the status of certificates across national boundaries. The detailed guidance provided by the British Council on educational qualifications, worldwide, is designed to assist British colleges and universities with the process of student selection.

Operation of colleges and programmes outside and across national boundaries increases opportunities to ensure cross-border recognition of qualification. American colleges and universities have taken the lead in this respect with local campuses in a number of countries. The spread of the UK's Open University to a number of European countries has similar outcomes.

An alternative approach is that advocated by Peters and Wills (1998) who argue that 'branding' in education has failed and, therefore, that seeking curriculum homogeneity at a national or international level is futile given institutional, organisational and cultural variation between providers, an argument in line with that posited by this chapter. Their remedy is to address institutional quality issues through ISO 9000 accreditation which, as a minimum, is designed to give consumers a degree of assurance that the delivery, support mechanisms and resourcing of educational provision within accredited institutions will be of a consistent standard. Given the increasing level of access to education via non-conventional delivery mechanisms, notably through remote access technology, such assurances could provide a valuable first level of quality guarantee although issues of content and relevance remain unaddressed by this proposal.

These recognition and transfer models are, in the main, concentrated at university and professional levels where trans-national mobility for education or within the labour market is most likely. Efforts to engineer the cross-border recognition of qualifications, for example within the European Union, have also concentrated at this level in order to facilitate a free movement of labour within a common market. By contrast, it would appear that education of a non-vocational nature and within the school system has been treated in the context of subsidiarity principles; in other words, outside the scope and competence of trans-national bodies.

Universality in educational experience is also fostered by widespread use of common teaching and learning resources in different countries. This is a result of more widespread translation of materials; the increasing practice of packaging learning materials in complete form including teachers' resources and assessment tests; the growth of English as the international communication language; lowering costs of resource production; and access to materials through technological gateways, notably the Internet. Common textbooks, Internet sources and CD-ROMs, inevitably impact on curricula, encouraging the adoption of common classes. In the tourism sector, the widespread international adoption of American Hotel and Motel Association (AHMA) materials and class texts, especially in the developing world, is an example of this process at work.

Education and economic growth

Responding to the diversity within education systems by means of institutional attempts to remove some of this variation in order to create homogeneity is driven by, primarily, economic imperatives although common culture and values derived through education may also be argued as a means of creating greater understanding between nations and as a contribution to peace. Economic arguments in favour of education point to the benefits which a highly educated workforce bring to the competitiveness of an economy, permitting businesses and the corporate nation to achieve more in the global marketplace with its goods and services than rival and competing countries and companies. The literature (for example sources cited in Ashton and Green (1996)), points to the cases of Germany, Japan and, increasingly, Singapore as countries where investment in education provides the basis for economic competitiveness and enhanced productivity. Public sector investment in education and training in many countries has, therefore, been motivated by a desire to compete by matching the skills levels of the workforce of the countries in question. This is true in both developed and developing economies, and in both there is the underlying political assumption that a better educated workforce is more readily employable and that, therefore, education can contribute to reducing unemployment in certain categories. The mandate of the Social Fund for Development (SFD) in Egypt is to support training initiatives which will enhance employment prospects in both the manufacturing and service economies. Likewise, European Social Fund investment in the educational systems of poorer countries of the European Union has been motivated by a belief that such support is an essential prerequisite to improve regional inward investment, reduce unemployment and in support of an effective common market within which all regions can compete without institutional or infrastructural barriers.

The case of Singapore (and to some extent, the Republic of Ireland) is frequently quoted to illustrate the value of education as a means of attracting high quality inward investment into the economy. In Singapore, in particular, this process and the very evident investment in education at all levels was undertaken through direct and focused state leadership with strategies that would be alien to Anglo-Saxon cultures. For example, the polytechnic sector in Singapore concentrates on technician level education to the highest possible standards but progression within the system from this starting point is unlikely except via the workplace. Until relatively recently, Irish investment in education provided many young people with a passport to emigration rather than contributing to the competitiveness of the local labour market, but this has altered as the country has assumed 'Celtic Tiger' status through exploitation of its educated workforce in knowledge-based sectors of the economy.

Ashton and Green provide a very sceptical analysis of the evidence for general direct links between investment in education and training and enhanced economic performance within developed economies and conclude that institutional and cultural factors play a major, if not dominant role, in differentiating the economic performance of countries. The economic arguments in favour of common education systems and qualifications as a necessity within the global economy are certainly open to debate. Some countries, notably the UK during the late 1980s, developed successful if short-term strategies based on attracting low-skills, low-cost investment, primarily in the service and assembly sectors within which education played a relatively limited role. Certainly there is evidence in a number of European countries that enhancing access and duration with respect to education creates expectations among school leavers and college graduates which are not satisfied by low skills, low status and routine work and that vacancies in these areas of the economy remain unfilled or attract immigrants, legal or illegal. This scenario is of particular importance in the context of tourism.

Informal learning and globalisation

The process of globalisation and harmonisation within education, however, is not only manifest in terms of institutional arrangements. Indeed, in many respects, the process of globalisation in education has much more profound impacts at an informal non-institutional level. This assessment recognises the importance of informal learning within the process of education. Such impacts include:

1   The effects on education and the educational expectations of students of an increasingly global communications and media environment, represented through global television channels (CNN), a common entertainment media and Internet technology. This permits students access to common learning resources, wherever they are located, in real-time conditions. It is possible, for example, for Australian students studying the politics of Ireland to access day-by-day information via *The Irish Times* in a way that the communications systems of just ten years ago did not permit on technological and cost grounds.

2   The reduction of local interests and local loyalties to be substituted by global alternatives, whether sporting (through support for and interest in 'global' stars and teams – Michael Johnson, the Superbowl final, Manchester United), in terms of 'universal' musical interests and following instead of the local and traditional or through the creation of global entertainment and leisure brands encompassing film, print media, computer games, toys and clothing accessories – Disney is the best example here. These loyalties create learning opportunities about the general but, correspondingly, reduce learning interest in local phenomena and the particular.

3   Almost universal access to global product and service brands at costs considerably lower than locally produced goods (the heart of globalisation) influences the education of consumers in terms of their expectations – the impact of McDonald's in many Asian countries on the acceptability of local street foods to young people – but also in terms of wider consumer behaviour. Global clothes products enable young people to learn about institutions and experiences with which they may otherwise have no contact.

4   Widening access to the experience of travel provides learning opportunities about other cultures, systems and beliefs, much facilitated by price reductions born of global economic treaties and political changes which, in turn, made them possible. Five decades ago, learning about

other cultures and countries was a remote experience, depending on book and similar sources and vulnerable to prejudiced interpretation. Today's alternative is direct access to the sights and sounds of the country and society in question. Travel, for example, has made a major contribution to diversifying food tastes, creating demand for global food products but also creating opportunity for innovation and the development of new food concepts, drawing on more than one culture and tradition.

Global interests, therefore, create universal learning opportunities at an informal level. The down-side of this process, however, is reduced interest in and, consequently, more limited learning about the local as young people travel less within their own countries, have more limited interest in local and national institutions (the reduced level of participation in the electoral process in many countries) and may, as a result, lose some of their sense of identity with their home community.

Tourism education and globalisation

Tourism education is a relatively recent phenomenon in most countries and has its origins rooted in both the traditions of vocational education and the structure and organisation of the tourism sector of the country in question. The relationship between tourism and education is complex and is faced by tensions and contradictions at both the levels of policy and implementation (Baum *et al.* 1997; Amoah 1999). Tourism education is also a field which generates considerable debate and tensions among academics and practitioners (Baum 1997; Airey 1995).

In most countries, tourism education has evolved out of a tradition of skills training and an emphasis on the acquisition of trade-level competencies in the accommodation and food services sectors (Baum 1995). Most European schools as well as a large number elsewhere commenced existence in providing training for chefs, waiting staff, receptionists and housekeepers, to the general neglect of both managers and the skills needs of the wider tourism sector. Some exceptions can be found from the early years of tourism education. Both Cornell University's School of Hotel and Restaurant Administration in the USA and The Scottish Hotel School (now part of the University of Strathclyde, Glasgow) started with a management education mission in 1926 and 1944 respectively. Tourism as a broader field of study frequently grew out of studies in accommodation and food service – this was true in The Scottish Hotel School in the late 1960s – but in some institutions tourism, as an academic field with limited vocational consideration, grew out of interest from geographers, anthropologists and those in management departments of universities, and these origins have coloured the educational approach adopted since. Newer entrants to tourism education have frequently followed similar paths to development although over a rather shorter period of time. Hong Kong's vocational schools provided the initial focus on craft education followed by the Hong Kong Polytechnic (now the Hong Kong Polytechnic University) with supervisory and then management programmes in first the hotels area and then in tourism. The Singapore Hotels Association's training arm, SHATEC, moved from craft origins to management education in a relatively brief period of time while, in the same country, the first university programme was in tourism (at Nanyang Technological University) followed some three years later by a parallel programme in hotel management at the same institution.

The model of education and training provided within the tourism sector itself exhibited similar origins to that of college and university learning, with an initial focus on providing basic skills to front-line technical staff, to the neglect of more senior and strategic learning objectives. It is only in recent years that companies such as Disney, McDonald's and Oberoi have established higher level management training and development institutes in support of their human resource development needs.

The main thrust of tourism education commenced and remains within the vocational education sector, whether colleges (specialist as in Switzerland or more general community colleges as in Australia, Canada and the UK), technological universities or through various training schemes in industry itself. While different national models exist for vocational education (Ashton and Green 1996), variation is reflected in location (trade school, college, industry), duration and status rather than in overall objectives. More recent initiatives have seen the introduction of tourism-related education into the mainstream secondary school curriculum, in both vocational and academic forms (STRU 1998).

There is no universally accepted curriculum for tourism education, reflecting attributes within education (as discussed above) combined with supply- and demand-side diversity within the industry and the form which tourism takes at a local as well as national level. The characteristics of a tourism curriculum at both trade and university level, for example, are influenced by, among other factors:

- the sponsoring department or faculty within which provision is located – business, geography, hotel and catering, parks and recreation;
- the origins of provision – whether it developed out of other, related subject areas (such as home economics) or grew as an independent field of study;
- the balance between vocational and academic, teaching and research priorities within the sponsoring institution;
- the experience and qualifications of faculty delivering the programmes;
- the sub-sectoral focus of provision, whether it is in tourism administration, hospitality, leisure, recreation or a combination of these;
- the role which the industry, private and public sector, plays in determining curriculum and supporting delivery;

- the level and quality of technical, learning and academic resourcing available to support the programme;
- the aspirations which major stakeholders, notably students, their parents, the industry and the local community, have for the programme and the university;
- the influence of national education and labour market policy as well as planning on curriculum, resourcing and related matters; and
- finally, in many developed countries, the characteristic of labour intensity which is found in many sub-sectors of tourism, notably hospitality, results in tourism being described as a low skills industry and this, in turn, impacts on the nature of educational provision and the level of investment it receives. Low skills levels are a feature of a weak labour market (Riley 1997) which in turn confers certain characteristics on the educational system with which it relates. However, it is arguable that the low skills argument is specific to developed tourism economies and has dubious relevance in many developing country situations (Baum 1997).

The same range of factors influencing diversity can be seen to exist when other forms of education and training for tourism are considered. As we have already suggested, tourism education is by no means unique in this respect but the fragmented nature of the industry, its wide geographic dispersal and, of particular importance, the labour intensity of much of its operations, give particular potency to the impact of this diversity and its consequences in relation to the globalisation debate. In short, internationally agreed programmes of study within the tourism sector are unusual and are really only to be found within specialist, technical areas such as those sponsored by the International Air Transport Association (IATA) or in the training of key technical airline personnel.

This is not to say that there have not been attempts to create greater homogeneity in the provision of tourism education. Accreditation initiatives by CHRIE in the USA and wider North American region have sought to establish specific curriculum and resource parameters for two and four year college provision in the hospitality sector, but up-take of this optional scheme has been by no means universal. The debate about a core curriculum for both hospitality and tourism has run over a long period in the UK, led by professional organisations such as the HCIMA and the Tourism Society. The arguments in favour of a core curriculum (Airey 1995) have been countered by those opposing the concept (Baum 1997).

It is not surprising, therefore, that attempts to develop trans-national tourism curricula and qualifications have likewise met with limited success. In Europe, the EU's vocational education arm, CEDEFOP, took the lead in the 1980s by seeking to establish agreement across all member countries with respect to the content of key tourism jobs as the first step towards establishing mutual recognition of qualifications. This move met with only limited success and was becalmed by excessive bureaucracy and the unwillingness of management, especially in small to medium sized businesses (SMEs), to recognise paper qualifications over practical experience and expertise. Attempts at the same time to create a widely accepted EuroDip failed because of a lack of interest among stakeholders, notably the tourism industry and college providers. There was also no evidence of real demand from the student consumer market for whom international vocational mobility in tourism, when not company sponsored, was generally to low skills and casual or seasonal positions. *Atlas,* as a trans-European tourism education network attempted to return to this issue without much success in the 1990s. Individual universities and colleges have made rather more progress in introducing common tourism programmes across national boundaries and the best examples of this are a number of the multi-campus American international universities which offer common programmes throughout the Americas, Europe and Asia.

At the level of in-company training, there is some evidence that the general training manuals which were developed in the 1970s in the hotel sector and applied without local adaptation within all units of major multinational companies have been replaced by rather more culturally and local-business sensitive approaches to staff training. The experience of the Disney Corporation in translating its training culture without compromise from the USA to France, one contributory factor in EuroDisney's (now Disneyland Paris) early difficulties, has been taken aboard by other companies internationalising their operations.

At a formal and institutional level, therefore, there is little evidence of a significant impact of globalisation on the formal organisation and delivery of tourism education. However, this does not mean that, in rather more informal terms, globalisation can be ignored. The impact can be seen at three levels relating to trends and developments within the international tourism industry; the use of resources in support of teaching and learning in tourism, human and other; and the influence which students and trainees themselves have on the education they receive.

1   The **tourism sector**, while retaining many of its small business characteristics in most countries, is increasingly subject to forces of globalisation as international acquisitions, mergers, out-of-country expansions and strategic alliances create larger and larger companies with interests in a growing number of countries. This is true in accommodation, air transport, travel and tour operations and the travel finance sector among others. The impact of this is to impose aspects of common culture across units and stations, demanding universal service delivery skills and the use of common systems and language. As a result, companies across the Star Airline Alliance or those operating as British Airways franchisees in Denmark, France, South Africa or the UK, increasingly demand certain core and common competencies, placing demands

upon the external education system as well as on in-house training provision. Accommodation units across the seventy-plus countries where Accor is represented follow common accounting systems and service standards with a similar impact on education and training. In a different form of global development, tourism companies are following the lead of manufacturing in locating certain back-office functions, notably accounting, reservations and IT, in places where labour is cheap and skills readily available. British Airways and Swissair, for example, have both done this by locating certain functions in India. Trends in these respects are likely to increase in the future with growing liberalisation in key tourism sectors, pressures to reduce costs with a global business and continuing consumer demand for familiar product and service standards wherever they travel. However, unless the challenges of multiculturalism in tourism are fully recognised in their broadest context, these moves may well founder (Baum 1996).

2   The **resourcing** of tourism education and training is also subject to global influences. Teachers are more mobile than they may have been in the past and bring with them the influence of their own education and professional experience elsewhere. A diversity of international faculty and student exchange programmes also contribute to dissemination of tourism education practice between institutions and countries. The globalisation of the educational publishing business also means that textbooks and other print and electronic resources are marketed worldwide. The same economic pressures which ensure that major car manufacturers continue to take more and more of the market also work to the advantage of major educational publishers who are able to offer 'standard' American or European texts at a far lower cost than local alternatives which may be available. The impact of the training volumes of the American Hotel and Motel Association (AHMA), especially on educational provision in developing countries, is a good example of specialist tourism publishers and their influence. There is also an increasing range of distance/open learning options for students to pursue in the tourism area: a good example is the Canadian programme developed by Mount St Vincent University in Halifax which, potentially, can serve a far wider marketplace than that in Nova Scotia for which it was designed. Internet access to publishers such as Amazon.com as well as direct to industry sources also contributes to tendencies which are seeing more and more students and teachers depending upon common teaching and learning resources in tourism education.

3   The **student** stakeholders in tourism education increasingly demand courses and case materials that have a strong international focus. Students as travel consumers have experienced the effects of globalisation and global access to travel products and services and expect their education to build upon this experience. Students participate in

exchange experiences in colleges and universities in other countries. They also see the value of an internationally focused education in helping to meet their career aspirations. Through use of the Internet, students are exposed to the international industry but also to the educational experience of others engaged in tourism education in colleges and universities throughout the world. They are able to benchmark their experience against the very best elsewhere.

Conclusions

It is evident, therefore, that the impact of the forces of globalisation on tourism education is mixed and fragmented. There is little evidence for strong pressure by the industry itself in favour of common provision as a response to globalisation although this may well appear in time through organisations such as the World Travel and Tourism Council (WTTC). Consumer demand, likewise, is not evidently well articulated. Any moves in this direction, to date, have been driven at an organisational level by colleges, professional bodies and quasi-governmental organisations at a trans-national level (CEDEFOP, for example) and these have been less than successful in most instances.

Tourism education cannot and will not remain immune to the forces of globalisation in the future, although the structure of the sector, its labour intensity and diversity may result in considerable 'lag' time before the sector's educational provision matches that in other sectors of the economy. The diversity of factors relating to change within tourism, within education, but above all, within the technological architecture which supports both areas, means that pressures towards globalisation will increasingly impact upon tourism education. The political and policy-related challenge, for tourism educators, is to retain control of the agenda which will drive this change and avoid being hostage to the fortune of developments in the wider learning and knowledge industry.

References

Airey, D. (1995) 'The future of tourism in higher education', *Tourism*, Spring: 22–3.
Amoah, V. (1999) 'Tourism education and training: an exploratory study of links to Policy formulation and implementation in the areas of tourism and education', unpublished DPhil thesis, University of Buckingham.
Ashton, D. and Green, F. (1996) *Education, Training and the Global Economy*, Cheltenham: Edward Elgar.
Baum, T. (1995) *Human Resource Management for the European Tourism and Hospitality Industry: A Strategic Approach*, London: Chapman and Hall.
Baum, T. (1996) 'Unskilled work and the hospitality industry: myth or reality?', *International Journal of Hospitality Management* 15(3).
Baum, T. (1997) 'Tourism education at the crossroads?', *Insights*, January, London: ETB.

Baum, T., Amoah, V. and Spivack, S. (1997) 'Policy dimensions of human resource management in the tourism and hospitality sector', *International Journal of Contemporary Hospitality Management*, 5(6): 221–9.

Davies, D. (1998a) 'Learning and work: an index of innovation and APEL', in (eds) R. Teare, D. Davies and E. Sandelands *The Virtual University: An Action Paradigm and Process for Workplace Learning*, London: Cassell.

Davies, D. (1998b) 'Lifelong learning competency in the twenty-first century – a prospectus', in R. Teare, D. Davies and E. Sandelands *The Virtual University: An Action Paradigm and Process for Workplace Learning*, London: Cassell.

Peters, J. and Wills, G. (1998), 'ISO 9000 as a global educational accreditation structure', in R. Teare, D. Davies and E. Sandelands *The Virtual University: An Action Paradigm and Process for Workplace Learning*, London: Cassell.

Riley, M. (1997) *Human Resource Management in the Hospitality and Tourism Industry* 2nd edn, Oxford: Butterworth-Heinemann.

Scottish Tourism Research Unit (1998) *International Benchmarking and Best Practice Study of Training and Education for Tourism*, Glasgow: STRU.

Teare, R., Davies, D. and Sandelands, E. (eds) (1998) *The Virtual University: An Action Paradigm and Process for Workplace Learning*, London: Cassell.

10 Globalisation, safety and national security

Gui Santana

Introduction

At the dawn of a new millennium, the tourism industry, the largest and most prosperous industry in the world, is facing a series of increasing security and safety issues that may potentially curtail its development in some countries or entire regions. Considering that a destination's image is one of the main, if not the major, determinant of travelling, perception of security-related issues by either the travelling public or investors can easily derail the evolutionary process of tourism in a destination. Although the area of security and safety in tourism is a fairly recent field of research, it is well recognised that tourism only thrives under certain conditions, the most important being a safe and secure environment (Mansfeld 1995; Wahab 1996; Vukonic 1997; Hall and O'Sullivan 1996; Richter and Waugh 1986; Sönmez and Graefe 1998; Buckley and Klemm 1993; Edgell 1990; Santana 1999; Teye 1986; Gartner and Shen 1992; Sönmez 1998).

While the process of globalisation has promoted constructive global and political alliances, it has concurrently advanced the activities of, and alliances among, transnational terrorists, organised crime enterprises, and extremist groups of every conceivable type. These 'threat alliances' have formed transnational cooperatives with immense global financial and human resources. Moreover, national, regional and international security issues are of quite a different nature from those that the world has been dealing with for decades. The post-Cold War era presents a different, diversified and more threatening range of security issues that places large civilian populations in direct danger. Today, new security issues have supplanted traditional ones (e.g., force balances, deterrence, etc.) and one of the main threats comes from the spread of conventional weapons. Concerns with the global spread of conventional weapons is now one of the major issues in world and regional security. Contemporary national and international security agendas are also concerned with ethnic conflicts, managing multilateral peace and security operations, negotiating arms control treaties, addressing new issues such as environmental changes and their challenges, demography, terrorism, human rights concerns, and aligning foreign policies with domestic economic and social concerns.

The process of globalisation is not only making travelling more accessible but is also facilitating the development of new destinations, increasing destination choices, and making it easier for tourists to shift from one destination to another. The pace and sophistication of the new global economy has alienated some countries or entire regions, especially those in developing and underdeveloped areas. Due to structural or political reasons many of these countries are not prepared to seize the opportunities presented by a globalised trade environment. In fact, most of the countries in developing regions have been unable to compete or engage in the lucrative international trade offered by the globalisation process and have therefore gained little, if any, benefit from it. The effects of globalisation on an increasing number of countries have been quite detrimental. Many are facing unprecedented levels of unemployment and hardship which have direct implications for local and regional political stability, security and safety, and consequently the tourism industry.

This chapter addresses pressing issues and consequences of the process of globalisation for the tourism industry in the light of safety and security, and examines the relationship between, and the effects of, security and safety related issues and tourism. First, and more specifically, an overview of international security issues and trends is provided. Then, the issues of tourism safety and the relationships between national security, terrorism, political instability and the tourism industry are discussed.

International security in the twenty-first century

The meaning of 'international security' has been redefined since the end of the Cold War. The global confrontation of two superpowers and of the two military alliances they led has ceased to exist and today there are new priorities and requirements in the field of international security. The threat of a global war has diminished dramatically and attention is now focused on the growing number of armed conflicts within, rather than between, states. During the 1990s the world witnessed an increase in civil wars, large-scale atrocities and even genocide. During the Cold War, notions such as 'strategic deterrence', 'regional security systems', 'arms control', etc., had operational meaning. This is no longer the case today. The meaning and understanding of 'international security' is now vague and by no means shared by the international community. This lack of consensus and understanding could prove disastrous since the threat of a large conflict still exists, although this threat is clearly smaller. Attention now, however, has shifted to local or regional security concerns, principally of a non-military nature.

Other issues have contributed to this process, such as the inadequacy of the United Nations as global guarantor of global security (Gasteyger 1998). While international treaties such as the Non-Proliferation Treaty and the Chemical Weapons Convention maintain a notion of world consciousness and 'international' security, important nations have challenged them or have simply not signed them. For example, the nuclear tests by India and Pakistan in early 1999 clearly demonstrated their disagreement with the Non-Proliferation Treaty. Their refusal to join the world community in the landmines convention was also a sign that the United States does not share the same definition of 'international security' as the rest of the world. Although the reasons for doing so are many, it is clear that their national security needs, or understanding of them, do not concur with that of international security. Moreover, 'globalisation', however it is defined, is making consensus more difficult rather than facilitating it.

As the East–West confrontation disappeared, changing for good the dynamics of global security, a new world order or orientation, as far as security is concerned, became mandatory. The world now is a pluralistic place, with a large array of ideologies, demanding a totally different and innovative approach. During, or even preceding the Cold War, a number of institutions were created for the purposes of fighting the Cold War or preventing a global war. Those institutions included the UN (United Nations), NATO (North Atlantic Treaty Organisation), the OECD (Organisation for Economic Cooperation and Development), ASEAN (Association of South East Asian Nations), and the Council of Europe, among others. While they served a clear purpose until the late 1980s, their functions seem to be inadequate for the new requirements imposed by an ever-changing and increasingly dynamic environment. Thus, their legitimacy is being questioned. The United Nations has been struggling to reform itself in order to adjust to the newly emerging tasks or to improve its performance when it comes to dealing with traditional ones. As for the other organisations, it is difficult to see any dedicated effort, or interest, in changing their organising principles.

While some of these organisations seek to justify themselves by enlarging their bases (such as by taking on new members) it is arguable whether this makes them more efficient or appropriate to tackle the new challenges. It is clear that the world today needs 'partnerships' for peace. This requires greater cooperation between international institutions. However, the promotion of a broad-based structure for security is an almost impossible task due to the very nature of these organisations. As with governments, international organisations are in constant need of justifying their utility and legitimacy and this makes them reluctant to share whatever responsibility they were given at their creation (Gasteyger 1998).

Among other pressing issues of international security, three legacies of the Cold War contribute significantly to shaping international security scenario. The unresolved disarmament programme and the proliferation of weapons of mass destruction, the enormous stockpiles and consequent proliferation of conventional weapons, and finally the appearance of volunteer armies. Admittedly, this assertion is rather simplistic. However, it does indicate the developments the world will have to confront in the years to come and they are major forces in shaping the international security scenario. The multiple nuclear tests by India and Pakistan in the late 1990s are a clear reminder of this. Although the number of conventional weapons has been significantly reduced by being destroyed in compliance with disarmament agreements or by becoming obsolete, large amounts of them have been transferred either freely or at a considerably reduced price to third world countries. One of the negative consequences of disarmament is that a large amount of these arms are now in the hands of rebels or private armies and not under government control. If, as will be discussed later, there is one threat to personal and national security, it is this uncontrollable and indiscriminate availability of conventional weapons. The appearance of volunteer armies is not a new phenomenon but their increase and range of activities (including terrorism) should be of major concern. In their early days, their recruiting methods were based on ideologies. Today however, volunteers increasingly join resistance movements for financial gains.

Paradoxically, the post-Cold War world is, in many respects, more unstable than the Cold War era itself, and is characterised by increased violence, increased proliferation of military technology and the potential for these trends to continue. An increasing economic disparity between developed and developing countries and its consequences (such as poverty, increasing state debts, population explosion and inadequate living standards, and so on), have contributed significantly to the instability of some regions. This has led to local wars and crises, ethnic and religious conflicts, and has resulted in regional arms races and in the proliferation of weapons of mass destruction. Consequently, it has become a major threat to local, regional, and international security. Instability and a state of chaos in some regions have also stimulated drug trafficking and international organised crime. Regions that have been affected, among others, are the Balkans, some regions of the former Soviet Union, some parts of Africa and Asia, some countries in Latin America, and the Middle East.

Other international security issues that became prominent in recent years, apart from the proliferation of conventional weapons and other 'military threats', include environmental degradation, resource depletion and economic disparity, demography, drugs, terrorism, etc. Clearly, international security today is a highly complex and amorphous subject that is being shaped by an increasingly dynamic environment. Having said that, some major forces or drives that have been influencing this process can be identified. Although a discussion on each of them would divert the focus of this chapter, a sample of the main driving forces has been summarised and is displayed in Table 10.1.

Table 10.1 A sample of major forces shaping international security

---

- The rise of the 'Market State'

Globalisation of economies and technology will continually change and shape the nature of governments and private sectors. The international financial market (and their actors) have accomplished what governments around the world have not even attempted: to bring down superficially successful but corrupt and illegitimate governments in Asia. Armies, borders, and passports are becoming less important.

- The fall of the Soviet Union and subsequent decline of Russia

The fall of the Soviet Union meant the predominance of one superpower, the United States. Despite its weak situation and lack of power to retaliate or react, the decline of Russia is of serious concern for regional and international security. One major issue is the continued leakage of nuclear capability to 'illegitimate' states.

- The inadequacy of international security bodies

There is a pressing need for new approaches to international security. Existing institutions are not adequate, structured, or prepared to confront the new security or 'threats' presented by the globalised world.

- The proliferation of conventional weapons

Large amounts of cheaply available conventional weapons. State producers are being forced to engage in export activities to maintain their military bases or for hard currencies.

- The attractiveness of weapons of mass destruction to weak countries

The pressing issue regarding WMD rests not only with states but more importantly with non-military groups, including terrorists. An example of these threats is the religious group Aum Shinrikyo which gassed the Tokyo subway with sarin in 1995. Fabricating chemical or biological agents is not difficult. If terrorists or other groups have not resorted to it in the past, this may be because conventional weapons or other means provided them with the result they sought. The future is still an open question.

- The rise of China

During the Cold War, China played a strategic pivotal role in the triangle with the Soviet Union and the United States. With the demise of the Soviet Union and an increasingly weak Russia, China emerges as an independent strategic player with no association with any given alliance or potential adversary. The great question is how China will behave in the next century. China's military expenditure was consistently around 10 per cent over the 1990s. It was accused of 'stealing' sensitive military technology from the US. It is not a secret, however, that China is extremely interested in technology weapons systems.

- Technological breakthrough

Information flow, apart from being the enabler of globalisation in that it makes distance and locations irrelevant, can also have implications for the future of some regions. It can shape people's expectations and force regimes to change. Governments, such as the Chinese, are less and less willing to control what people hear and read. The isolation cost seems to be too high since it goes hand in hand with poverty. Instantaneous news requires governments to be ready to react and respond to their actions (or lack of them).

Table 10.1 (continued)

• A rise in organised crime

Criminal activities are increasing all over the world. Domestic 'ills', such as crime, are a major threat to personal, national, and regional security since they increasingly cross national borders and the existing structures for dealing with them are increasingly mismatched. Policing, for example, has clearly defined jurisdiction and is geographically contained. New technologies, such as telecommunications, are also being employed by criminals and in many instances the law is unable to keep pace with technological advances and use.

• Demography

Population growth is another international security issue. Apart from the industrialised nations the rest of the world is going to experience a population explosion. The biggest challenge will be to accommodate the agricultural population of low-income countries. Countries with high fertility rates are associated with a higher number of armed conflicts, either external or internal. Another demographic indication of security is the ratio between the population of trained professional soldiers and the civilian population. Some regions, such as the sub-Saharan African nations, have a ratio of trained soldiers and trained teachers of 15 to 1. In Sweden this ratio is 1 to 4. One may conclude that the majority of conflicts are likely to occur in the developing world and parts of Europe.

Sources: Nazarkin (1998); Treverton *et al.* (1998), Center for Defense Information (1999).

Globalisation and national security – a brief overview

Globalisation can be broadly defined as the process of softening geopolitical boundaries, favouring relatively unrestricted travel, commerce and communication of all types. The process of globalisation has, directly or indirectly, involved all sectors of the global economy and no country can avoid it. While transnational companies and a few countries benefit enormously from the gains of 'playing' on a global scale, some nations experience just the opposite. A study conducted by Boxberg and Klimenta (1999) reveals a sinister side of globalisation, with endemic unemployment in a growing number of countries, criminal exploitation of the finite natural resources of emerging and underdeveloped countries, a concentration of financial resources, and weakening of national state. Financing of welfare has become an impossible task in both developed and underdeveloped nations and this poses a great threat to the welfare of us all.

This irreversible process has also favoured the activities of groups that promote terror and has permitted others the access to destructive technologies that are today threatening national and regional security. Moreover, in the aftermath of the Cold War, states and other groups began to have unlimited access to conventional weapons, weapons of mass destruction, and communications technologies. According to the Center for

Defense Information (1999) the number of armed conflicts in the world on 1 January 1999, totalled twenty-three, with two additions during 1998. Indeed, the number of conflicts is on the increase. Another indication is the number of United Nations peace operations, which mushroomed from five in the 1980s to over thirty in the 1990s (Treverton *et al.* 1998). Old and new conflicts ensure that virtually all parts of the world are involved or engaged in one way or another in high security situations (as protagonists of conflicts or playing roles such as peace mediators, etc.). Another feature today is that many settled disputes now run the risk of being re-started.

New or renewed conflicts in Africa and Europe have contributed to the increased levels of worldwide-armed conflicts. Among these are the allied attack on the Yugoslav Republic in 1999, the 'Kosovo War', and renewed fights in Angola, between government forces and UNITA (Union for Total Independence of Angola) which put a definite end to the fragile 1994 Lusaka peace accord. To the north on the African continent, the Democratic Republic of Congo was involved in disputes with Rwanda and Uganda which supported rebels who opposed the ruling regime. Table 10.2 illustrates some ongoing major conflicts around the world. Table 10.3 presents a list of political violence or conflicts that have been suspended but run the risk of being resumed.

Civil wars have become increasingly brutal. Events such as those that occurred in Burundi, Rwanda, Angola and Somalia in the 1990s are just some examples of sheer brutality that shocked the world. Other deadly civil wars were seen in Algeria, Sierra Leone, Sri Lanka, Afghanistan, and so forth. Ongoing prominent political conflicts include those in the Middle East, between Israel and Hamas and Hizbollah – continued struggle in southern Lebanon, the dispute of India and Pakistan over Kashmir, the Kurds against Turkey and old disputes between Iran and Iraq. In South America, Marxist rebels and drug lords still keep the Colombian government under siege. Some of those brutal episodes were only possible due to the increasing access that ruling regimes and/or opposition groups are gaining to weapons which have become readily, and sometimes cheaply available, following the end of the Cold War. In fact, some conflicts, such as in Rwanda, were partially fuelled by the fact that one party was engaged in arms purchases. The relationship between acquiring arms and the escalation of disputes to war is a positive one and significant.

The global arms trade and conventional weapons

When addressing the 'Conference on Disarmament in Geneva', in 1997, the United Nations Secretary-General Kofi Annan stated that 'Our challenge today is to build on our hope and optimism at the end of the Cold War, and not allow real progress in international security to be undermined by the new conventional arms races at regional and subregional levels.' The interest in this new threat crystallised with the 1990–1 Iraqi

Table 10.2 Major current conflicts – the world at war

Conflict: Main warring parties	Year began	Cause(s)
Middle East		
Israel/Hamas and Hizbollah	1975	Religious and territorial
Iraq government (Sunni) Shi'ite	1991	Religious
Kurdish factions/govts. of Iran, Iraq and Turkey	1961	Independence
Asia		
Afghanistan: Taliban/Other factions	1978	Ethnic and religious
Cambodia govt./Khmer Rouge and Royalist	1979	Political
India govt./Jammu and Kashmir Liberation Front	1989	Ethnic and religious
India govt./Punjab	1982	Religious
India and Pakistan	1948	Ethnic and religious
Indonesia govt./Revolutionary Front for East Timor	1975	Independence
Indonesia govt./Irian Jaya and Aceh	1969	Autonomy and religious
Philippines govt./New People's Army, National Liberation Front	1969	Ideological and religious
Sri Lanka govt./Tamil Eelam	1978	Ethnic and religious
Africa		
Algeria govt./Islamic Salvation Front (FIS), Arms Islamic Group (GIA)	1991	Religious vs. Secular rule
Angola govt./UNITA	1975	Economic and ethnic
Burundi: Tutsi vs. Hutu	1988	Ethnic
Democratic Republic of Congo govt./ Rwanda, Uganda and indigenous rebels	1997	Ethnic
Rwanda govt. (Tutsi) Hutu	1990	Ethnic
Sierra Leone govt./Revolutionary United Front, National Provisional Ruling Council	1989	Ethnic
Somalia: factions	1978	Ethnic
Sudan govt./Sudanese People's Liberation Army	1983	Ethnic and religious
Europe		
Yugoslavia govt./Kosovo Liberation Army (KLA)	1998	Autonomy and ethnic
Latin America		
Colombia govt./National Liberation Army (ELN), Revolutionary Armed Forces of Colombia (FARC)	1987	Drug trade and ideology
Peru govt./Sendero Luminoso	1981	Ideology and drug trade

invasion of Kuwait after the realisation that the major arms suppliers had helped to create the arsenal they faced on the battlefield.

In the past three decades the global arms market has traded more than a trillion dollars worth of arms. Overall international arms sales declined sharply from an estimated high of US$71 billion in 1985 to a low of

Table 10.3 Political conflicts or violence that run the risk of being resumed – a fragile world

Parties in conflict	Duration	Cause(s)
Asia		
Armenia and Azerbaijan	1990–94	Nagorno-Karabakh
Myanmar (Burma) govt./various factions	1942	Ethnic and drug trade
Peoples Republic of China/Uighur	1996	Independence
Tajikistan/Popular Democratic Army	1992–97	Religious
Africa		
Cameroon and Nigeria	1994–96	Bakassi Island
Chad govt./Muslim Separatist	1965	Religious
Eritrea and Ethiopia	1998	Territory
Kenya govt./Kikuyu	1991	Ethnic
Liberia govt. National Patriotic Front	1989	Ethnic and economic
Europe		
Serbs, Croats and Bosnian Muslims	1990–96	Final status of Bosnia-Herzegovina
Russia/Chechnya	1994–96	Independence
Republic of Georgia/Abkhazia and South Osset	1992–93	Independence
Moldova/Trans-Dneister Region	1991	Ethnic and economic
United Kingdom/IRA and other factions	1969–97	Ethnic and religious
Middle East		
Israel/Palestine	1948	Independence
Iraq/Desert Storm Coalition development	1991	Prevent WMD
Americas		
Guatemala govt./Nat'l Revolutionary Unity (URNG)	1968–96	Ethnic
Haiti	1991–94	Economic
Mexico govt./Zapatista and Popular Revolutionary Army	1983 and 1993	Ethnic and religious

US$20 billion in 1992, despite the increased availability of weapons. This trend suffered a reversal in 1994 when the international arms market reached the figure of US$25.4 billion (Hull and Markov 1996). In 1996, global arms trade expanded by 8 per cent in real terms to US$39.9 billion and most of this growth was due to increased demand from East Asia and the Middle East (Global Futures Bulletin 1997).

While virtually all international efforts in arms control and non-proliferation (multilateral and bilateral) since 1945 have focused on weapons of mass destruction (WMD), which resulted in the Chemical Weapons Convention and indefinite extension of the Nuclear Non-Proliferation Treaty, all the wars and violent conflicts in the world during the same period have employed conventional weapons. Most of these wars and conflicts are taking place in the developing world (as illustrated in Tables 10.2 and 10.3). According to Booth (1991), conventional weapons have been responsible for virtually all the twenty to forty million war-related deaths since 1945 (1991 figures). Tragically, up to three-quarters of these deaths have been of civilians.

Conventional arms were broadly defined by Krause *et al.* (1995) as including all weapons and military technologies (including dual-use technologies whose primary application is military) that fall below the threshold of weapons of mass destruction (which are understood to encompass nuclear, chemical and biological weapons). The authors also consider under the category of conventional weapons delivery systems used for weapons of mass destruction (such as ballistic missiles or combat aircraft) and major weapons platforms, as well as land mines, small arms, light weapons, and other non-lethal military equipment, for example, transport vehicles.

Conventional proliferation, which is defined as 'the diffusion of weapons, associated technologies or expertise that produces an adverse effect on local, regional or global security and stability' (Mutimer 1994: 10), has been fuelled by a number of factors. The end of the Cold War has had a major impact on global trade in conventional armaments, just as it has on most facets of national security and defence. One of the main characteristics of this era is that the nature of demand for weapons has shifted from the context of rivalry between superpowers and their associated client states to providing for national defence within the context of regional security needs. In general, these changes have influenced the general level of world demand for arms, which is in decline. However, countries continue seeking to acquire sophisticated weapons. Although a lengthy discussion on the topic is well beyond the scope of this chapter, it is worth mentioning, albeit briefly, two important reasons for this proliferation. For a number of consecutive years the military forces in the former Eastern and Western blocs have been experiencing a massive downsizing which has meant that arms industries throughout the world are struggling to survive. In this scenario, states are being forced to promote exports in order to maintain their defence industrial base. Moreover, concerns with domestic welfare in industrialised nations, coupled with a long recession at the beginning of the 1990s, has also meant that the economic and employment benefits from military exports cannot be overlooked. Thus, while the impetus of arms activities in the recent past was mainly strategic and political in nature, it is now based solely on economic advantage. In many cases, the notion of selling arms to strengthen a given regime has been replaced by the urgent need for hard currency or strategic commodities such as oil. In this sense, the arms trade is supply-driven. As wars and conflicts multiply all over the world, added to the quest for international power and prestige sought by some states, there has naturally been an increase in the arms trade figures, making it a demand-driven activity as well.

While it is totally legitimate for a nation-state to build up its defence system (right of self-defence), conventional proliferation can also generate a series of non-desirable outcomes. Considering that conventional proliferation is taking place mainly in developing regions, arms transfers can exacerbate internal conflicts, obstructing progress in democratisation, and perpetuating authoritarian regimes. It can also instigate a regional inter-state arms race and military expenditure which can potentially lead to conflicts and even war. Arms acquisitions and expenditure in militarisation can also consume vital financial resources that could be used for social and economic development.

The discussion in this chapter so far has focused on general aspects of international and national security drawing on the overall international environment following the end of the Cold War. The discussion above is clearly an over-simplified account of an extremely complex and dynamic issue. It has been developed here, nevertheless, with the sole purpose of illustrating some aspects of the macro security scenario. The following sections look at how security and safety affects and influences tourism development and prosperity.

Tourism and safety

Tourism is described as an effective instrument for promoting peace and understanding between nations and people. Due to its rewarding economic impact, tourism has also become one of the best-adopted development strategies in many countries or regions around the world. For most developing countries tourism represents unrealised developmental potential and many of these nations, in an effort to raise their general level of prosperity, have embraced tourism as a way of accomplishing economic development. In spite of all the glossy and growing figures in international tourism, and the status and recognition it currently enjoys, the tourism industry cannot shield itself from the sinister power of armed conflicts and the detrimental effects of terrorism and crimes against it. In fact in some circumstances, it could be argued that the industry's very success has made it an attractive target for terrorism and other forms of violence and abuse. The tourism industry is very sensitive to any security-related issue, be it of a political nature (instability, upheavals, military coups, civil unrest, wars, guerrilla, terrorism, declaration of martial law, ethnic cleansing, riots, etc.) or with respect to risk to personal safety.

When tourism is the issue, the meaning of security naturally encompasses 'personal security'. However, national security is not necessarily sufficient to ensure personal security, since it is primarily designed to protect the national territorial integrity and sovereignty of the state from external aggression. A close observation of the overall security situation in the recent past indicates that while security has increased at a state level it has dramatically decreased for the population. The combination of developments since the end of the Cold War and those produced by the process of globalisation has resulted in a growing number of civil wars, large-scale conflicts, genocides, violent crimes, drug related activities, terrorism, diseases, environmental pillage, and so on. While the first obligation of a nation-state is to provide a generally safe environment for its citizens, national security capability or make-up have failed to address or have been unable to safeguard people's physical, psychological and emotional well-being. In most cases they have proved insufficient to guarantee safety. The inability of states to deal with rising levels of crime and the increased number of conventional weapons in our society illustrate this point well. The spread of conventional weapons is a major security issue at all levels today. Conventional weapons in the hands of civilians, away from combat situations, have caused havoc and trauma all over the world. In the United States, after a series of 'accidents', the government is now contemplating new laws to prohibit the commerce or even private ownership of weapons. In Brazil, the state of Rio de Janeiro passed a resolution in June 1999 prohibiting civilians to own firearms. Rio de Janeiro has for over two decades been literally under siege from drug barons and organised crime. Personal security must also be extended to issues such as economic vulnerability and health. The benefits of globalisation in those two areas in particular have been countered by the negative effects of some of its mechanisms. In an interdependent world with open markets and world trade, an economic crisis in one corner of the globe most certainly affects negatively people's destiny in another, especially on the lives of the most vulnerable. The increased mobility of people and materials has direct implications on health issues (pollutants, infectious disease, etc.).

The security and safety of travellers has been brought into focus in many destinations in recent years. Internationally, terrorism, and particularly the kind that is specifically aimed at tourists, has attracted a great deal of media coverage and scrutiny and has caused the world tourism industry billions of dollars in lost revenues. For example, following the 1992–3 terrorist attacks against international tourists in Egypt, the country's tourism industry lost more than US$2 billion in tourism revenue (Associated Press 1993). The attack on a tourist bus (loaded with German tourists) in September 1997 and the tourist massacre in Luxor had far-reaching implications and will cost Egypt dearly for years to come. In Cuba, in August-September 1997, a series of bomb attacks at tourist hotels resulted in thousands of holiday cancellations. This was a devastating blow to an economy hungry for foreign currency. Considering the general pattern of reaction by the public to safety related issues, it can be assumed that every time a bomb scare warning is given, the tourism industry loses a few million dollars.

The first days of 1999 were marked by the international headlines of an international tourist kidnapping followed by murder in Yemen. On 28 December 1998, armed militants kidnapped a group of foreign tourists (two US citizens, twelve British nationals, and two Australians). On the following day three Britons and one Australian were killed, and one American was seriously wounded following a Yemeni security forces attempt to rescue the victims. A few months later, another politically motivated crime was committed against international tourists in Uganda, where American and British tourists were singled out and brutally murdered. An increasingly wide range of other high profile safety and security related issues that affected the international tourism industry have been making the headlines. They range from fatal airplane disasters (transport) to pollution and contamination (environment), from food poisoning (hygiene) to tourist rape (personal safety), from social unrest (political instability) to tourist victimisation (socio-economic impacts), from terrorist attack (deviant behaviour) to natural disasters ('act of God'), and so forth. These events have not only drawn attention to the fact that tourism is not always a safe option, but have also demonstrated that the industry has limited capacity and ability to deal effectively with safety issues both in the prevention of disasters and crises and the management of them when they occur.

People travel for many different reasons. Regardless of their purpose of travel, their decision to travel is based on several psychological traits, such as desire for adventure, peace/tranquillity, comfort, education/experience, and so forth. However, none of those desires have the capacity to overrule one of the most basic human desires (or needs) – safety. It is important to emphasise that safety in this case refers not only to the real, factual safety but, more importantly, to perceived safety. A revision of the growing amount of literature analysing the relationship between tourist flow and expenditure and politically motivated violence, as well as other types of situation that convey the impression of insecurity at a destination, indicates that they all severely inhibit tourism (Sönmez 1998; Gartner and Shen 1992; Mansfeld 1995; Wahab 1996; Aziz 1995; Cassedy 1992; Brayshaw 1995; Vukonic 1997; Sönmez and Graefe 1998; Teye 1986; Buckley and Klemm 1993; Hall and O'Sullivan, 1996; Ryan 1993; Scott 1988; Goodrich 1991).

The process of globalisation means that travel is increasingly accessible and official statistics from most regions of the world show a positive trend in the number of international arrivals and tourism receipts. This process also means that there are more destinations available to the travelling public and, naturally, competition is increasingly fierce between regions and products. At the same time, terrorist groups, political guerrilla groups, etc., have seen in the tourism industry the opportunity they need for directing the attention of the international community to their causes. Their methods vary as much as their causes. Blowing up aircraft full of innocent civilians, hostage taking, kidnapping, gunning down tourists, and targeting tourism facilities, are just a small sample of the criminal acts directed at the industry in recent years. Some countries, or even entire regions, have been taken as hostages by organised political, and other groups (narcotics, arms dealers, etc.). This is not, however, the privilege of developing or 'third world countries' or regions. Within the main developed blocs in the world, these kinds of activities involving tourists are still happening and in some cases intensifying. Given that tourist destinations are easily substituted products, crisis-hit regions are usually left with the difficult, long, and always expensive task of rebuilding their image and regaining the trust of travellers.

Risk perception and tourism demand

The terrorist attack on tourists in Luxor, Egypt (1997), political turmoil and ethnic cleansing in Indonesia (since 1997), persistent armed conflict in the Balkans, tourists being kidnapped and murdered in Uganda by Rwandan Hutu rebels (1999), the terrorist bombing of Cape Town's 'Planet Hollywood' (1998), tourist kidnappings in Venezuela, Sri Lanka, India, Peru and Colombia, a surge in street crime in Brazil, drug related crimes in many parts of the globe, the terrorist sarin gas attack in Tokyo's subway (1995) which killed twelve persons and affected more than 5,000, and so on, are all situations that had a profound negative effect on potential visitors' perceptions of those destinations. Moreover, as far as tourism is concerned security issues always have a 'spill-over effect'. That is, tourists tend to associate a security 'incident' with an entire region. For example, during the Gulf War North-American tourists avoided the entire region, including Europe, and preferred Caribbean destinations (Wall 1996; Sönmez 1998). Tourists are deterred from travelling to perfectly safe places because they assume the entire region to be risky (Vukonic 1997; Mansfeld 1995; Sönmez 1998; Richter and Waugh 1986; Wall 1996).

However, it is important to observe that tourists are not discouraged from travelling altogether. Indeed, Wahab (1996: 176) argues that 'When some chain of events deters tourists from visiting a certain destination, other destinations, whether proximate or far away, will benefit.' Since the dismantling of the former Yugoslavia (a war which made the entire region unattractive) other countries such as Portugal, Greece, Spain, Cyprus, Italy, and Turkey experienced an increase in tourist arrivals (Vukonic 1997; Wahab 1996). Therefore, it can be said that there is a clear correlation between an increase in violence in one place and an increase in tourism demand elsewhere. Whenever a security scare arises (real or perceived), destination substitution is an inevitable and logical consequence, even if the substitution means a change in the primary purpose of travel. A cultural destination, such as Egypt, which is clearly not easily substituted, may 'compel' tourists to look for other alternatives such as scenic or beach-based destinations. Overall, the effect on worldwide tourism is often small. Having said that, some segments are more sensitive than others to this phenomenon of substitution. Business travellers or tourists visiting family or relatives have more limited ability to relocate than do pleasure travellers (Sönmez and Graefe 1998; Wall 1996). This, of course, also depends on the level of risk presented to potential travellers.

Considering that in tourism perception is reality, the implications of any security/safety issue can prove disastrous to a destination. A study of three

major tourism offices dealing with major crises which affected their destination (Hong Kong, after the Tiananmen Square student massacre in
China; Fiji, after a bloodless military coup in 1987; and San Francisco,
after the Loma Prieta earthquake in 1989) revealed that a crisis can very
quickly cripple the travel industry (Cassedy 1992). In these cases the crises
were ones of perception and it was observed that they could be just as
devastating, if not more so, than crises that actually cause physical damage.
In each of the cases, the crises were the perception of stability and danger
by the travelling public. The industry suffered because people lost confidence in the destination as an attractive and safe place to visit and stay
in. Safety 'misperception' can indeed be a dangerous issue for destinations. The tourism industry in Florida underwent a phase of decline in
visitor arrivals following a series of crimes against tourists. The negative
effect on Florida's tourism industry, it is argued, was caused by distorted
and extensive media coverage of incidents of crime against tourists and
an overestimation of tourist's real risk and victimisation (Crystal 1993).
Other evidence suggests that the real risks for tourists were substantially
smaller than public perception would suggest (Brayshaw 1995). Despite
the arguments, Florida, as a destination, experienced a real crisis.

The case of Florida is not an isolated one. It is common for destination
officials to complain that the media over-react and tend to give more promi-
nence to incidents than they warrant. While it can be argued that the likelihood of being involved in acts of violence and other incidents at home are
statistically much greater than the chances of being mugged on a London
street or murdered in Egypt, the reality, however, is that tourists travel in
order to experience travelling attributes such as relaxation, pleasure, peace,
tranquillity, enjoyment, comfort, etc. These are all concepts conveyed by
the concept of tourism. Anything that might suggest a move away from
achieving them is not part of most travellers' equation. The main problems
are that not only are tourists' logical base for fear often poor but also that
their perception of risk usually has little to do with logic. During the Gulf
War, Egypt was not only safe to travel to (and about 2,000 kilometres away
from the conflict) but it was also unusually uncrowded. However, tourists
perceived the risk as too high and stayed away.

The media plays an important role in travel patterns since it has the
power to shape the public's perception of a destination or issue. With
advances in telecommunications technology, many events today are covered
live, or are covered by amateur video cameras and shown later to a worldwide audience. Usually, graphic scenes or pictures are presented which have
profound effect on the public. Moreover, a destination, even long after the
occurrence of an incident, may still be remembered for a negative episode
and therefore send the public a reminder (even when no longer justified) of
relative or partial safety. Some crises are perpetuated by the media in the
public's mind. This is mostly due to invidious comparisons, perhaps due to
our insatiable appetite for superlatives. Typically, the news media make

comparisons in their stories such as 'the worst terrorist attack against tourists since Luxor', thereby keeping the crisis alive.

Although travel always involves a certain degree of risk, and is a function of the decision travel choice, not much research has been conducted in this area. One of the few studies in risk perception in the tourism industry was conducted by Roehl and Fesenmaier (1992). In their study, which looked at the relationship between risk perception and pleasure travel, it was suggested that 'a choice involves risk when the consequences associated with the decision are uncertain and some outcomes are more desirable than others' (1992: 17). Roehl and Fesenmaier also identified that information searching can be a risk reduction strategy. According to the authors, however, studies into risk perceptions have produced mixed results due to the difficulty in operationalising the phenomenon. Depending on how risk is operationalised it produces different results when trying to establish the relationships between risk attitudes and behaviour, making it in turn difficult to observe consistencies in this relationship. Roehl and Fesenmaier (1992: 17), after investigating a number of studies into risk, concluded that 'risk perceptions are situation-specific and therefore should be evaluated using measures appropriated to the context of interest'. Sönmez and Graefe investigated how key factors affected the travel decision process and used risk perception as a framework for their study. The results revealed that risk perception levels do indeed have a direct influence on international destination choice.

> When the consequences of a travel decision are uncertain, that decision is perceived as risky. Decision-makers proceed by comparing the benefits and costs of destination choices in order to select the one that promises the most benefits for the least cost.
>
> (Sönmez and Graefe 1998: 122)

Considering the susceptibility and vulnerability of the tourism industry to crises, this research field warrants more attention by both academics and practitioners.

Terrorism: definitions, patterns and causes

Definitions of terrorism have not gained universal acceptance. The US Department of State has provided some definitions of terms that are generally used and applied when terrorism is discussed. For the US Department of State the term 'terrorism' means 'premeditated, politically motivated violence perpetrated against non-combatant targets by subnational groups or clandestine agents, usually intended to influence an audience.' The term 'international terrorism' has been defined as 'terrorism involving citizens or the territory of more than one country', and the term 'terrorism group' as 'any group practising, or that has significant subgroups that practice, international terrorism' (1999).

Although the amount of international terrorism has been falling consistently since the mid-1980s, the threat of terrorist acts is increasing. The reduction in the number of attacks can be largely attributed to progress in law enforcement measures and diplomatic efforts around the world. The total number of terrorist attacks in 1998 was 273, a reduction from the 304 attacks registered in 1997, and the lowest annual total figure since 1971. However, the number of fatal victims in 1998 was 741, a record high, and the number of injured reached 5,952 (US Department of State 1999). The most prominent attacks in 1998 were the bombings in August of the US Embassies in Nairobi, Kenya and Dar es Salaam, Tanzania. The attack in Kenya helped to increase significantly the number of fatal victims and injuries. The US Embassy in Nairobi was located in a highly congested area and as a result of the explosion 291 people died and about 5,000 were wounded.

Terrorism attacks peaking in the 1980s, with a record high number of attacks in 1987 (666 terrorist attacks), called attention to the fact that tourism could be negatively affected. The aftermath of the Gulf War and the fear of possible 'global terrorism' in the early 1990s, as well as the increasing amount of political turmoil in many parts of the world, contributed to the growing attention the topic has been receiving. The number of publications dealing with violence-related topics grew significantly and whole conferences and working groups were devoted to debating the effects of terrorism on the industry in general, influences on tourism patterns, tourism vulnerabilities, and possible solutions to continued terrorist threats.

Terrorism is not a new phenomenon and has been used as a political weapon in virtually all parts of the globe. Sendero Luminoso (Peru), the IRA (Northern Ireland), the ETA (Basque Separatist Group – Spain), the Red Brigade (Italy), the Fundi (Algeria), the Abu Nidal Organisation (Iraq), the Hizbollah (Lebanon), and the Khmer Rouge (Cambodia), are just a few examples of some well-established groups. The causes of terrorism, or what leads to the organisation of terrorism groups, are many and diversified. Terrorism root causes have been attributed to religion fanaticism, political instability, chronic economic problems, famine and disease, environmental problems, demography (over-population), lack of opportunities, civil unrest, wars and guerrillas, among others. All these issues can contribute or trigger the organisation of terrorist groups and action. The globalisation process, as discussed in previous sections, provides the right conditions and environment for terrorism to flourish and become more deadly. Globalisation has excluded more nations than it has included in the way of development and prosperity. This has led to unprecedented levels of unemployment and economic hardship in many parts of the world and a sense of discontent and resentment towards some nations and peoples.

The tourism–terrorism relationship

Terrorist organisations target tourist facilities and tourists in an attempt to achieve ideological objectives and to strengthen claims to political legitimacy by making the incumbent government appear weak. Terrorist attacks on the tourism industry are often justified or validated by their perpetrators on the basis that the tourists or tourist facilities represent a threat to their culture, value system, religious convictions or because they represents capitalism. The Leader of Sendero Luminoso, a Peruvian terrorist group that deliberately targeted the tourism industry between 1989 and 1991, justified the group's actions on the grounds that they supported a Maoist philosophy that condemns capitalism. According to the group's logic, tourism was regarded as a symbol of capitalism and tourists as representatives of capitalist regimes, or oppressors, and an attack on tourism was an attack on the government since tourism is a state sponsored industry. The results of this terrorist campaign in Peru were disastrous for the industry, which experienced a reduction in international arrivals from 350,000 in 1989 to 33,000 in 1991 (Wahab 1996).

As described by Buckley and Klemm (1993: 184): 'Terrorism is a political weapon which can be used to greater effect in the postwar world because of the mass media, whose coverage gives the terrorist organisation an illusion of power and efficiency which is out of proportion to its real zeal.' Indeed, it can be argued that one of the prime objectives of terrorists is to make their causes heard by a wide audience. In this respect, international tourists are one of the main 'mediums'. Karber (1971 in Sönmez 1998) explains terrorism 'as a symbolic act' and that it can be 'analysed much like other mediums of communications' (Sönmez 1998: 418). Therefore, the components of the communication process can be explained as: the terrorist (transmitter of message); the terrorist's target (intended recipient of message); the terrorist act itself (message); and the reaction of the message recipient (feedback).

In the last fifteen years a number of studies have theoretically explored or shed light on the relationship between terrorism and tourism (Richter and Waugh 1986; Aziz 1995; Wahab 1996; Sönmez 1998; Buckley and Klemm 1993; Mansfeld 1995, among others). It is interesting to observe that tourism and terrorism have some issues in common: both make use of travel and communications technologies, both cross geographical boundaries, and both involve citizens of different countries. It appears that there is a consensus among experts that tourist destinations, tourists, and tourist facilities make logical targets for terrorists. Tourists are usually 'easy prey' since in most cases they are in totally unfamiliar environments, do not speak the language of the host country, are unfamiliar with customs and behaviour (and so are unable to detect suspicious behaviour), are unarmed, are very easily identified (they 'flock' in groups and act in very predictable patterns or are singled out when off the beaten track by their 'foreign-looking' appearance), and police forces usually keep a low profile so as not to cause unnecessary concern among the tourists or even to avoid 'intruding' on tourists' experiences, leaving tourists totally unguarded. Tourist locations provide the perfect environment for terrorists since there is usually good transport infrastructure (facilitating the movement of terrorists as well as their apparatus) and facilities where they can make money transactions (bring in and take out foreign currencies) without raising suspicion.

As the economic significance of tourism activities gains relevance and states become more dependent on it, added to the symbolic image tourism represents, tourist facilities and tourists become logical and practical targets for terrorist attacks (Richter and Waugh 1984). In addition, tourism itself can be viewed as a political issue and as such becomes susceptible to terrorist attacks. The literature on the topic also suggests that the communications gap between tourists and host communities, cultural misperceptions, and socio-economic differences that separate the two groups may cause resentment and lead to violence (Wall 1996; Aziz 1995; Wahab 1996; Ryan 1993; Hall and O'Sullivan 1996; Buckley and Klemm 1993).

Effects of terrorism on the tourism industry

Terrorism can not only jeopardise the future of the tourism industry (local, national or regional) but it can also threaten the industry's legitimacy. In respect to the latter, tourism development is often a matter of national policy (this is especially true in developing nations). When terrorism strikes, the industry inevitably becomes a matter of political significance and debate since it involves national security measures (and resources) that have implications on the whole population. The effectiveness and use of tourism as a development strategy is easily questioned or discredited by those who oppose it.

Whenever a tourist destination is affected by an act of terrorism, the consequences are many and diversified. Invariably, the net result is that the industry's profitability is negatively affected due to both travellers' reactions to the event and the enormous amount of effort the industry has to exercise to lure tourists back. The occasional terrorist attack, the type that is localised and low key, usually targeting locals, has less implications for the tourism industry since its repercussions are much smaller than a persistent pattern of attacks aimed at and directly intended to disrupt the industry or those involving international travellers. The series of terrorist attacks in Egypt is an example of the latter. Since the tourism industry is of great economic importance for that country and enjoys a high profile, it became a target for those seeking political influence. In this case, the consequences were enormous for the country. In an attempt to maintain tourist flow and occupancy rates, both the government and the private sector engaged in promotional campaigns. Typically, transportation prices to the destination are drastically cut and the accommodation sector reduces its prices sharply. Simultaneously, a large amount of effort (human and financial) is devoted to improving security and fending off an unfavourable security image with international media campaigns. Therefore, on the one hand it has a drastic reduction in profitability and on the other hand a sharp increase in operational costs.

Terrorism can affect the industry in many ways and a long list could be drawn from the potential implications for each individual sector within the industry. It would depend, naturally, on the nature and severity of the situation. However, one can generalise some impacts:

- Employment in destination countries and businesses in generating countries may be hurt (tour operators, travel agencies, etc.).
- Neighbouring countries may lose inbound tourists (generalisation effect).
- Airlines suffer from low demand and when they are involved (victim of terrorism) there may be legal disputes over compensation or damage claims for victims of incidents, or awards for loss of life.
- Due to its economic significance, tourism is immensely important politically for both developed and developing nations. Whenever international terrorism occurs, its impacts are seldom localised and it has repercussions on the whole country, or region, as well as on political relations between countries.

Although short-term reaction to terrorist attacks is relatively easy to predict, the long-term effects depend on a large array of issues and is more difficult to anticipate. For example, the long-term effect of adverse events on the tourism industry depends on how the event is perceived by the public and the tourism trade, the strength of the industry, the kind of product (how easily it can be substituted), the level of government commitment to the industry, the determination of law enforcement agencies to contain and/or eliminate threat, and so on. Although it seems that potential tourists have a relatively short memory, the implications for investment in the industry may be a different matter altogether. Investors are usually more cautious and would prefer 'safer' locations and/or guarantees on their investments.

The future of terrorism

The number of terrorist attacks has been experiencing a downward trend for several years. This is the result of counter-terrorism measures adopted in many parts of the world and reflects both diplomatic and law enforcement efforts. Terrorist groups have been discredited and intelligence and security measures have made it harder for them to operate. A range of coordinated interstate actions against terrorism are under way such as vigorous diplomacy, law enforcement, treaty negotiations, development of new technology, improvement in security, travel restriction for suspected terrorists, sharing of information, closing down of illegal activities, disruption of terrorists' training, blocking of assets, and extradition treaties (Albright, 1999). Concerted actions against Libya, such as a series of embargoes which isolated the country for several years, prompted the Libyan government to allow the suspected terrorist of the Pan Am 103 explosion (Lockerbie, Scotland) to face trial in The Netherlands.

The US response to the bombing of the Embassies in Africa by attacking terrorist facilities in Afghanistan and Sudan sent a clear message of the American government's determination to use any means, including military, to protect their interests and to disrupt terrorist networks and activities. US$5 million was announced as the reward for information leading to the arrest and conviction of any of the suspects of the Embassies bombings (US Department of State 1999).

The international community is increasingly engaged in the effort to condemn, isolate, and pressurise states that sponsor terrorism by applying a broad range of sanctions. Sanctions are more effective if they are imposed multilaterally and there is still work to be done in this respect. It is not uncommon to see unilateral and isolated sanctions which, although important, are not as effective as a collective effort. In 1998 seven countries were identified as sponsors, in different ways, of terrorism activities: Cuba, Iran, Iraq, Libya, North Korea, Sudan and Syria (US Department of State 1999). Although state sponsored terrorism has diminished in recent years, its threat is still too great to be ignored. The US Department of State (1999) made it clear that the majority of the most violent terrorist attacks would not have been possible without such state sponsorship. State sponsorship involves providing safe haven, financial support, weapons, training, logistic support, use of diplomatic facilities, etc.

While it is difficult to predict the future of terrorism, it is becoming increasingly transnational, with loosely affiliated organisations which operate independently of state sponsors. Those groups are extremely mobile and operate on a global basis, have at their disposal and utilise sophisticated technology (including weapons of mass destruction, i.e. chemical, biological, and nuclear weapons), train their members in various countries, and are able to raise large amounts of financial resources. To maintain their operations, organisations that are less state dependent resort to criminal activities such as drug smuggling, kidnapping and extortion.

Terrorist organisations in the future will employ innovative methods of attack, attacks will be more frequent and indiscriminate, and terrorists will continue to attack 'soft' targets. As telecommunications technology develops, the world will witness more reports of such attacks. As a consequence (and in some respects this is already a reality), travel experiences might be significantly altered or influenced due to increasing security procedures, exposure to searches and questioning, constant delays and disruption, constant military presence, and intensified use of surveillance technology such as cameras, etc.

Although preventive measures and other efforts are succeeding in diminishing the number of terrorist attacks, those that do occur are, unfortunately, increasingly powerful and deadly. Terrorists now have access to technology that is incredibly destructive.

Political instability

Political instability, wherever it occurs, and especially when accompanied by acts of violence or hostility of any kind, is likely to have a detrimental effect on the tourism industry. The development of a sound tourist industry depends greatly on political stability (Mansfeld 1995; Sönmez 1998; Teye 1988; Hall and O'Sullivan 1996). From a recent history perspective, there are many examples of politically unstable situations that led to the near destruction of destinations' positive image and consequently their tourism industries. The most prominent cases were the dismantlement of the former Yugoslavia and the war that engulfed the whole region, and the 4 June 1989 troubles in China, when virtually the entire world witnessed live coverage of the brutal clash of pro-democracy protesters with Chinese governmental forces in Tiananmen Square. In the case of China, tourism was hard hit with large numbers of international tourist cancellations, followed by international condemnation of the crackdown and disproportionate use of force against non-violent civilian protest, and sanctions imposed by Western governments on corporations that conducted business in China (Gartner and Shen 1992). Overall, the Chinese tourism industry experienced hardship until 1991, when it showed signs of recovery.

International and regional wars, military coups, racial, religious and ethnic conflicts, riots, political unrest, strikes, student uprisings, etc., usually emanate from political instability and threaten the safety of involved countries or entire regions, such as the events in the Balkans and the Middle East, and the continued dispute over Kashmir between India and Pakistan. Political instability is described by Hall and O'Sullivan (1996: 106) as:

> A situation in which conditions and mechanisms of governance and rule are challenged as to their political legitimacy by elements operating from outside of the normal operations of the political system. When challenge occurs from within a political system and the system is able to adapt and change to meet demands on it, it can be said to be stable. When forces for change are unable to be satisfied from within a political system and then use such non-legitimate activities as protest, violence or even civil war to seek change, then a political system can be described as being unstable.

While a terrorism act is an immediate, sudden event, which invariably involves a great number of agents in responding to the incident and large media coverage and scrutiny, violence and hostilities resulting from political instability is often a result of long disputes and struggles, usually of public domain, and rarely comes as a surprise. It is relatively easy to predict; foreigners are advised to leave the troubled region, and in this sense does not receive the intense and immediate media scrutiny and sensationalism associated with terrorism. The effect on tourism, however, is not different. Political instability also curtails tourism and can create long-lasting barriers to international tourism (Bar-On 1996; Richter and Waugh 1986; Lea and Small 1988; Mansfeld 1995; Millman 1989; Vukonic 1997). It is important to emphasise, however, that terrorism may also result from political instability (Hall and O'Sullivan 1996; Sönmez 1998; Vukonic 1997). Politically unstable areas intimidate tourism demand and obstruct development for more prolonged period of times than do terrorism affected areas.

While terrorism attacks pose a direct threat to tourists, warfare, military coups, strikes, protests, etc. may hinder development and inhibit tourist demand but do not in themselves constitute a direct threat to tourists. In the case of terrorism, tourists may be the very target of terrorist action. The nature of the threat posed to tourists, then, is quite different from that of political instability. Domestic tensions rarely involve tourists and tourist involvements are usually by coincidence rather than by design.

Since we live in an interrelated and interdependent world, political instability in one country is likely to affect governmental policies of other nations in the immediate vicinity or region and even of countries on distant continents. Governments, when reacting to political instability in a third country, usually adopt policies that influence, in one away or another, tourist flow to that country or region. Governments can, by means of their foreign policy, discourage tourist visitation to a country or entire region by influencing their perception of potential destinations, or by objectively discouraging travel through their travel warning advisory services. The United States, through their US Department of State, the United Kingdom, through the Foreign Office, and most other Western countries, provide a travel advisory service concerning safety and security, advising business and pleasure travellers as to the safety or potential risks associated with travelling to a certain region or country. This information is used by the tourism industry (airlines, travel agencies, tour operators, etc.) and by the individual traveller. In extreme cases, governments may prohibit tourist visitation to a destination, as was the position of the United States of America's policy in relation to American tourists visiting Cuba in the 1960s (Sharpley and Sharpley 1995; Hall and O'Sullivan 1996). Investment on tourism may also be blocked by governmental policies which have a direct impact on the development of the tourist industry, delaying development or even threatening its future.

The impact of political instability in major established tourist destinations, such as that in China, Mexico, Yugoslavia, Fiji and many parts of the African continent, the effect of the Gulf War in European tourism, frequent problems in Israel, and so forth, generated a great number of studies (Mansfeld 1995; Scott 1988; Goodrich 1991; Hall and O'Sullivan 1996; Millman 1989; Wall 1996; Vukonic 1997; Aziz 1995, Mihalic 1996; Pitts 1996, among others). In each case, the tourism industry suffered significantly and the road to recovery is still difficult in many situations. While there is a general consensus that tourists always return following the cessation of conflicts or threats, the effects of political instability can change for ever the nature of the industry and what constitutes it. For example, following the end of the conflicts and hostilities in the Balkans, Croatia was unable to attract its former affluent Western market. Germans, Austrian, British and French tourists, the main market prior to the war, showed no sign of returning (Ivandic and Radinic 1997; Vukonic 1997). Instead of high spending tourists, the industry started to receive a new kind of visitor from countries such as the Czech Republic, Slovenia, Slovakia and Hungary. This had tremendous implications for the industry as a whole. One explanation for this might be the perception of former tourism markets towards the general level of safety and security of the region. For example, when British troops are stationed as peace-keepers in a country, British citizens tend to assume that the destination is unsafe and naturally seek alternative places in which to enjoy their holidays. Ironically, this occurs when destinations are most in need of hard currencies and the benefits generated by tourism.

Crisis management

Clearly, the tourism industry is one of the most vulnerable and most susceptible industries to crises. Ironically, the ability of the industry to react or prevent crisis is very limited. While a great deal of attention is given to the impact of crises on the industry, very few attempts have been made to provide effective tools for the prevention and management of crisis within the industry. While other industries are benefiting from the development and employment of crisis management tools, the tourism industry is still only reactive to divergent situations. Sönmez *et al.* (1994) in Sönmez and Tarlow (1997) defined tourism crisis as:

> Any occurrence which can threaten the normal operation and conduct of tourism related businesses; damage a tourist destination's overall reputation for safety, attractiveness, and comfort by negatively affecting visitor's perception of that destination; and, in turn, cause a downturn in the local travel and tourism economy and interrupt the continuity of business operations for the local travel and tourism industry, by the reduction of tourist arrivals and expenditure.

Therefore, tourism officials and management should have in place all the necessary mechanisms to deal with emergencies and crises, both prior to their occurrence (such as prevention mechanisms) and to effectively manage those that will inevitably happen. Organisations responsible for tourism must be proactive with regards to safety and national security. One reason for this is that it is impossible to learn about crisis management in the heat of a crisis and, even worse, it is virtually impossible to respond adequately to the demands of a major crisis without having previously established support mechanisms.

Conclusion

The tourism industry is highly susceptible to the perception of political instability and concerns with the associated risks to personal safety. At its most extreme manifestation (outbreak of violence, hostilities, war, terrorist attacks against tourists or tourist facilities, etc.), political instability and persistent terrorist action can destroy established tourism sectors in a very short period of time, as in the case of the former Yugoslavia, or severely disrupt the industry as in the case of Egypt, following a string of terrorist actions which deliberately targeted tourists or tourist enterprises. At the less extreme end, but with equally devastating results, short-term political disruption or adverse media publicity can destroy years of growth and development in tourism, such as the Tiananmen Square episode (4 June 1989), or hamper development altogether, as in Brazil, until the travelling public's memory of the event dies out or the root causes of the disruption are unequivocally controlled or eradicated.

With the new millennium it seems that the number of countries facing or experiencing conflicts are, unfortunately, increasing rather than decreasing. National, regional, and international security today is of quite a different nature from that of the Cold War era when two superpowers and their alliances dictated world order. The vast majority of conflicts today, and to a certain extent the potential ones, are within individual countries (domestic problems) or at most confined to a region, rather than interstate in nature. However, the threat of interstate conflict has not disappeared and should not by any means be disregarded. The legacies of the Cold War, especially the indiscriminate and uncontrollable availability of conventional weapons, the increasing number and range of activities of volunteer armies, the unresolved disarmament programme, and the proliferation of weapons of mass destruction, have all contributed significantly to the escalation of armed conflicts and will have to be dealt with for generations to come. If progress is to be made towards a wider and more widely shared concept of international security in this century, the global community will simultaneously have to accomplish three demanding tasks. The first is to cope with the legacies of the preceding generations mentioned above; the second is to decide who the responsible actors on the world stage should be; and the third is to promote and strengthen those institutions that will prevent the kinds of wars that occurred in the twentieth century.

A logical consequence of this new security environment is that civilians now are at the centre of existing and potential conflicts. With advances in modern technology and the increasing proliferation of weaponry, any armed conflicts today invariably exact a horrific toll on civilians.

The effects of globalisation, while clearly benefiting some regions, have had different impacts on others, creating problems for national security. Moreover, the globalisation process has favoured the incrementation of activities and alliances among organised crime enterprises, transnational terrorism groups, and extremist groups of all types. The globalised environment has also permitted those groups that promote terror access to incredibly destructive technologies. These organisations today have formidable power and influence.

National security today should encompass a more diversified base of aspects, of a non-military nature, such as environmental degradation, economic disparity, demography, irrational exploitation of natural resources, drugs, proliferation of crime, etc. These issues are major forces that shape and influence the concept of national security. They all directly affect the tourism industry. Moreover, the nature of tourism operations and activities, and the facilities (both natural and man-made) used for promoting customers'/visitors' experience/satisfaction, expose the sector to safety vulnerability not encountered in other economic activities. As far as tourism is concerned, safety is not simply restricted to the macro aspects. There have recently been concerns with some elements of the travel process after a number of high profile incidents and disasters which attracted large media attention and scrutiny. Among these were a number of aircraft disasters, tourist victimisation, food contamination, tourist kidnappings, fatal accidents in leisure facilities, unethical conduct of travel agencies or tour operators, and many other incidents, which indicate that international travel can be a risky business. In addition, safety is by no means concerned just with the physical well-being of travellers. Yet, most of the attention to this topic to date reflects just that or relates to damage to tourism infrastructure and facilities. Not much has been said about the emotional and psychological toll that a negative travel experience causes (or may cause) to the individual traveller.

While large-scale natural disasters can significantly affect tourism patterns in a destination, man-made threats (terrorism, war, political instability, crime, etc.) seem to have a stronger, more negative, and longer-lasting influence on tourism demand. Unlike natural disasters, man-made crises tend to receive severer condemnation from the potential travelling population. One reason for this attitude from the public is that, in principle, all man-made 'disasters' are preventable.

Finally, despite advances in many areas, the world today seems to be more unstable than during the Cold War and is characterised by increased violence and by the potential for this trend to continue. Given the economic and social importance of tourism, and its high susceptibility and vulnerability to crisis, both government and tourism organisations should be prepared for the unexpected. Some regions are more prone to disasters and instability than others. Those regions, therefore, need to address safety and security issues with a strong emphasis on planning. The world has witnessed time and time again persisting safety problems and their consequences which sadly have been mainly human and environmental. Destinations that have acquired a reputation as unsafe travel zones have found it difficult to maintain their tourism industries. It is important to understand that people's decision to travel is based mostly upon the perception they have of a destination and of the travel process. Crisis management planning and training must be implemented if destinations are to face the new security and safety challenges of the twenty-first century.

References

Albright, M. (1999) Secretary of State Madeleine Albright statement on release of *Patterns of Global Terrorism 1998*, 30 April 1999, Washington, DC, as released by the Office of the Spokesman – US Department of State.

Annan, K. (1997) *United Nations Secretary-General statement to the Conference on Disarmament*, Geneva 30 January. United Nations Press Release SG/SM/6151 DCF/284, 1997.

Associated Press (1993) 'Tourism falls prey of terrorism', *Gainsesville Sun*, 6 May 1993.

Aziz, H. (1995) 'Understanding attacks on tourists in Egypt', *Tourism Management*, 16 (2): 91–5.

Bar-On, R. (1996) 'Measuring the effects on tourism of violence and of promoting following violent acts', in A. Pizam and Y. Mansfeld (eds) *Tourism, Crime and International Security Issues*, New York: John Wiley.

Booth, K. (ed.) (1991) *New Thinking about Strategic and International Security*, London: HarperCollins.

Boxberg, G. and Klimenta, H. (1999) *As Dez Mentiras da Globalização*, São Paulo: Aquariana.

Brayshaw, D. (1995) 'Negative publicity about tourism destinations – a Florida case study', *Travel & Tourism Analyst* 5: 62–71.

Buckley, P. and Klemm, M. (1993) 'The decline of tourism in Northern Ireland', *Tourism Management* 14(2): 184–94.

Cassedy, K. (1992) 'Preparedness in the face of crisis: an examination of crisis management planning in the travel and tourism industry', *World Travel and Tourism Review* 2: 169–74.

Center for Defense Information (1999) *Center for Defense Information Finds Wars on the Increase, Press Release*, 7 January.

Crystal, S. (1993) 'Welcome to Downtown, USA', *Meetings and Conventions* 28(3): 42–59.

Edgell, D. L. (1990) *International Tourism Policy*, New York: Van Nostrand Reinhold.

Gartner, W. C. and Shen, J. (1992) 'The impact of Tiananmen Square on China's Tourism Image', *Journal of Tourism Research* 30(4): 47–52.

Gasteyger, C. (1998) 'Towards the 21st century: trends in post-Cold War inter-national security policy', *Workshop 4, 3rd International Security Forum and 1st Conference of the PfP Consortium of Defense Academies and Security Studies*, 19–21 October 1998, Kongresshaus Zurich, Switzerland, Online Publications http://www.isn.ethz.ch/securityforum/Online_Publications/WS4/Gasteyger.htm (02/07/99).

Global Futures Bulletin (1997) 'Global arms trade increases', *Global Futures Bulletin*, V.47 http://www.cdi.org/ArmsTradeDatabase/Arms_Trade_Patterns_and_Trends/ANALYSIS/Global_Arms_Trade_Increases.txt (15 June 1999).

Goodrich, J. (1991) 'An American study of tourism marketing: impact of the Persian Gulf War', *Journal of Travel Research*, Fall: 37–41.

Hull, A. and Markov, D. (1996) *The Changing Nature of the International Arms Market*, Institute for Defense Analysis, IDA.

Hall, C. and O'Sullivan, V. (1996) 'Tourism, political stability and violence', in A. Pizam and Y. Mansfeld (eds) *Tourism, Crime and International Security Issues*, New York: John Wiley.

Ivandic, N. and Radinic, A. (1997) 'War and tourism in Croatia – Consequences and the road to recovery'. Paper presented at the International Conference on *War, Terrorism, Tourism: Times of Crisis and Recovery*, Dubrovnik, Croatia, 25–27 September 1997.

Krause, K., Epps, K., Weston, B. and Mutimer, D. (1995) *Constraining Conventional Proliferation: A Role for Canada*, Center for International and Strategic Studies.

Lea, J. and Small, J. (1988) 'Cyclones, riots and coups: tourist industry responses in the South Pacific', in B. Faulkner and M. Fagence (eds), *Frontiers in Australian Tourism*, Canberra: Bureau of Tourism Research.

Mansfeld, Y. (1995) 'Wars, tourism and the "Middle East" Factor', *Proceedings of the First Global Research and Travel Trade Conference on Security and Risks in Travel and Tourism*, Östersund, Sweden.

Mihalic, T. (1996) 'Tourism and warfare – the case of Slovenia', in A. Pizam and Y. Mansfeld (eds) *Tourism, Crime and International Security Issues*, New York: John Wiley.

Millman (1989) 'Pleasure seeking vs the "Greening" of the World Tourism', *Tourism Management* 10(4): 132–43.

Mutimer, D. (ed.) (1994) *Control But Verify: Verification and the New Non-Proliferation Agenda*, Toronto, York Center for International and Strategic Studies.

Nazarkin, Y. (1998) 'Arms control and disarmament: analysis and prospects, towards the 21st century: trends in post-Cold War international security policy', *Workshop 5D, Arms Control and Disarmament, 3rd International Security Forum and 1st Conference of the PfP Consortium of Defense Academies and Security Studies*, 19–21 October 1998, Kongresshaus Zurich, Switzerland, Online Publications http://www.isn.ethz.ch/securityforum/Online_Publications/WS5/W5_5D/Nazarkin1.htm (02/07/99).

Pitts, W. (1996) 'Uprising in Chiapas, Mexico: Zapata lives – tourism falters', in A. Pizam and Y. Mansfeld (eds) *Tourism, Crime and International Security Issues*, New York: John Wiley.

Richter, L. K. and Waugh Jr, W. L. (1986) 'Terrorism and tourism as logical companions', *Tourism Management* 7(4): 230–8.

Roehl, W. and Fesenmaier, D. (1992) 'Risk perception and pleasure travel: an exploratory analysis', *Journal of Travel Research* 30: 17–26.

Ryan, C. (1993) 'Crime, violence, terrorism and tourism – an accidental or intrinsic relationship?', *Tourism Management* 14(3): 173–83.

Santana, G. (1999) 'Tourism: towards a model for crisis management', *Turizam, Special Issue – Tourism and Violence: Crisis and Recovery* 47(1): 4–12.

Scott, R. (1988) 'Managing crisis in tourism: a case study of Fiji', *Travel and Tourism Analyst* 6: 57–71.

Sharpley, R. and Sharpley, J. (1995) 'Travel advice – security or politics?', *Proceedings of the First Global Research and Travel Trade Conference on Security and Risks in Travel and Tourism*, Östersund, Sweden.

Sönmez, S. (1998) 'Tourism, terrorism, and political instability', *Annals of Tourism Research* 25(2): 416–56.

Sönmez, S. and Graefe, A. (1998) 'Influence of terrorism risk on foreign tourism decisions', *Annals of Tourism Research* 25(1): 112–44.

Sönmez, S. and Tarlow, P. (1997) 'Managing tourism crises resulting from terrorism and crime', paper presented at the International Conference on *War, Terrorism, Tourism: Times of Crisis and Recovery*, Dubrovnik, Croatia, 25–7 September 1997.

Teye, V. B. (1986) 'Liberation wars and tourism development in Africa: the case of Zambia', *Annals of Tourism Research* 13: 589–608.

Treverton, G. F., Heuven, M. and Manning, A. (1998) 'Towards the 21st century: trends in post-Cold War international security policy', *Workshop 4, 3rd International Security Forum and 1st Conference of the PfP Consortium of Defense Academies and Security Studies*, 19–21 October 1998, Kongresshaus Zurich, Switzerland, Online Publications http://www.isn.ethz.ch/securityforum/Onl. . ./WS4/Treverton/Table%20of %20Contents.htm (02/07/99).

US Department of State (1999) *Patterns of Global Terrorism: 1998*, Washington DC: US Department of State.

Vukonic, B. (1997) *Tourism in the Whirlwind of War*, Zagreb: Golden Marketing.

Wahab, S. (1996) 'Tourism and terrorism: synthesis of the problem with emphasis on Egypt', in A. Pizam and Y. Mansfeld (eds) *Tourism, Crime and International Security Issues*, New York: John Wiley.

Wall, G. (1996) 'Terrorism and tourism: an overview and an Irish example', in A. Pizam and Y. Mansfeld (eds) *Tourism, Crime and International Security Issues*, New York: John Wiley.

11 Globalisation, total quality management and service in tourism destination organisations

Darren Lee-Ross and Nick Johns

Introduction

Since the early 1970s the marketplace for tourism and hospitality services has become increasingly international and competitive. This process of 'globalisation' has mirrored macro-trends of escalating worldwide trade and strategic international alliances between nation-states and major corporations, particularly in developing nations (Hall 1997; WTO 2000). Globalisation may be defined in a number of ways, for example:

> The extension of traded goods into global markets. Or, The expansion of exports and imports and the import of capital equipment.
>
> (Manning 1996: 2)

> [Where] political, economic and technical developments are transforming the world into a conglomeration of interconnected and interdependent nations, economies and peoples.
>
> (Clark and Arbel 1993: 84)

The rationale for globalisation usually centres upon a supposed positive relationship between increased trade, competition and wealth generation for all. However, some commentators draw attention to the negative aspects of globalisation including a downward thrust on wage levels and working conditions for less-skilled workers of all nations (Ross 1996; Sylvan 1996; Nunnenkamp 1995).

The factors usually cited for the globalisation of tourism demand include low prices for air travel, low accommodation costs, economic prosperity in developed and emerging nations, greater free time, technological changes, improved health among retirees, reduced international tensions and interest in other cultures (Baum 1995). Some authors suggest (quite reasonably) that the tourism industry has always been globally focused with people travelling to and from diverse countries and regions to experience the unique tourism product (Clark and Arbel 1993; Vellas and Bécherel 1995). Clearly this is an oversimplification as destinations may be unique but the services offered in hospitality organisations may not be because many hotels, restaurants, amusement parks and so on, are owned and operated by 'foreign' companies and the service will be standardised accordingly, for example, Marriot, Sheraton, Motel 6, Hilton, KFC, McDonald's, Disneyland Paris (Clark and Arbel 1993). This is also true of the hospitality sector because it is a key element of tourism and the global infrastructure. However, globalisation trends have become more pronounced because of issues such as increasing international ownership and franchising of hotel and restaurant chains, joint international finance syndicates, saturation of home markets, and global political changes.

Similar to the caution advised by Ross and Sylvan, Tribe (1995) considers that globalisation of tourism and hospitality may also have negative outcomes. For example, increased travel could mean a regrettable outcome of a standardised cosmopolitan world, or a homogenisation of tourism products (Ritchie 1991). This trend for worldwide 'tourism culture' could, amongst other things, make travel to 'authentic' destinations superfluous.

Naisbitt's (1993) forecast is not so gloomy. He points out that host societies and indigenous cultures of tourist destinations tend to strengthen efforts to maintain their unique character. By doing this, small and medium-sized hotels and restaurants, for example, will strive to maximise *local* competitive advantage rather than relying on a standardised international service product (as is often the case for international corporations). Whether existing and future tourist destinations become populated by homogeneous, chain-type service organisations or by smaller 'unaffiliated' enterprises is difficult to forecast. However, Ritchie's (1991) prediction seems plausible when he refers to this development as a paradox, where cultural diversity will thrive in a sea of homogenisation.

To maximise the chances of success in an increasingly competitive and global marketplace, tourist destinations must develop their product more effectively than anywhere else. In other words, they must achieve competitive advantage. Some destinations are blessed with awe-inspiring natural and man-made attractions which will always attract tourists such as the Great Barrier Reef, rainforests and various antiquities; other regions are not so lucky and have to reinvent themselves. For example, partnerships between public and private sectors of the tourism industry are redefining certain UK tourist destinations to satisfy new and emerging consumer tastes and preferences, such as development of the 'Black Country Museum' near Birmingham, and a new focus on the Norfolk Broads as a substitute for the former 'bucket and spade' product of Great Yarmouth.

The globalisation of tourism has also resulted in changes of consumer preferences. Dalen's (1989) typology provides a useful point of reference for these changing consumer tastes by dividing tourists into four categories: these are 'modern materialist'; 'modern idealist'; 'traditional idealist'; and 'traditional materialist'. Dalen identifies an increasing number of modern idealists who demand, amongst other things, high levels of service quality,

value for money, and holidays which are customised to their individual preferences. Other researchers have also identified a similar trend (Brackenbury
1991; Middleton 1991; and Poon 1993).

Unlike tourist attractions, destination organisations often achieve competitive advantage by effective management of service provision. In particular, small to medium-sized enterprises (SMEs) find it difficult to take
advantage of the globalisation process because they cannot achieve sufficient cost-control or sustain competition based on price. Instead, they must
use and manage quality improvement strategies to differentiate themselves.

Employees (especially front-line) have a key impact upon 'added value'
and total service quality instead of being simply 'providers' or 'deliverers'
of a product. Managers must adopt a philosophy of total quality management (TQM) where customers and employees are viewed as an essential
part of the hospitality product during the service encounter. This chapter
does not seek to further discuss globalisation, nor comment on the development of global tourist destinations or demand; these areas are dealt with
in other chapters. Instead, issues of quality and service provision crucial
for achieving competitive advantage amongst global hospitality organisations are discussed. The chapter is divided into three sections. The first
explains the concept and development of TQM; the second explains the
nature of quality and quality provision of the hospitality product. The
importance and impact of the service encounter upon the hospitality
product is discussed in the third section. Finally, some recommendations
for managing the service encounter are made.

Total Quality Management

During the latter half of the 1980s the acronym TQM entered both business and academic vernaculars. Some consider it little more than a 'buzz'
phrase, covering business activities designed to improve product quality
(Dahlgaard 1999). Moores (1996: 35) suggests, 'There are as many definitions of TQM as the number of individuals to whom the request is put'.
Similarly, Johns (1993) believes that distinctions between TQM and other
quality improvement techniques/approaches such as quality circles and
quality assurance are largely philosophical, rather than practical.

It may therefore be possible to view TQM as an eclectic of quality
improvement processes and procedures designed to improve the provision
of goods and services. However, the difference between these components
and TQM is first, it has a customer and employee imperative (Schalkwyk
1998) and second, it embraces the idea of 'holism'. In other words, the
customer is sovereign, employees are empowered, and both drive a
'systemic' TQM philosophy. In simple terms, TQM allows departments, activities and procedures operating within an organisation to be
viewed as sub-systems with their own customer–supplier relationships. The
objective of TQM is to optimise these relationships at every system

Table 11.1 Principles and components of TQM

- Management commitment/leadership
- Continuous improvement as a result of a focus on quality
- Strong customer focus
- Customer driven organisation
- Total involvement/commitment/responsibility from all in organisation
- Actions based on facts using statistical process control and statistical performance measurement
- Focus on employees
- Learning, training and education
- Cultural changes
- Partnership with suppliers, customers and society
- Holistic approach

Source: Adapted from Dahlgaard (1999)

boundary, that is, where one department, activity and/or set of procedures finishes and another begins. Here, systems boundaries exist largely at the behest of the manager or are defined by the product/service.

After reviewing the literature, Dahlgaard (1999) provides a comprehensive list of elements essential for TQM. These are shown in Table 11.1.

Total quality movement

Historically, the concept and main consideration of quality was related to manufacturers and their conformance to specified standards. Slowly, Juran's (1979) term 'fitness for use' became important because it recognised the dynamics of quality specifications. Gradually, the TQM philosophy evolved and came to mean a meeting of customer requirements (Shimizu and Akao 1996). Later this developed into 'satisfying' customers and finally, to 'delighting' customers. Thus the quality movement went from detective backward-looking quality control to preventive forward-looking quality assurance, from looking at results towards processes and interrelationships.

The main feature of quality control was that products must already exist in order to be tested. To overcome the limitations of retrospectivity, an approach emphasising process monitoring, control and documentation was developed. Quality assurance came into being with the aim of systematically eliminating faults during processing. Thus defects do not occur in the product in the first place (see Oakland 1989).

Total quality management philosophy now also focuses on service processes and procedures. These developments have also meant a shift in human motivational aspects; for example, rewards are now viewed in terms of empowerment and social aspects rather than simply material dimensions. In short, two new areas have been introduced into TQM, these are customer and employee satisfaction.

The concept of quality

Most business texts (and for that matter, most managers) consider private sector organisations to have a range of objectives including increasing profitability and market share. Moreover, these may only be attained by achieving competitive advantage through the provision of high quality goods and services. However, the quality of a product has two dimensions, for example a car may be red in colour. Colour is a characteristic or attribute of the product. In this sense, it is one of the product's 'qualities'. The other quality aspect of this car is something the CEDT (1992: 390) terms as its 'degree of excellence'.

It is the second notion of quality which is difficult to define objectively. For example, a Rolls-Royce, Hyundai, McDonald's meal and a fine-dining experience may all be perceived to be of excellent quality. Clearly therefore, customer expectations and perceptions are a key consideration in the management of quality. This dimension is classified by Juran (1979: 2.2) as 'fitness for purpose'. Indeed, this is acknowledged tacitly by a number of authors who agree that quality is defined by customers constantly comparing and refining their perceptions and expectations of service experiences (for example, Grönroos 1988; Peters 1989; and Oberoi and Hales 1990). In destination organisations such as hotels and restaurants, the expectations and perceptions of customers are crucial because they are involved in the performance of the service. Moreover, customer demands are individual and each service encounter has the potential to be different invoking a variety of expectations against which the perceived service is measured.

There also is another difficulty in that the appreciation of quality is a 'non-thinking process' (Pirsig 1974: 200). In other words, consumers perceive a product as a unified whole rather than differentiating between tangible and intangible dimensions (Sasser Olsen *et al.* 1978; Jones 1983). The problem for service providers is the identification of key quality attributes (a number of researchers have done so, see Lewis 1984 for a review; Martin 1986; Grönroos 1988; Parasuraman *et al.* 1988, Johns and Lee-Ross 1998). Those attributes which are tangible are easier to identify and measure than intangibles.

Table 11.2 shows Johns and Wheeler's (1991) tangible ('production') and intangible ('service') components of hospitality processes in their typography of hospitality management strategies.

Effective quality management systems should contain elements of quality control and quality assurance. In hospitality organisations, quality control emphasises measurable, tangible aspects of the post-operational product (for example, clean kitchens, sufficient staff and provision of tea-making facilities) which facilitate a basic service. Quality assurance is pre-operational and designed to prevent poor performance. However, the term pre-operational, in a hospitality service sense, is different to production-oriented manufac-

Table 11.2 Management strategies for 'production' and 'service' components of hospitality processes

Types of processes	Examples	Process features	Productivity enhanced by	Quality enhanced by
'Production'	Food, prepared room	Customer sees product but not process. Some 'stock piling' possible	Production line specifications, use of capital, control of labour, waste and energy	Design, specs., standardisation, statistical sampling
'Service'	Reception, F and B service which is 'instant' and totally perishable	Customer sees and participates in service process	Maximising sales per staff member, maximising marketing advantage per staff member	Maximising quality of service per staff member

Source: Adapted from Johns and Wheeler (1991).

turing quality assurance. Although some aspects such as 'standardised procedures' may be applied to food production operations in a hotel, 'packaging' and 'storage' are clearly inappropriate. This is because the output is intangible, the product is perishable, the delivery system is complex and highly integrated, the customer takes part in the service process, and each customer has individual expectations and perceptions, especially during the service encounter.

The positioning of the last characteristic of service provision by no means indicates its relative level of importance. Indeed, a number of authors argue that the customer's overall perception of service quality stands or falls at the customer/employee interface (for example, Lundberg 1991; McGuire 1999). This is because the service interaction is distinguished by perceptions, preconceptions and social relationships, rather than an expectation of material quality characteristics from the tangible hospitality product. Thus service operations require a management system which reflects the evolution of TQM. That is, a system which empowers employees during customer transactions. The following section discusses the role and impact of the service encounter upon customers' total quality impression of the hospitality product.

Serving customers: aspects and dynamics

The economic importance of service industries in many countries has driven a plethora of centrally driven quality-based initiatives including the European Foundation for Quality Management; the British Quality Foundation; the BS EN ISO 9000 Series; the Australian Quality Council;

and the Quality Assurance Institute of America. In addition, expanding globalisation in the tourism industry has increased competition and the supply of 'sophisticated' customers. Tourism and hospitality organisations have responded by relying more upon their employees to exceed client expectations and thus achieve competitive advantage. Many of these firms have operationalised this facet of TQM by focusing upon employee performance at the customer interface, recognising the crucial nature of the 'service encounter' (MacVicar and Brown 1994 – Moat House Hotels; Dodwell and Simmons 1994 – Scott's Hotels; Breen and Liddy 1998 – Ramada Franchise; Bouldner *et al.* 1997 – Adventure Theme Park).

The service experience or encounter may be considered as a role performance that customers and providers have to enact because they are central to service provision. The definition of 'encounter' ('meet in conflict' CEDT 1992: 161) applied to hospitality services suggests a potential for conflict. Although the meaning cannot be taken literally, it implies that the service encounter has the potential to ruin an otherwise excellent hospitality experience. The service situation is difficult to supervise because a ubiquitous manager is both impractical and not particularly desirable in busy organisations. Thus, service provision relies solely on the employee; managers do not have direct control over their worker's performance during the encounter (Jones and Lockwood 1989).

However, service encounters can be managed indirectly by considering other issues, like organisation culture, system design, selection and training. Thus, front-line employees can perform adequately at the customer interface so long as they are supported, there is an appropriate service delivery system and adequate training is provided. Clearly this is a complex mix of variables and interactions which managers of tourism and hospitality destination organisations must understand if they are to achieve competitive advantage effectively.

This section continues by focusing on the role of the service encounter in destination organisations. The nature and main features of social interactions are first explored followed by an outline of service encounters and how to manage them. Other issues of encounter management are then considered including service delivery and system design, 'scripting' and organisational culture and selection and training strategies.

Social interactions

Service encounters are complex because social contact between actors carries a variety of expectations. For example, employees providing a service must feel satisfied with their performance almost as a justification for their career choice. They also need to satisfy the immediate demands of the customer. Customers seek immediate satisfaction related to the interaction purpose such as choosing a particular experience in an adventure park, currency exchange, or ordering a drink in a hotel bar. In addition, they need to be treated respectfully in a polite and 'appropriate' manner. Horney (1996) classifies these expectations as procedural (tangible) and convivial (intangible) aspects of service delivery. Recognising and under-standing these two overall dimensions of a service product is necessary for managing the encounter. Grönroos (1990) considers service encounters as a three-phase consumption cycle beginning with a joining phase (activated by the customer), an intensive consumption phase (where the core service is consumed), and a detachment phase. The length of the cycle varies for different services. For example, the consumption cycle for insurance services would be long-term, whereas tourism and hospitality services have a short, periodic cycle. However, the complexity of the situation is exacerbated because each one has the potential to be different from the last.

The role played by customers during the service encounter also has a significant impact on the outcome of the exchange. A normal pattern contains several stimuli and responses communicated both verbally and non-verbally, with actors providing cues for each other to react in certain ways. Usually, cues are determined by the current situation and stereo-typing from experience of previous service encounters. Perceptions of social and economic status, personal characteristics and appearance also impact upon the relationship between participants. In addition, expectations and behaviour are particular to specific occasions and this may cause changes in the service encounter dynamic (Farber-Canziani 1996). For example, a tourist enjoying the clubhouse facilities of St Andrew's golf course in Scotland, UK will have different service encounter expectations to that of their local 'pitch and putt' course.

Despite the impact of these dynamics upon the service encounter, psycho-logical research (for a summary see Eysenck and Eysenck 1985) shows that people possess a more or less stable set of characteristics which predispose them to react in certain ways in a variety of encounters. These include culture; introversion and extroversion; stability and neuroticism; aggression and passivity; self-esteem; and desire for approval.

Service encounters

A number of key features distinguish service encounters from other social interactions (see Czepiel *et al.* 1985). These features modify the properties of social interactions depending upon the specifics of the hospitality organ-isation-based situation. Generally, the narrow and purposeful focus of the encounter is understood by participants and all will have expectations modified by the tasks and functions of the job. For example, customers enjoying a pizza meal would not expect a formal, 'silver-served' (by com-petent, unobtrusive but polite waiting staff) fine dining experience. Perceptions of status may also shift or become suspended. Outside the work-place, the service provider may have a 'higher' status than the customer but will nonetheless be expected to defer during the encounter. Task-related information also helps characterise exchanges because it focuses on encounter-based goals of which actors are aware. For example, when booking a holiday with a travel agent, the activities involved are restricted by the nature and content of the service.

Service encounters have a significant impact on participants and play a key role in the subsequent total quality perception of the customer and the employee. 'Successful' encounters motivate employees and help improve job performance and job satisfaction. In their quest for competitive advantage, global destination organisations must manage these encounters effectively so that they become longer-term service 'relationships'.

Encounter management

It is something of an irony that the key element of service provision cannot be controlled directly by managers. Employees assume the role of 'ambassadors' for their organisations, influencing overall customer perceptions of quality. In an effort to assert indirect control, a number of service-related techniques of 'blueprinting', 'flow-charting' and other quality assurance systems may be applied (Comen 1989). For example, blueprinting appreciates the overall service process as a series of interconnected stages. This allows certain areas or service elements, such as the encounter, to be identified, modified or customised to customer's individual requirements. This technique allows managers to increase their range of options at any stage of the service process. Scrutiny is essential for all service processes but is especially important for those which are complex and convoluted. Thus, a complicated service system gives rise to a sophisticated service encounter with greater provider and customer expectations. For example, the Gleneagles hotel in the UK offers an extensive 'package' including use of leisure facilities and a championship golf course. Whereas a Travelodge offers a limited range of accommodation and catering services. Managers require a detailed blueprint of the process and Gleneagles' employees need a comprehensive knowledge of facilities and how to interact with clients whose purpose of visit and social status may be different to those of Travelodge clients.

Another common indirect approach to managing the service encounter is known as 'scripting'. This is where employees are trained to respond in a prescribed manner at the service interface. Simply, the role of hotel hall porter, tour guide, receptionist and so on is conceptualised as the typical series of tasks and behaviours associated with that position. A script usually includes standardised routines for greeting customers, enquiring, demonstrating empathy and positivity. There are a few potential problems associated with the scripting procedure however. Managers need to be flexible when designing scripts because each encounter is unique (although there is usually some commonality). The procedure must be planned carefully and employees encouraged to share in the developmental process, otherwise scripted responses will not be their own. Simple repetition of learned 'lines' at predetermined points of the service encounter may give a shallow, unoriginal and uncaring impression to customers. In addition, employees may become bored and lose their self-esteem.

Managers may also rely on methods which improve employee observation, diagnostic and improvisation skills. In short, managers may empower employees to deal with each individual situation as it arises. This approach requires a training support system to be successful; otherwise there is a danger of employee incompetence and inexperience leading to poorer levels of service and increased worker stress. Limitations of these procedures are detailed by Farber-Canziani (1996).

Other strategies which minimise or eliminate the need for service encounters include using automation or self-service operations. For example, many hotels use drinks and snack vending machines and incorporate buffet style self-service food systems. The layout and physical design of facilities in destination organisations is also considered to have an impact on providers and recipients during the service encounter (Czepiel *et al.* 1985).

Selection

Perhaps one of the most controllable issues for managing the service encounter is selecting the right person for the right job. Ideally, people should be selected who already have a predisposition toward providing an organisation-determined quality service. This idea is based on the notion that there are differences in people's skills and personality characteristics which can be matched to specific jobs. A number of techniques may be used for this purpose including 'situational interviews' which are based on identified 'critical incidents' occurring during performance of the job. The incidents are turned into descriptions which are then transposed into interview questions asking how applicants would deal with particular situations. Answers are transposed by interviewer(s) into a Likert-type response on a behaviourally anchored rating scale. Other approaches focus on personality traits suggesting that sociability, conscientiousness and so on, reflect innate competencies related to service encounters. The resulting 'service orientation' index constructed from such traits indicates an interviewee's level of suitability for the job (Lewis and Entwhistle 1990).

These techniques have been criticised because of their apparent inability to predict behaviour at work. Therefore, it is advisable to use them as part of a composite approach to employee selection so that biases specific to a technique may be counteracted by strengths of another.

A recently developed diagnostic tool known as the Service Predisposition Instrument (SPI) (Lee-Ross 1999) may also be used as part of an organisation's screening procedure for people operating at the service interface. The instrument uses empirically-determined service dimensions to ascertain suitability of respondents. The service dimensions are shown in Table 11.3.

Table 11.3 Seven service dimensions

Disposition
Affinity/understanding
Competence/ability
Deference
Extra
Individual consideration
Communication

The SPI consists of twenty-one question statements developed around these dimensions and are scored on a Likert-type scale. The questions are based on the popular 'three-component' view of attitudes as recommended by a number of authors (for example, Rosenberg and Hovland 1960; and Ajzen and Fishbein 1980). The three classes are:

- cognitive – perceptual responses and verbal statements of belief;
- affective – sympathetic nervous responses and statements of affect/emotion; and
- conative – overt actions and statements concerning behaviour.

Using simple calculations comparisons can be made between candidates who have completed the SPI with a pre-established set of job norm indices. This reveals the extent to which individuals are predisposed to operating effectively at the customer interface. Managers are then in a position to judge whether first, the candidate is suitable for a front-line position, and second, how much training is required.

Training

If training procedures are to be effective they must mirror TQM philosophy. That is commitment at all levels; involvement of everyone; opinions of staff at all levels; and chief executives willing to participate in the programme. According to Jones and Lockwood (1989) effective training can be divided into three areas:

- The first is 'traditional' where staff are shown the correct way of dealing with customers based on previously identified standards. Staff are then encouraged to adopt this approach through role-play and video techniques. Although this approach is useful, it may not be suitable for all situations because it engenders inflexibility.
- The second concerns the use of quality circles where staff are encouraged to consider how they can improve the service given to the customer. To be effective, this approach needs everyone's support and the provision of resources by which ideas generated can be tried and tested.

- Finally, encouragement plays a vital role in any training process. Consequently, staff develop their (own) customer service skills using incentives where they are judged against previously identified standards of performance.

Lewis and Entwhistle (1990) consider both 'evangelical' and 'exploratory' approaches as effective techniques for training staff. The former relies on a charismatic trainer generating excitement and enthusiasm amongst large numbers of trainees. The latter is a subtler and more gradual approach creating a climate where trainees can participate in decision-making, grow, learn and become committed to the organisation.

Training is a key element in the development of 'customer care programmes'. Forward-looking destination organisations usually begin the process with a statement of what they want to achieve and where they 'want to be'. In other words, the starting point for these programmes is a policy or vision statement. An overview of a training schedule is shown in Figure 11.1.

Culture

Customer care, and the provision of excellent service at the customer interface needs more than a simple shift of organisational focus; rather, it requires the creation of a service-oriented culture. This may be especially true in developing countries where employees have little experience of receiving and providing a quality service. Chon (1999: 53) considers that '. . . developing an organisation's culture that focuses on serving the customer is essential in providing quality customer service'. Expressions of organisational culture range from complex and convoluted deposits of

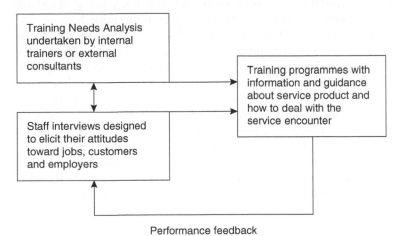

Performance feedback

Figure 11.1 Schedule for a customer care training programme.

values, beliefs, ideals and rituals, to simply 'the way things get done here'. In smaller tourism and hospitality organisations, culture may be ill-defined and a function of owners and managers. Larger organisations however, should have clearer aims, objectives and procedures to achieve and thus satisfy their cultural aspirations. In order to develop an appropriate service culture, managers and employees need training and education to understand the concept itself and the key change triggers. The main elements cited by Jones and Lockwood (1989: 123) include:

- values – sense of direction and identity guiding day-to-day behaviour;
- heroes – successful individuals because of their adherence to organisational values;
- rites and rituals – communicate what is expected of employees; and
- cultural network – informal organisation through which communication of the culture takes place.

These are key elements and should assist managers to establish an appropriate organisational culture. The 'correct' organisational climate impacts on customers. That is, the organisation establishes a climate which affects the way employees behave during the service encounter and, therefore, the customers' perceptions of the organisation and its service.

Conclusion

The process of globalisation appears to be unstoppable with international mergers, acquisitions and strategic alliances becoming the norm rather than the exception. Unfortunately, increased trading will not benefit everyone, particularly low-skilled workers in new and emerging economies. Unless global corporations become more philanthropic, living standards and social structures in certain countries may be undermined.

The tourism and hospitality industry already has a global focus. However, due to dynamic trading and technological conditions, there have been changes in product supply in response to shifting consumer preferences. In an increasingly competitive trading environment it is important for destinations to reinvent or differentiate themselves successfully, especially if they become homogenised as predicted by some researchers. In particular, destination organisations will need to seek competitive advantage by providing a quality service product using the philosophy of TQM. Total Quality Management is based on a systems approach, whereby organisations and processes are viewed in a holistic manner. In other words, all activities, processes and procedures are interdependent and issues for service delivery include:

- use of blueprints and flow charts to identify key service areas, eliminate potential problems and recovery from service failure;

- participation of customers in the service encounter;
- indirect management of the service encounter;
- indirect management of customers through layout planning; and
- adequate selection and training support systems.

A defining characteristic of hospitality organisations is the kind of service they deliver. Service quality stands or falls by the outcome of the service encounter. Service excellence can only be achieved by commitment starting at the top of the organisation being translated by managers providing leadership by example. Clearly this is only workable with appropriate support systems which focus on service quality. It is vital that managers and employees understand the 'service encounter'. It is the essence of the service experience because customers derive an initial 'total quality' impression of the organisation from this interaction. In addition, the ensuing behaviour at the service interface has a lasting impression upon the customer and is likely to influence subsequent exchanges. The service encounter is also the point at which managers lose much of their direct control over employee performance. It is therefore essential that managers strive to influence employee conduct wherever possible. There are a number of ways this may be achieved including examination of the service delivery system, use of appropriate selection and training procedures, and an understanding of organisational culture. However, achievement of excellence during the service encounter requires scrutiny of the total service process, managerial commitment and enthusiasm from everyone in the organisation.

References

Ajzen, I. and Fishbein, M. (1980) *Understanding Attitudes and Predicting Social Behaviour*, New Jersey: Prentice-Hall.

Baum, T. (1995) 'Trends in international tourism', *Insights*, March, London: ETB.

Bouldner, A., Baker, D. A. and Fesenmaier, D. R. (1997) 'Effects of service climate on managers' and employees' rating of visitors' service quality expectations', *Journal of Travel Research* 36(1): 15–22.

Brackenbury, M. (1991) 'The preferences of tourists in the European market', in *Strategy for the Development of Tourism in Europe Towards the Year 2000*, Heraclion.

Breen, P. and Liddy, J. (1998) 'The Ramada revolution: the birth of a service culture in a franchise organisation', *National Productivity Review* 17(3): 45–52.

CEDT (1992) *Collins English Dictionary and Thesaurus*, HarperCollins, London.

Chon, K. (1999) 'Quality in hospitality and tourism services', *Australian Journal of Hospitality Management* 6(2): 51–3.

Clark, J. J. and Arbel, A. (1993) 'Producing global managers: the need for a new academic paradigm', *Cornell Hotel and Restaurant Administration Quarterly* 34(4): 83–93.

Comen, T. (1989) 'Making quality assurance work for you', *Cornell Hotel and Restaurant Administration Quarterly*, November: 22–9.

Czepiel, J. A., Solomon, M. R. and Surprenant, C. F. (1985) *The Service Encounter: Managing Employee/Customer Interactions in Service Businesses*, Massachusetts, Lexington Books.

Dahlgaard, S. M. P. (1999) 'The evolution patterns of quality management: some reflections on the quality movement', *Total Quality Management* 10(4/5): 473–80.

Dalen, E. (1989) 'Research into values and consumer trends in Norway', *Tourism Management* 10(3): 183–6.

Dodwell, S. and Simmons, P. (1994) 'Trials and tribulations in the pursuit of quality improvement', *International Journal of Contemporary Hospitality Management* 6(2/3): 14–18.

Eysenck, H. J. and Eysenck, M. W. (1985) *Personality and Individual Differences*, New York: Plenum Press.

Farber-Canziani, B. (1996) 'Integrating quality management and customer service: the service diagnostics training system', in M. D. Olsen, R. Teare and E. Gummesson, *Service Quality in Hospitality Organisations*, London: Cassell.

Grönroos, C. (1988) 'Service quality, the six criteria of good perceived quality service', *Review of Business* 9: 1–9.

Grönroos, C. (1990) *Service Management and Marketing: Managing the moments of truth in service competition*, Massachusetts: Lexington Books.

Hall, C. M. (1997) *Tourism in the Pacific Rim: Developments, Impacts and Markets*, South Melbourne: Longman.

Horney, N. (1996) 'Quality and the role of human resources', in M. D. Olsen, R. Teare and E. Gummesson, *Service Quality in Hospitality Organisations*, London: Cassell.

Johns, N. (1993) 'Quality management in the hospitality industry part 3: recent developments', *International Journal of Contemporary Hospitality Management*, 5(1): 10–15.

Johns, N. and Lee-Ross, D. (1998) 'A study of service quality in small hotels and guesthouses', *Progress in Tourism and Hospitality Research* 3: 351–63.

Johns, N. and Wheeler, K. L. (1991) 'Productivity and performance measurement and monitoring', in R. Teare and A. Boer (eds), *Strategic Hospitality Management*: 45–71.

Jones, P. (1983) 'The restaurant, a place for quality control and product maintenance', *International Journal of Hospitality Management* 2(2): 93–100.

Jones, P. and Lockwood, A. (1989) *The Management of Hotel Operations: An Innovative Approach to the Study of Hotel Management*, London: Cassell.

Juran, J. (1979) *Quality Control Handbook*, 3rd ed., New York: McGraw-Hill.

Lee-Ross, D. (1999) 'A comparison of service predispositions between NHS nurses and hospitality workers', *International Journal of Health Care Quality Assurance* 12(3): 92–7.

Lewis, B. R. and Entwhistle, T. W. (1990) 'Managing the service encounter: a focus on the employee', *The International Journal of Service Industry Management* 1(3): 41–51.

Lewis, R. C. (1984) 'Isolating differences in hotel attributes', *Cornell Hotel and Restaurant Administration Quarterly*, November: 64–77.

Lundberg, G. C. (1991) 'Productivity enhancement through managing the service encounter', *Hospitality Research Journal* 14(3): 63–71.

MacVicar, A. and Brown, G. (1994) 'Investors in people at the Moat House International, Glasgow', *International Journal of Contemporary Hospitality Management* 6(2): 53–60.

Manning, C. (1996) 'The great globalisation debate with Peter Thompson', http://abc.net.au/rn/talks/bbing/stories/s10617.htm: 1–18.

Martin, W. B. (1986) 'Defining what quality service is for you', *Cornell Hotel and Restaurant Administration Quarterly*, February: 32–8.

McGuire, L. (1999) *Australian Services Marketing and Management*, South Yarra: Macmillan.

Middleton, V. T. C. (1991) 'Whither the package tour?', *Tourism Management* 12(3): 185–92.

Moores, B. (1996) 'A glossary of terms encountered in quality and customer service', *International Journal of Health Care Quality Assurance* 9(5): 24.

Naisbitt, J. (1993) Megatrends, *Het Spektrum*, Wijnegem.

Nunnenkamp, P. (1995) 'Verschärfte Weltmarktkonkurrenz, Lohndruck und begrenzte wirtschaftspolitische Handlungsspielraume', in *Aussenwirtschaft*, 50 Jahrgang, Heft 4.

Oakland, J. (1989) *Total Quality Management*, London: Butterworth-Heinemann.

Oberoi, V. and Hales, C. (1990) 'Assessing the quality of the conference hotel service product: towards an empirically-based model', *Service Industries Journal* 10(4): 700–21.

Parasuraman, A., Zeithaml, V. A. and Berry, L. L. (1988) 'Servqual: a multiple-item scale for measuring customer perceptions of service quality', *Journal of Retailing* 64, Spring, 12–37.

Peters, T. (1989) *Thriving on Chaos*, London: Pan Books.

Poon, A. (1993) *Tourism Technology and Competitive Strategies*, Wallingford: CABI.

Pirsig, R.M. (1974) *Zen and the Art of Motorcycle Maintenance*, London: Corgi.

Ritchie, J. R. B. (1991) 'Global tourism policy issues: an agenda for the 1990s', *World Travel and Tourism Review* 1: 152.

Rosenberg, M. J. and Hovland, C. I. (1960) 'Cognitive, affective and behavioral components of attitudes', in C. I. Hovland and M. J. Rosenberg (eds), *Attitude Organization and Change*, New Haven: Yale University Press.

Ross, E. (1996) 'The great globalisation debate with Peter Thompson', http://abc.net.au/rn/talks/bbing/stories/s10617.htm: 1–18.

Sasser, W.E., Olsen, R. P. and Wyckoff, D. D. (1978) *Management of Service Operations*, New York: Allyn & Bacon.

Schalkwyk, J. C. V. (1998) 'Total quality management and the performance measurement barrier', *The Total Quality Management Magazine* 10(2): 124–31.

Shimizu, S. and Akao, Y. (1996) 'The latest trends of quality management in Japan', *ICQ Proceedings*, Yokohama: JUSE.

Sylvan, L. (1996) 'The great globalisation debate with Peter Thompson', http://abc.net.au/rn/talks/bbing/stories/s10617.htm: 1–18.

Tribe, J. (1995) *The Economics of Leisure and Tourism: Environments, Markets and Impacts*, Sydney: Butterworth-Heinemann.

Vellas, F. and Bécherel, L. (1995) *International Tourism*, London: Macmillan.

WTO (2000) World Tourism Organization, http://www.world-tourism.org

Part IV

Globalisation of tourism's approach to sustainability

12 Environmental policies and management systems related to the global tourism industry

*Donald E. Hawkins
and Christopher Holtz*

Introduction

Environmental quality is essential to the viability of the tourism industry. This chapter focuses on major environment-related forces influencing the tourism industry in the twenty-first century, and is organised under the following headings:

1 Global environmental policies.
4 Sustainable tourism initiatives at the national and regional level.
3 The use of environmental management systems.
4 The growth of eco-labels.
5 The important role of public/private sector cooperation and collaboration.

Global environmental policies and the tourism industry

In the last decade, global environmental policies have made substantial progress in institutional development, international cooperation, public participation and private sector action. National governments and the private sector have developed stronger legal frameworks, market-based environmental incentive instruments, environmentally sound technologies, and cleaner production processes. Consequently, several countries report significant progress in controlling environmental pollution and slowing the rate of resource degradation as well as reducing the intensity of resource use (UNEP 1997).

The tourism industry, arguably the world's largest, bears a great responsibility in the effort to move towards sustainable development. The economic importance of tourism is undeniable. A study sponsored by the World Travel & Tourism Council (WTTC) and conducted by Wharton Econometric Forecasting Associates (WEFA) found that the tourism industry generates 11.7 per cent of global gross domestic product (GDP) and nearly 200 million jobs worldwide. These figures are forecast to total

11.7 per cent global GDP and 255 million jobs in 2010 (United Nations Economic and Social Council 1999).

While the economic development potential of tourism is substantial, there is also strong evidence of the negative environmental impact of tourism development. Nevertheless, tourism may be a more sustainable option for economic development because, unlike other natural resource based industries, it is based on enjoyment and appreciation of local culture, built heritage, and the natural environment which provides a powerful economic incentive to conserve these valuable assets (United Nations Economic and Social Council 1999).

The potential for tourism development to be a sustainable – that is for it to *meet the needs of the present without compromising the ability of future generations to meet their own needs* – has resulted in support for environmentally sustainable tourism development from both the public and private sectors at the national and international levels.

The United Nations and its associated organisations, the World Bank, and regional development banks are attempting to 'green' their loan programmes and assistance mechanisms. They are also supporting the development of environmentally sustainable tourism directly and indirectly through a variety of means, including sustainable tourism product identification, infrastructure development, environmentally sound hotel financing, and ecotourism development in protected areas.

The Commission on Sustainable Development (CSD) – an inter-governmental forum to coordinate and monitor the progress of Agenda 21's implementation – has produced a number of policy recommendations for stakeholders involved in the sustainable tourism development following their annual meetings in 1999. These recommendations were broken down into private sector, public sector, non-governmental organisation, and international community policy challenges. The following section summarises these recommendations (United Nations Economic and Social Council 1999).

Private sector policy challenges

The key challenges facing the tourism industry are to:

- promote wider implementation of environmental management, particularly in the many small and medium enterprises that form the backbone of the tourism industry, and spread initiatives to all sectors of the tourism industry;
- use more widely environmentally sound technologies, in particular to reduce emissions of CO and other greenhouse gases and ozone-depleting substances, as set out in two international agreements;
- address the key issues of siting and more eco-efficient design of tourism facilities;

- raise the awareness of tourism clients of the environment and social implications of their holidays, and of opportunities for their responsible behaviour;
- develop a better dialogue with the local communities in travel destinations, and promote the involvement of local stakeholders in tourism ventures;
- work with governments and other stakeholders to improve the overall environmental quality of destinations; and
- report publicly on environmental performance.

Public sector policy challenges

Governments need to further develop and implement the legislative and policy frameworks for sustainable development. In particular, they need to:

- ratify, if they have not already done so, and work towards the effective implementation of, international and regional environmental conventions;
- integrate more fully tourism development into the overall plans for sustainable development and develop participatory approaches;
- develop more widely land use planning, and protect the coastline through building restrictions (for example, legislation in France, Spain, Denmark and Egypt forbids building within a defined distance from the coast);
- identify and adopt the most appropriate mix of regulation and economic instruments, and, in many cases develop economic instruments to address environmental issues; and
- work towards the effective enforcement of regulations and standards.

Governments need to raise awareness, build capacity and promote effective action for sustainable tourism. This requires that they strive to:

- improve the understanding of the benefits and burdens of tourism in environmental, social and economic terms, for the areas under their jurisdiction;
- strengthen capacity for the management and control of tourism in their sphere of responsibility, and establish and maintain procedures for cooperation and coordination with neighbouring authorities, and with relevant state authorities;
- provide support through pilot projects and capacity development programmes, including capacity development at the local government level;
- ensure the participation of all stakeholders affected by or involved in tourism and its development, especially indigenous and local communities;

- ensure that tourism makes a positive contribution to economic development, and that the economic benefits of tourism are equitably shared;
- encourage and catalyse industry initiatives for sustainable tourism across all sectors of tourism, including accommodation, land, air and sea transportation, tour operators, travel agents, attractions sectors, etc.; and
- promote changes in consumer behaviour in both tourist-originating countries and destinations towards more sustainable forms of tourism.

Governments will also need to develop monitoring of progress towards sustainable tourism. It is important to develop activities to monitor, control and mitigate adverse effects that may arise from tourism activities and development.

Non-governmental organisations policy challenges

The key environmental policy challenges that face non-governmental organisations are to:

- more specifically voice their views in tourism policies and strategies;
- contribute to the development and implementation of environmental standards for tourism;
- develop or participate in raising awareness and education activities for sensitising tourists towards improving guest consumption patterns; and
- assist in monitoring tourism activities and development and progress towards more sustainable tourism.

International community policy challenges

The key challenges facing the international community are to:

- assist and support governments in the development of national strategies or master plans for the sustainable development of tourism, and of environmental land use and building regulations and standards for tourism;
- raise awareness and build capacity of all stakeholders by providing information on best practices for sustainable tourism;
- encourage the private sector to develop and apply codes and guidelines, and environmental management systems, and promote the development of the use of environmental reporting by companies in the various branches of the tourism sector;
- assist in assessing the environmental effectiveness of existing voluntary initiatives in the various branches of the tourism sector, and make recommendations accordingly;

- promote the transfer of environmentally sound technologies (ESTs), practices and management tools adapted for the tourism sector, and disseminate information on ESTs to governments and the tourism industry;
- work with other stakeholders to establish and disseminate lessons from best practices projects on sustainable tourism;
- provide support through provision of information and capacity development programmes, particularly on the costs and benefits of tourism development, the use of economic incentives to promote sustainable tourism, and on destination management; and
- assist in the establishment of monitoring of progress towards sustainable tourism.

The role of the international community and global environmental initiatives

Some other notable international environmental institutions that support sustainable tourism initiatives include:

The World Bank Group – Despite its formal distancing from the tourism sector in the 1970s, the Bank's focus on economic development and its private sector capacity, particularly in the International Finance Corporation, makes it technically well placed to address the pressures of a global tourism industry. Its now strong environmental capacity also makes it well placed to address the impacts of the tourism sector on biodiversity (Vorhies 1999).

The United Nations Development Programme (UNDP) – a United Nations organisation whose mission is to help countries in their efforts to achieve sustainable human development by assisting them to build their capacity to design and carry out development programmes in poverty eradication, employment creation and sustainable livelihoods, the empowerment of women and the protection and regeneration of the environment, giving first priority to poverty eradication.

The United Nations Environmental Programme (UNEP) – the environmental voice of the United Nations, responsible for environmental policy development, scientific analysis, monitoring, and assessment.

The Global Environment Facility (GEF) – a financial mechanism that addresses the incremental costs that developing countries face in responding to selected global environmental problems. The World Bank, UNEP and UNDP implement GEF projects.

New international environmental conventions and agreements are being adopted, older treaties are being improved, and new approaches to international policy are being developed and implemented. Four important international environmental conventions and treaties that are particularly relevant to the tourism industry include the following:

Rio Declaration on Environment and Development (Agenda 21)

The plan of action adopted by governments in 1992 in Rio de Janeiro provides the global consensus on the road map towards sustainable development. Agenda 21 is grouped around a series of themes – comprising 40 chapters and 115 separate programme areas, each of which represents an important component in the overall strategy towards global sustainable development. The Agenda identifies three core tools to be used in achieving sustainable development goals:

1　Introduction of new, or strengthening of existing, regulations to ensure the protection of human health and the environment.
2　Use of free market mechanisms through which the prices of goods and services will reflect the environmental costs of resource inputs and process outputs.
3　Industry-led voluntary programmes that deliver environmentally responsible products and services.

In 1996, WTTC, the World Tourism Organisation and the Earth Council joined together to launch 'Agenda 21 for the Travel & Tourism Industry: Towards Environmentally Sustainable Development', making the tourism industry the first industrial sector to develop an industry specific action plan based on Agenda 21. WTTC has now introduced an addition to this programme – The Alliance for Sustainable Tourism – which invites public and private sector tourism organisations to record their Agenda 21 based activities on a central Internet site and encourage cooperation with other local partners (United Nations Economic and Social Council, 1999).

The Convention on International Trade in Endangered Species of Wild Fauna and Flora (CITES)

CITES is an international convention banning commercial international trade in an agreed list of endangered species and by regulating and monitoring trade in others that might become endangered. The international wildlife trade, worth billions of dollars annually, has caused massive declines in the numbers of many species of animals and plants. The scale of over-exploitation for trade aroused such concern for the survival of species that an international treaty was drawn up in 1973 to protect wildlife against such over-exploitation and to prevent international trade from threatening species with extinction (World Conservation Monitoring 1998).

The United Nations Framework Convention on Climate Change

In the 1992 United Nations Framework Convention on Climate Change (UNFCCC), finalised for the Earth Summit in Rio de Janeiro, Brazil, the world's nations agreed on voluntary actions to reduce greenhouse gas emissions. Negotiations on the Kyoto Protocol to the UNFCCC were completed on 11 December 1997, committing the industrialised nations to specified legally binding reductions in emissions of six 'greenhouse gases'. During negotiations that preceded the December 1997 meeting in Kyoto, Japan, little progress was made, and the most difficult issues were not resolved until the final days – and hours – of the Conference. There was wide disparity among key players especially on three items:

1   The amount of binding reductions in greenhouse gases to be required, and the gases to be included in these requirements.
2   Whether developing countries should be part of the requirements for greenhouse gas limitations.
3   Whether to allow emissions trading and joint implementation, which allow credit to be given for emissions reductions to a country that provides funding or investments in other countries that bring about the actual reductions in those other countries or locations where they may be cheaper to attain.

The convention on biological diversity

A convention adopted as part of the 1992 UN Conference on the Environment and Development with the goals of:

* maintaining biodiversity;
* using its elements sustainably; and
* sharing in a balanced and fair way the advantages springing from the exploitation of genetic resources.

The Convention on Biological Diversity (CBD) is the major and most visionary global biodiversity agreement. Now ratified by almost all the countries in the world – with the notable exception of the United States, the CBD is attempting to develop a global framework for the management of biodiversity. It is striving to meet the trinity of biodiversity objectives – conservation, sustainable use and equitable benefit sharing – through globally agreed policies and procedures for managing biodiversity. These policies and procedures are also intended to support the overall goal of sustainable development and the corollary objective of poverty alleviation. But given the tradition of biodiversity management, the CBD has understandably yet to address the pressures of globalise commerce directly.

There is, however, an increasing recognition that the private sector must be an active player in managing biodiversity, but this recognition has yet to be articulated into clear roles and responsibilities for the private sector and global market processes (Vorhies 1999).

Critical weaknesses in the global environmental policy structure

While global policy structures and international environmental solidarity are growing in strength, they remain too weak to make significant progress a worldwide reality. As a result, the gap between what has been done thus far and what is needed is widening. From a global perspective, the environment has continued to degrade during the past decade, significant environmental problems remain, and the outlook is, unfortunately, pessimistic.

Internationally and nationally, the funds and political situation are not sufficient to halt continuing global environmental degradation or address the most pressing environmental issues, even though the technologies and knowledge are available to do so. The recognition of environmental issues as necessarily long-term and cumulative, with serious global and security implications, exists but the will to act remains limited. The continued preoccupation with immediate local and national issues and a general lack of sustained interest in global and long-term environmental issues remain major impediments to environmental progress internationally (UNEP 1997).

According to the CSD's report on Tourism and the Environment, there are several important emerging issues with regard to tourism and environmental protection that must be addressed in order to overcome these impediments. These include:

Developing partnership. For sustainable tourism, the involvement and commitment of all stakeholders is essential. However, public, private and academic sector partnerships are still underdeveloped and therefore need to be encouraged.

Involvement of the banking and insurance sectors. Banks and insurance companies could greatly expedite the progress of sustainable tourism by incorporating environmental and social criteria into assessment procedures for loans, investments, and insurance. They could help to finance environmentally sound technologies and provide incentives for sustainable tourism. This approach has worked well in other contexts. Widespread involvement of the banking and insurance sectors should be sought.

Use of economic instruments. The tourism industry consumes increasingly scarce natural resources. The costing of energy and water in particular could expedite greatly eco-efficiency in the tourism industry and raise revenue for the improved management of those resources. Governments should consider the development and widespread use of economic instruments for sustainable tourism.

Involvement of tourism boards. Often, marketing strategies and messages are not in line with the principles of sustainable tourism. There is a need to better involve tourism boards in sustainable tourism efforts.

Capacity-building of local government. In many countries, local governments have important responsibilities regarding tourism development. Capacity-building programmes should be implemented to help them understand those responsibilities, develop integrated and participatory approaches, and define and implement policies for sustainable tourism.

Greater focus on transport. There is a continued development of long-haul travel. Economic, technological and management approaches should be developed to reduce emissions, waste and pollution resulting from tourism transportation. Changing consumption patterns should also be considered.

Emerging types of tourism. Tourism is rapidly diversifying. Emerging forms of tourism should also develop according to sustainability criteria. The increase of cruises and the current trend towards mega-ships necessitate that the cruise ship industry develop a socially and environmentally responsible approach.

Improve monitoring. Careful monitoring of impacts and results, as well as the adoption of corrective measures, are conditions for sustainable tourism. All stakeholders at all levels should thus develop monitoring programmes. As previously stated, the private sector should develop monitoring and public reporting of its activities. Local and central governments should develop monitoring tools, such as indicators, and should incorporate the results into their decision-making process. Where appropriate, participatory approaches should be used. Monitoring is currently uncommon and that should be made a priority (United Nations Economic and Social Council 1999).

Sustainable tourism at the national and regional level

While the global policies and international donor organisation priorities discussed above are important because they tend to act as positive drivers toward environmental sustainability, particularly in the developing world, it is the national and regional policies within individual countries that are the key to sustainable tourism development strategies.

Some examples are shown in Figures 12.1, 12.2 and 12.3.

Calvia is the most important tourist destination in the Balearic Islands and one of the most visited in the Mediterranean with 120,000 tourist units and 1,600,000 visitors per year. Its 60 km of coastline with natural areas of great quality and diversity is the primary attraction for a destination that represents the typical sun and beach tourism of the northwestern Mediterranean.

Calvia experienced a tourism boom in the 1960s and its physical development lacked any comprehensive planning, in favour of a short-term approach driven by unlimited construction growth, cheap labour, and the unsustainable use of natural resources. In the early 1990s, Calvia reassessed its tourism development and growth strategy and the Town Council made a decision to implement a series of programmes and actions to promote sustainable tourism. Their goal was to transfer the guidelines that emerged from the Rio Conference and the 5th Environmental Programme of the European Union into a Local Agenda 21 Plan of Action.

Seven key criteria inspired Calvia's Local Agenda 21 Programme:
• An integrated conception of local development that harmonised financial, social, cultural, and ecological elements.
• Consideration of the basic concepts of sustainability in tourist destinations.
• Priority consideration of the time factor which projects the problems and opportunities from the present into the future.
• An overall vision of local and regional space and a position of solidarity against global environmental problems.
• Transparent planning and implementation through a municipal Citizens Forum.
• The coordination of the planning of the Local Agenda 21 Programme with municipal actions.
• The production of final action plans that can be specified in lines of action, initiatives and working programmes.

Figure 12.1 Local Agenda 21 for Travel and Tourism: Calvia, Mallorca.

Environmental management systems and the tourism industry

Within the last two decades the concept of quality management systems emerged in an effort to gain consistent performance in meeting specified standards (initially in military equipment procurement and operations). The best-known QMS in the commercial world is the ISO 9000 standards of the International Standards Organisation. The implementation of a QMS is intended to provide consumers with an assurance that a company's products and services will be of consistent quality.

As the importance of environmentally-friendly private sector operations grew (as reflected generally in the industry sponsored sustainable development principles described previously), the conceptual model of QMS was applied to industry operations that impacted on the environment. (An EMS can be defined as a management system that incorporates management commitment, organisational structure, operational practices and procedures, and resources into a documented and implemented environmental policy.) The implementation of an EMS represents the basis by

In 1998, the US Environmental Protection Agency (EPA), the International Institute of Tourism Studies at the George Washington University (IITS) and the World Travel and Tourism Council (WTTC) joined forces to work together towards **the development and implementation of more sustainable business practices within the travel and tourism industry in the United States.**

Ultimately, the EPA expects that a set of voluntary environmental principles will be developed and implemented within the sub-industries that make up the travel and tourism industry.

Specifically they hope to:
- Identify critical environmental issues associated with travel and tourism development and operations in the US; access what issues are already being addressed and how these efforts may be mirrored, bolstered or learned from. Establish a foundation of mutual interest from which collaborative solutions to complex issues associated with travel and tourism development can be identified.

Results to benefit both the environment and the industry include:
- Increased capacity of private and public industry stakeholders to network and work together in a collaborative fashion.
- Better understanding of the critical environmental issues associated with tourism and recreation development.
- Reduction of inadequate solid waste treatment, less negative transportation/air quality impacts, more wastewater recycling, and less inadequate energy consumption.
- Enhancement of collaboration between the public and private sector tourism stakeholders, stronger arguments for sustainable tourism product development.

The first stage of the project includes:
- A Sustainable Tourism Roundtable meeting with 'captains' of the travel and tourism industry to establish what are the current concerns of the industry with regard to environmentally friendly business practices. From the results of these meetings, the long-terms goals of this project will be established and the possibility of establishing a permanent 'forum' on environmental management for the hotel and recreation sector will be considered. See below for a brief description of the Sustainable Tourism Roundtable outcomes.
- The establishment of a draft Green Globe/Green Seal standard for environmental management of lodging facilities. The planning of an advanced information system, which will include a virtual library of tourism and recreation environmental management resources.

The second phase of this project will be determined as a result of the Sustainable Tourism Roundtable conducted in December 1998, along with subsequent strategic thrusts, action plans and resource commitments developed through a proposed permanent **'Sustainable Tourism Leadership Group'.**

Figure 12.2 The Environmental Protection Agency leading a national sustainable tourism initiative in the United States.

which an organisation can exercise control over its impact on the environment by systematically gathering and coordinating knowledge about those impacts. Implementing an EMS demonstrates a strong commitment to Agenda 21 principles.

Green Globe's Destination Programme has worked or is currently working in four countries through funding provided from private sector, government and international donor sources.

The first tourism destination to have completed the Green Globe destination process is Jersey in the Channel Islands of the United Kingdom. Funding for this project was provided through a partnership of public and private sector stakeholders. Major achievements of the Green Globe Destination project in Jersey include:

- Installation of a state-of-the-art ultra-violet water treatment plant to improve effluent discharge quality.
- Incentive programmes to preserve historic buildings.
- Construction of an information/interpretation centre that provides environmental and cultural information to visitors.

Green Globe is also working in the coastal resort of Vilamoura, Portugal, with the support of master developer Lusotour, S.A. Examples of environmental improvements made as a result of Green Globe's involvement include:

- Increasing recycling points within the destination.
- Producing environmental guidance pamphlets for boat users.
- Installing water quality monitoring equipment in the marina facility.
- Introducing walking and cycling networks to reduce automobile traffic.

Green Globe's most extensive Destination Programme engagement, and first to receive international donor support, is in the Philippines where UNDP, UNESCO and the Government of the Philippines are supporting work in three destinations:

- Camiguin Island in Mindanao.
- Ifugao Rice Terraces (a World Heritage Site) in Luzon.
- Ulugan Bay in Palawan.

Green Globe has completed the *Strategic Environmental Assessments* and *Action Planning* phases for Camiguin Island and the Ifugao Rice Terraces and is currently beginning the *Implementation* phase of the Green Globe Destination process.

Figure 12.3 Green Globe regional sustainable tourism development projects.

There are in fact many common features shared by QMS and EMS. These include:

- documented policy statements;
- appropriately organised management structures with documented responsibilities;
- documented operational and process controls;
- documented quantifiable performance targets;
- record keeping systems;
- self-auditing programmes;
- awareness and training programmes; and
- management reviews.

Typically, a private sector EMS most often conforms with the International Standards Organisation ISO 14001 standard, although the European Union's Eco-Management and Audit Scheme (EMAS) is a valuable reference because of its rigorous parameters.

It is important to note that all of the preceding material regarding QMS/EMS and ISO14001 is actually intended for use by private businesses or perhaps certain public sector agencies. It is not designed for use with a tourism destination that comprises many different types of organisations from the public and private sector; but it does lay a foundation of terminology and processes that will be used in discussing EMS for tourism destinations. The application of the EMS process to tourism destinations is an emerging area of interest, and while there is not yet a substantial amount of direct anecdotal or documentary evidence to examine we can look to recent work by groups like the World Travel & Tourism Council's Green Globe Alliance and others for case studies that illustrate the potentials and pitfalls of the destination-based EMS process.

The rationale behind applying the EMS process to tourism destinations is simple and logical. The natural environment is an extremely valuable resource for most tourism destinations and the aggregate impact from many different sectors of the travel and tourism industry – transportation, accommodations, and tour operations – tends to have a negative environmental effect. Coupled with this assumption is the fact that the public sector is responsible for many functions that should minimise negative environmental impacts – waste management, land use planning, transportation infrastructure, biodiversity conservation, etc. – but often fails to meet this challenge adequately.

Proponents of the destination-based EMS believe that truly sustainable destination management requires a public/private sector partnership in the form of a cooperative management structure that deals proactively with environmental issues. With this in mind, we will be looking closely at methodologies and processes that can be used in the creation of a strategically designed EMS with broad public/private stakeholder support.

The benefits of such an EMS will include:

- providing a systematic framework for public and private sector cooperation on environmental issues;
- improving compliance with regulatory requirements and industry codes of conduct;
- reducing public and private sector operation costs as greater energy/resource savings are achieved;
- increasing competitive advantage in the market for 'green' tourism destinations; and
- creating a practical mechanism for pursuing the Agenda 21 principles of sustainable development.

Property-based environmental management systems

The creation of property-based EMSs to guide site audits and monitoring processes and in which to anchor certification processes is a new phenomenon in the hospitality industry and, to date, one without a defined standard. This is generating a mounting confusion within the industry over what environmental standards, which criteria, and whose certification programme to use. The advent of EMSs into tourism is based on the success of EMS operations in other industries, the perceived market benefits of independent certification of environmental standards for individual tourism enterprises, and a demonstrated growth in demand for environmentally friendly or 'green' tourism destinations in major outbound markets. EMSs are desirable because their adoption can reduce operation costs through energy and resource savings, improve internal management methods, reduce liability/risk from environmental deterioration, improve a property's image in the area of environmental performance and compliance with regulatory requirements, and open opportunities for profit in the emerging market for 'green' tourism destinations. The overall benefit of a properly designed and administered EMS, then, is its ability to provide credible and objective assurance to inbound markets that the environmental conditions in a particular property or destination are of a higher quality than those of competing locations (Holtz 1998). (See Figure 12.4.)

The growth of eco-labels in the tourism industry

The creation of environmental management systems for the tourism industry is a new phenomenon that is looking to build on three important trends:

1   The success of EMS operations in other industries.
2   The perceived potential market benefits of independent certification of environmental standards for individual tourism business.
3   The demonstrated growth in demand for environmentally friendly or 'green' tourism destinations in major outbound markets.

A number of organisations are currently certifying tourism providers as 'environmentally friendly', including Green Globe, Green Seal and HVS Ecotel. The primary benefit, to date, of conforming with the various EMS standards developed for the tourism industry is the cost savings achieved through various operational efficiencies. It is anticipated, however, that as awareness of environmental certification within the tourism industry is raised, there will be an increased competitive advantage to being independently certified as environmentally friendly. The hope is that as consumer consciousness of environmental issues increases, more people will choose tourism service providers and, indeed, destinations, based on their

The Environmental Audits for Sustainable Tourism (EAST) project in Jamaica, funded by the US Agency for International Development (USAID), is a programme of audits within a corporate environmental management system aimed at the tourism and hospitality sector.

The project's objectives are:

- To develop greater awareness and understanding of the benefits of environmental management systems and audits among the tourism and hospitality sector.
- To upgrade the technical skills of Jamaicans who are expected to conduct audits and advise on environmental systems.
- To assist a select representative number of tourism-related establishments in carrying out environmental audits.
- To help finance selected audit recommendations, on a cost-sharing basis, in order to demonstrate the financial benefits of the systematic application of environmentally friendly practices.

The EAST programme partnered with Green Globe, a UK-based environmental management programme for the travel and tourism industry, to secure tangible marketing benefits, in the form of international publicity, for hotels and resorts participating in the environmental audits programme. Green Globe is currently developing a worldwide environmental certification standard for the hospitality industry and four Jamaican hotels received this certification at World Travel Market 1998 under the EAST project.

The EAST programme facilitated the development of a formal environmental management system (EMS) within the participating hotels and resorts through which management and staff could exercise clear operational control over their impact on the environment. EMS implementation has also resulted in significant operational cost savings by increasing the efficiency of resource use. The participating hotels and resorts received technical assistance through the EAST programme to conduct an audit of the environmental impacts of their operations and develop an EMS for implementation. The areas addressed in the environmental audit included:

- Water use
- Energy use
- Solid waste generation
- Generation of pollutants
- Use of hazardous substances
- Generation of emissions
- Damage to the ecosystem

The findings of the audits resulted in a collection of recommendations that were organised into an EMS that included:

- An environmental policy that clearly communicates Mocking Bird Hill's commitment to maintaining the social, cultural and physical environment.
- An action plan to guide the Mocking Bird Hill's actions and expenditure of resources.
- The implementation of actions that impact upon all of the Mocking Bird Hill's operations relative to the environment, including awareness and training, product purchasing, staff procedures, incentives programmes and community outreach.
- Corrective action or monitoring to ensure that the EMS performs as expected and is reviewed periodically by senior management.

Figure 12.4 Hotel and Resort Environmental Management Systems in Jamaica.

environmental performance. As indicated previously, though, there is not yet quantitative evidence to indicate that consumers make these types of choices based on their perception of environmental factors.

While there is not currently a defined standard for an EMS that deals with a destination as a whole, organisations like Green Globe, Green Seal and others are examining the best method to institute such a standard. Destinations are looking into adopting an EMS system for many of the same reasons individual businesses choose to create an EMS:

- reduction in operation costs as greater energy/resource savings are achieved;
- improved internal management methods;
- reduction in liability/risk from environmental deterioration;
- improvement of a destination's image in the area of environmental performance; and compliance with regulatory requirements; and
- desire to profit in the market for 'green' tourism destinations.

Ecolabelling and certification programmes

The tourism industry is dependent on the environment for its sustainability and makes extensive use of the natural and cultural resources in its area of operation. The industry's prosperity is thus dependent on the conservation and responsible use of the environment. Several organisations, including government organisations, not-for-profit industry organisations and non-governmental organisations, have addressed the issue pertaining to environmental conservation and best practices within the tourism industry by introducing ecolabelling and green certification schemes. Each certification programme defines criteria and standards that enhance efficiency and reduce overuse and wastage. Each scheme is unique in that the certification period varies and may range from one to three years. Some schemes such as the PATA Green Leaf require the industry operator to merely sign the PATA Code, whereas others have more detailed application procedures. Evaluation methods also vary from scheme to scheme. Though there is abundant information on the criteria required to participate in these schemes, there is a shortage of information on the evaluation mechanisms and duration period of each scheme. The following section discusses the individual certification programmes in greater detail. The majority of this information was obtained through personal communication or taken from the United Nations Environment Programme (UNEP) report entitled 'Ecolabels in the Tourism Industry' (UNEP, 1998).

PATA Green Leaf (Asia Pacific)

This is a green certification scheme developed by the Pacific Asia Travel Association, which is an industry association. It was launched in 1995 and requires that participants officially accept and abide by the PATA principles of conduct listed below. The PATA Code urges Association and Chapter members and their industry partners to:

Adopt the necessary practices to conserve the environment, including the use of renewable resources in a sustainable manner and the conservation of non-renewable resources.

Contribute to the conservation of any habitat of flora and fauna, and of any site whether natural or cultural, which may be affected by tourism.

Encourage relevant authorities to identify areas worthy of conservation and to determine the level of development, if any, which would ensure those areas are conserved.

Ensure that community attitudes, cultural values and concerns, including local customs and beliefs, are taken into account in the planning of all tourism-related projects.

Ensure that assessment procedures recognise the cumulative as well as the individual affects of all developments on the environment.

Comply with all international conventions in relation to the environment.

Comply with all national, state and local environmental laws.

Encourage those involved in tourism to comply with local, regional and national planning policies and to participate in the planning process.

Provide the opportunity for the wider community to take part in discussions and consultations on tourism planning issues insofar as they affect the tourism industry and the community.

Acknowledge responsibility for the environmental impacts of all tourism-related projects and activities and undertake all necessary changes to those practices.

Foster environmentally responsible practices including waste management, recycling and energy use.

Foster in both management and staff, of all tourism-related projects and activities, an awareness of environmental and conservation principles.

Support the inclusion of professional conservation principles in tourism education, training and planning.

Encourage an understanding by all those involved in tourism of each community's customs, cultural values, beliefs and traditions and how they are related to the environment;

Enhance the appreciation and understanding by tourists of the environment through the provision of accurate information and appropriate interpretation; and *establish* detailed environmental policies and/or guidelines for the various sectors of the tourism industry.

Tyrolean Environmental Seal of Quality (Austria and Italy)

Tirol Werbung and Suditirol – public authorities operating in the area of accommodation and catering – promote this scheme. It was launched in 1994 and sets mandatory criteria pertaining to waste prevention, waste utilisation, energy, soil and transportation for businesses operating in the lodging industry. These include the hotel trade, the catering trade, private lodgings and farm holidays, camping sites and alpine refuges. The standards include but are not limited to the following:

WASTE PREVENTION

No portion packages in the catering and sanitary areas;
no sales of beverage cans;
refundable deposits instead of dispensers with disposable containers;
no dispensers for beverages in disposable containers;
no disposable tableware or cutlery;
use of recycled paper or chlorine-free paper in the office, advertising and sanitary areas;
drawing up of refuse concept guidelines.

TRANSPORT

Providing, hiring or arranging for guest bicycles (where terrain permits);
advising visitors of best public transport connections (rail and bus) for arrival and departures;
transfer service for guests arriving by public transport;
facilities for storing winter sports equipment over the summer or between holidays.

Green Globe (International)

The World Travel and Tourism Council, which is an industry association, promote the Green Globe certification programme, which focuses on all industries within the tourism sector. It was launched in 1994 and requires that the participating agents comply with the Green Globe minimum standard requirements, which are as follows:

WASTE MINIMISATION, REUSE AND RECYCLING

The company shall undertake a detailed assessment of the source and content of the waste produced.

The company shall ensure that all treatment of all waste is conducted in accordance with best industry practice and according to legislative requirements.

The company shall identify opportunities to reduce, reuse and recycle waste, and develop an action plan for implementing appropriate action.

ENERGY EFFICIENCY, CONSERVATION AND MANAGEMENT

The company shall undertake a detailed assessment of energy use throughout the company, establish the type of energy required for all activities, and monitor and review use on a regular basis.

The company shall set targets for reducing energy use throughout the company.

The company shall ensure that energy efficiency is a key consideration in the purchasing of new or replacement equipment and in the design of new buildings or facilities.

The company shall encourage energy efficiency among staff, residents, guests and business partners

The company shall research alternative, environmentally benign methods of energy generation, such as solar, wind or biomass power.

MANAGEMENT OF FRESHWATER RESOURCES

The company shall undertake a detailed assessment of water use throughout the company and monitor water use on a regular basis, installing sub-meters where necessary.

The company shall set targets for reducing water consumption and ensure that its water requirements do not adversely affect the water supplies for nearby communities.

The company shall identify opportunities to reduce and reuse water and take action accordingly.

The company shall minimise the wastage of water by undertaking regular maintenance checks.

The company shall where possible, install water saving devices in new and existing buildings.

The company shall ensure that water efficiency is a key consideration when purchasing new and replacement equipment, and is built into the design of new buildings and facilities.

WASTEWATER MANAGEMENT

The company shall dispose of waste water responsibly, by ensuring that all effluent is treated to match existing minimum standards for the area.

The company shall establish emergency procedures to ensure that the aquatic environment is protected from disasters within the facility.

The company shall avoid products containing potentially hazardous substances that may eventually find their way into the water system.

ENVIRONMENTALLY-SENSITIVE PURCHASING POLICY

The company shall have a purchasing policy to ensure environmental considerations are taken into account in purchasing decisions and products with a lower environmental impact (not entailing excessive cost) are selected.

The Audubon Cooperative Sanctuary System (International)

Audubon International promotes this programme. The scheme was launched in 1991 and follows a checklist detailing the measures that have to be adopted by industry professionals with respect to water conservation in order to compete for the association's golf course award. The checklist includes:

operating irrigation system for maximum efficiency;
planning to install new irrigation system to increase efficiency;
checking irrigation system for proper distribution;
noting leaks quickly and making timely repairs;
incorporating evapotranspiration rates or weather data;
setting watering priorities;
avoiding watering at peak evaporation times;
reducing irrigated turf areas where possible;
re-capturing and re-using irrigation water;
incorporating water conservation landscaping and/or drought-tolerant plants;
choosing turf species well suited to climate and soils;
using mulches in gardens to reduce water loss;
selecting water sources for least impact on local supplies; and
other water conservation methods.

Blue Flag (Europe)

This certification scheme is promoted by the Foundation for Environmental Education in Europe, which is a non-governmental organisation and is aimed primarily at the preservation and responsible use of beaches in Europe. It was launched in 1985.

Essential Blue Flag criteria for environmental information and education include:

• Prompt public warnings must be given if the beach or part of it is expected to become, or has become, grossly polluted or otherwise unsafe. This requirement includes unsafe conditions due to the discharge of storm water.

- Information on protected sites and rare or protected species in the area must be publicly displayed and included in tourist information, except where such information might endanger the site or specimens. Information must include a public code of conduct.

The responsible authority must:

- display on or close to the beach updated information on water quality and the location of sampling points;
- display the Blue Flag criteria (no information available) as close as possible to the location of the Blue Flag itself;
- remove the Blue Flag if any of the essential requirements are no longer satisfied; and
- the responsible authority must be able to demonstrate at least five educational activities related to the coastal environment.

We are an environmentally friendly operation (Germany)

This certification scheme is promoted by the Deutscher Hotel and Gaststatten Verbans DEHOGA (Hotel and Restaurant Association of Germany), which is an industry association operating in the field of lodging and catering. This scheme was launched in 1993 and outlines some specific do's and don'ts for the hotel and catering industry.

FIRST STEPS

Keep an environment file.
Appoint an environment protection officer.
Train the staff.
Invest with an ecological eye.
Inform your guests.

SAVING ENERGY

Check heating regularly; use alternative sources of heat.
Supply hot water centrally.
Recycle heat and monitor power curves.
Use energy saving bulbs.
Fit time switches and infra-red sensors.
Avoid electric hand-dryers; don't leave televisions on stand-by.

WASTE AND WASTEWATER

Establish how hard your own water is.
Fit flow regulators for bathroom sinks and showers.
Reduce WC flush capacity, fit a double flush.
Inspect all points of water consumption regularly.
Let guests decide when to change towels and bed linen.
Don't bombard the environment with washing powder.
Keep 90 degrees and pre-wash cycles to a minimum.
Avoid fabric conditioners.
Choose mild cleaning agents.
Avoid disinfectants.
Avoid toilet ducks and odour neutralisers.
Avoid aggressive toilet and drain cleaners.

CUTTING BACK WASTE

Avoid individually wrapped portions of food.
Avoid individually wrapped bath and shower products.
Avoid canned drinks and disposable bottles.
Take advantage of bulk sizes and returnable containers.
Avoid disposable items wherever possible.
Minimise plastic packaging and segregate materials.
Keep paper consumption down, use recycled paper.

SEGREGATING WASTE

Sort rubbish into paper and card, glass, recoverable materials, compost,
special and hazardous waste, residual waste.
Return packaging to suppliers.
Collect and dispose of organic waste separately.
Filter grease and oil.

AIR

Improve indoor air.
Avoid office equipment which pollutes the environment.

THE LOCAL ENVIRONMENT

Take care of grounds and gardens.
Buy ecological farm produce.
Purchase products from your own region.
Offer staff and guests a bonus for using public transport.

Committed to Green (Europe)

The European Golf Association's Ecology Unit developed this certification scheme. It was launched in 1997 and is aimed at the enhancement of environmental management systems used in the maintenance of golf courses without compromising on the quality of turf grass and playing conditions.

Committed to Green will help the environment by actively encouraging and supporting:

CONSERVATION

Biodiversity.
Landscape quality.
Cultural heritage.

APPLICATION

Sustainable development principles in the location.
Design.
Construction of new golf courses.

CONSERVATION

Water resources.
Maintenance of high quality standards in surface and ground waters.

SAFE AND APPROPRIATE USAGE

Storage and disposal of fertilisers and pesticides.

ENERGY

Conservation.
Minimising waste.

AWARENESS

Knowledge of environmental management principles and techniques among golf course managers.
Public awareness of golf's role in the wider environment.

The scheme also recognises and supports golf clubs that have made significant achievements across the categories listed below by bestowing the 'Committed to Green Award for Environmental Excellence'.

- nature conservation;
- landscape and cultural heritage;
- water resource management;
- turf grass management;
- waste management, energy efficiency and purchasing policies;
- education and the working environment; and
- communications and public awareness.

Ecotel (International)

This scheme has been developed by HVS Eco Services, the environmental consulting division of HVS dedicated exclusively to the hospitality industry. It was launched in 1994. Hotels that are able to demonstrate a heightened level of environmental sensitivity are awarded the Ecotel seal. The criteria on the basis of which HVS awards this seal include the following:

SOLID WASTE MANAGEMENT

Solid waste reduction, reuse, and recycling strategies will be evaluated. Waste disposal service and equipment suppliers will be selected and the procurement of waste storage facilities will be addressed.

ENERGY EFFICIENCY

A detailed energy audit will be conducted by certified engineers to identify energy and cost-saving tactics, isolate the most viable rebate programmes and to establish standard operating procedures.

WATER CONSERVATION

Conservation and preservation efforts, consumption patterns and rates, and related equipment will be analysed. Standard operating procedures will be established.

PROJECT COORDINATION AND EMPLOYEE EDUCATION

On-site programme implementation and education seminars will be performed to facilitate programme start-up, efficiency and payback.

COMMUNITY INVOLVEMENT

Relationships between the client hotel and community organisations will be established.

LEGISLATIVE REVIEW

Federal, state and local environmental regulations will be addressed to ensure that the property complies with all applicable environmental laws.

SPECIAL EVENTS

Distinctive, memorable environmental and cultural events such as Earth Day celebrations, charitable fund-raisers and guest lectures will be designed and coordinated.

British Airways Tourism for Tomorrow Awards (International)

These awards have been developed by British Airways and are directed at tour operators, individual hotels and chains, national parks and heritage sites and other activities associated with tourism in order to promote the responsible use of the environment by these agencies. The awards were launched in 1992 and are distributed annually and the criteria to compete for this award include the following.

CULTURAL AND SOCIAL

Including relationship to the community, provision of education, training and welfare services.

BUILT HERITAGE

Including the preservation and/or the renovation of buildings and cultural sites.

NATURAL ENVIRONMENT

Including conservation and/or habitat management.

MANAGEMENT OF VISITOR NUMBERS

Including present and future targets.

POLLUTION, WASTE AND ENVIRONMENTAL IMPACT

Ways in which these have been minimised including clean technology.

ENVIRONMENTAL COMMUNICATION

How your project has been communicated.

INNOVATION

What is new and unusual about the project.

LEADERSHIP

What makes the project a role model?

SUSTAINABLE TOURISM

How the project meets the long-term goal of 'development that meets the needs of the present generation without compromising the ability of future generations to meet their own needs'.

Code of Practice for Ecotourism Operators (Regional)

This code has been developed by the Ecotourism Association of Australia, a not for profit organisation, in 1991 to develop ethics and standards for ecotourism and to facilitate understanding and interaction between the tourist, host communities, the tourism industries and government and conservation groups. The code is as follows:

Strengthen the conservation effort for, and enhance the natural integrity of the places visited.

Respect the sensitivities of other cultures.

Be efficient in the use of natural resources, e.g. water, energy.

Ensure waste disposal has minimal environmental and aesthetic impacts.

Develop a recycling programme.

Support principals (i.e. hotels, carriers, etc. who have a conservation ethic).

Keep abreast of current political and environmental issues, particularly of the local area.

Network with other stakeholders (particularly those in the local area) to keep each other informed of developments and to encourage the use of the code of practice.

Endeavour to use distribution networks (e.g. catalogues) and retail outlets to raise environmental awareness by distributing guidelines to consumers.

Support ecotourism education/training for guides and managers.

Employ tour guides well versed and respectful of local cultures and environments.

Give clients appropriate verbal and written educational material and guidance with respect to the natural and cultural history of the places visited.

Use locally produced goods that benefit the local community, but do not buy goods made from threatened or endangered species.

Never intentionally disturb or encourage the disturbance of wildlife or wildlife habitats.

Keep vehicles to designated roads and trails.
Abide by the rules and regulations applying in natural areas.
Commit to the principle of best practice.
Comply with Australian safety standards.
Ensure truth in advertising.
Maximise the quality of experience for hosts and guests.

Conclusion: Public and private sector cooperation needed to develop a sustainable tourism industry

Traditional 'command and control' governmental approaches are not currently capable of developing the greatest value from investments in sustainable tourism. A special report by the World Tourism Organisation (WTO) emphasised the importance of the changing government role in tourism and the need to reexamine traditional activities undertaken by the public sector (WTO 1997). These changes have been particularly evident in Europe and North America where government owned assets have been divested and privatised. Governments have also created public-private partnerships to market their countries as international travel destinations.

Successful sustainable tourism initiatives require the active and concerted involvement of the public and private sectors. This involvement can and must occur at a variety of levels. At the most general level, we observe that the core resources that attract tourists are national patrimony – the cultural and resources that make a country, region or people distinctive or even unique. As national patrimony, these resources tend to be owned and managed (or at least directed) by the government. Tourism also requires the active participation of private sector entrepreneurs to develop services that make enjoyment of those assets possible (such as hotels, restaurants, tour companies, transportation providers). However, these services are largely meaningless if there is no infrastructure to permit tourists to enter an area or support the entrepreneurs. Airports, roads, potable water and other basic services are generally provided by the public sector.

At a more operational level, public-private partnerships can take many forms: such as concessions or outsourcing contracts to manage public assets; joint ventures to develop tourism attractions; agreements to develop and operate paid infrastructure such as roads, ports and water systems; user fee systems to support common objectives such as resource protection; and other simpler forms of cooperation such as agreements between governments and the private sector to support shared objectives with human and financial resources. The principal objective of these mechanisms is to leverage investment capital. The governments provide the core assets, necessary investment conditions and the 'licence' to invest in activities that have a much greater impact than the government alone is capable of achieving.

For example, these mechanisms are used to help advance tourism development in local areas near major protected areas. Declining government budgets and other financial difficulties have led to weak governmental mechanisms to care for national parks. Local governments are requesting the private sector and non-governmental organisations to assume management responsibility for most of the nation's parks. This form of public-private partnership allows for greater management agility, lowers government budget obligations and provides the parks with more flexibility to enter into agreements with donors, tour operators and the local hospitality sectors. However, the only thing public about the parks is the land itself; the public sector is not necessarily participating in the management of these areas. In many countries, government rules greatly restrict the options for generating park revenue and entering into alliances with the private sector. While not necessarily the ideal long-term strategy, the current arrangement offers numerous advantages over a purely public park system.

Additional public–private cooperation will be needed to protect the natural and cultural assets that attract visitors. National governments will need to work with the private sector to identify common objectives and opportunities for producing revenue to protect the parks and protected areas of the region, ensuring the sustainability of these core attractions into the future. User fees, room taxes, environmental performance bonds for new tourism development and a variety of other mechanisms could be considered. In addition to providing revenue, these types of mechanisms provide assurances to tourists that the destination is serious about the quality of its attractions.

Effective public–private partnerships can accomplish other desirable sustainable development goals, such as providing greater opportunities for smaller enterprises to participate in tourism development, and reducing 'leakage' of tourism revenues out of the country. Overall, tourism development decision-making could be improved by enhanced collaboration of the private sector in land use planning and environmental impact assessment and more transparent use of permitting and zoning.

References

Commission on Sustainable Development (April 1999) *The Global Importance of Tourism – Background Paper #1*, prepared by the World Travel & Tourism Council and the International Hotel and Restaurant Association, WTTC, London.

Holtz, Christopher (1998) *Sustainable Tourism Best Practices*, Washington, DC: The George Washington University.

United Nations Economic and Social Council (1999) *A Report of the Secretary General: Tourism and Sustainable Tourism Addendum on Tourism and Environmental Protection*, Commission on Sustainable Development 7th Session, UNEP, Paris.

United Nations Environmental Programme (1997) *Global Environmental Outlook – 1 Executive Summary: Global State of the Environment Report*, UNEP, Paris.

United Nations Environment Programme (1998) *Ecolabels in the Tourism Industry*, United Nations Environment Program Industry and Environment, Paris.

Vorhies, Frank (1999) *An Essay on Biodiversity and Globalisation*, IUCN-The World Conservation Organisation, IUCN.

World Conservation Monitoring (1998) *What is CITES?* WCM, Washington.

World Tourism Organisation (1997) *Towards New Forms of Public-Private Sector Partnership: The Changing Role, Structure and Activities of National Tourism Administrations*, WTO: Madrid.

13 Brief Encounters

Culture, tourism and the local-global nexus

Peter Burns

Introduction

For the purpose of analysis, the world can be understood by looking at the structures that frame the relationships between nation-states and global markets. Strange (1988), in the context of her work on the international political economy (IPE), identified these structures as: security; production; finance; and knowledge. In all of this, the key question is, as Strange asks, *cui bono?* (who benefits?). Balaam and Veseth (1996: 101) describe why this seemingly simple question is fundamental: 'Asking this question forces us to go beyond description to analysis. To identify not only the structure and how it works, but its relationship to other structures and their role in the international political economy [an understanding which] therefore becomes a matter of holding in your mind a set of complex relationships and considering their collective implications'.

The complexity referred to above is also present in globalisation, and is a term that in many ways can be aligned with a neo-liberalist, neo-conservative, *laissez-faire* economics. It is used by neo-liberals to describe how the world functions as a global whole. Both the roots and framing ideology to this are to be found in the Weberian formulation that the world should move towards adopting the features of a representative government, bureaucracy, a capitalist economy, the Protestant ethic and a scientific methodology. Neo-liberalism is epitomised by Thatcher and Reagan both of whom advocated minimal state interference in free global markets, with the role of the state being defined by security and defence issues. Their mantra was: markets are more important than states.

Taking this as a generalised starting point, this chapter will explore the links between tourism and culture within the context of a global-local nexus. The first part examines some theoretical and definitional issues arising out of globalisation. The second section investigates the meaning of culture as it is located within tourism, while the third part draws the first two together to discuss the impact of tourism as a global phenomenon on culture at a local level. The chapter ends on a forward-looking note by suggesting a number of research and policy issues that arise from the discussion.

Globalisation

The term 'globalisation' is rooted in the study of international relations and 'modernisation', itself a contentious term in that some reject its predilection for concentrating on that which is quantifiable – such as employment, literacy, GNP – while neglecting quality of life factors such as happiness (Robertson 1992: 11). The key words accompanying globalisation as an economic phenomenon have been 'privatisation' and 'deregulation'. Curiously though, it is a proponent of the marketplace, Vaclav Havel, who, during and immediately following the collapse of communism in the Eastern bloc, illustrates one of its greatest ironies: 'The market economy is as natural and matter-of-fact to me as air. After all, it is a system of human activity that has been tried and found to work over centuries. It is the system that best corresponds to human nature. But precisely because it is so down-to-earth, it is not, and cannot constitute, a worldview, a philosophy, or an ideology. Even less does it contain the meaning of life. It seems both ridiculous and dangerous when ... the market economy suddenly becomes a cult, a collection of dogmas, uncompromisingly defended and more important, even, that what the economic system is intended to serve – that is, life itself' (Havel 1992, cited in Balaam and Veseth 1996: 56).

Prior to the cult of the market, it was probably the global warming debates which illustrated for the public at large that there was more to global theory than global products 'borne of a high-tech, fast-moving society, frequently allied with the motive to maximise profit' (Boniface and Fowler 1993: 3).

Globalisation's main themes explore the ways in which economic relations between nations and regions are framed by cultural coding and subject to cultural contingencies (Robertson 1992). It differs substantially from the discredited convergence theory, i.e. the school of thought proposed by Tinbergen (1961, cited in Lavigne 1995) in which capitalist and socialist systems would borrow the best from each other until a *rapprochement* (a sort of market socialism) was achieved. This theory did not predict increasing levels of cold war hostility, surrogate wars (such as in Ethiopia/Eritrea and in Laos/Cambodia/Viet Nam) and the Reagan/Thatcher-inspired arms race that led eventually to the sudden death of communism.

In more recent times convergence has taken on a different meaning. The proposition that the world, pressured as it is by global products and globalised consumerism, is becoming the same – 'less differentiated' as Brewster and Tyson (1991: 1) term it; the 'homogenised world system' perceived by Inkeles (1981, cited in Robertson 1992) or Levitt's (1983) 'converging commonality'. Alongside these converging trends are paradoxical divergences. Most significantly by power moving towards metropolitan centres of capital and knowledge (economic divergence) coupled with a cultural shift whereby 'societies are converging in some respects (mainly economic and technological), diverging in others (mainly social relational) and, in a special sense, staying the same in yet others' (Robertson, 1992: 11).

Robertson (1992: 166) indicates that globalisation might be characterised as 'compression of the world', furthering Marshall McLuhan's earlier and optimistic concept of 'the global village' (1960). Dunning (1993) also identifies this trend: 'As the cross-border interchange of people, goods, assets, ideas and cultures becomes the norm, rather than the exception, so our planet is beginning to take on the characteristics of a global village' (1993: 315). This is one view, though I would argue that it is a monocentric, elitist one. For the most part, the tendency towards a 'global village' (even when used as metaphor for instant communications) is almost solely the territory of like-minded elites. It is they who have access to the electronic super-highway and who communicate with each other across the globe surrounded by seas of poverty, inhabited by those who don't communicate outside their own reference groups. So, even with the electronic revolution, there are still parts of the globe which remain 'uninformed and lacking in "adequate" and "accurate" knowledge of the world at large and of societies other than their own (indeed of their own societies)' (Robertson 1992: 184).

Culture, globalisation and tourism

On the supply side of tourism's equation, a particular problem arises. Part of tourism's product will include localised culture and people. This invokes an argument about whether culture and people should be enmeshed in corporate marketing strategies (Burns and Holden 1995) or provide the backdrop for relaxation and recreation. Paradoxically, on the consumption side, customer reaction against the Fordist, 'industrialisation' of services (cf. Ritzer, 1993) is emerging. Finally, with growing sophistication in market segmentation and the rise of both consumerism and green awareness, there is recognition that tourists themselves are not a particularly homogeneous group.

The complexity of tourism's social and economic dynamic, both as act and as impact, means that it should not be perceived as an integrated, harmonious and cohesive 'whole'. The intention of this section is to focus on the particular relationship between culture and tourism: the cultural dynamics, systems and structures that make meaning between visitors and the visited possible. This approach is important for at least three reasons:

- Culture (especially culture that is understood to be unique or unusual by actors including marketing specialists and planners) can be seen as a commercial resource – an attraction.
- Such understanding might help deflect or ameliorate negative change to a host culture occurring through the act of receiving tourists.
- Tourism literature rarely acknowledges the world as a system of relations wherein the properties of a 'thing' (in this case, culture) derives meaning from its internal and external relations.

These three items are reflected in Wood's (1993) seemingly structuralist perspective on the issue of culture, tourism and impacts. His analysis of culture and tourism relies on identifying and understanding systems, in discussing tourist and development discourses. He argues that: 'The central questions to be asked are about process, and about the complex ways tourism enters and becomes part of an already on-going process of symbolic meaning and appropriation' (Wood, 1993: 66).

I would suggest that insofar as looking for an analytical tool for tourism, Wood's 'process' and 'systems' discussed by Burns (1999a) are synonymous. While Wood (1993) gives a generalised account of cultural impact systems, Greenwood (1989) offers a more specific sense of the cultural problematic encountered by those who study tourism:

> Logically, anything that is for sale must have been produced by combining the factors of production (land, labor, or capital [and enterprise?]). This offers no problem when the subject is razor blades, transistor radios, or hotel accommodations. It is not so clear when the buyers are attracted to a place by some feature of local culture, such as an exotic festival.
>
> (Greenwood 1989: 172)

Underpinning Greenwood's insights on local culture is the notion that place and space are inexorable elements of culture that cannot be separated from the natural environment where it develops. If Greenwood's central concern, the commoditisation of culture for tourism, is to be addressed then a deeper analysis becomes essential. Proponents of *Tourism First* (cf. Burns 1999b) tend to see culture from a supply-side point of view framed by the notion of culture-as-attraction. Thus, while attractions may vary (an obvious point) for many destinations cultural elements will almost certainly be included as part of the 'product mix' (Ritchie and Zins 1978: 257). The extent to which these components of culture are adapted by the local population and offered to tourists for consumption is likely to be framed by at least two factors.

- First, the relative difference and thus the relative novelty between cultural components of the visitors and the visited.
- *Second*, by the type and number of visitors.

The next section examines various definitions of culture within the broad context of tourism.

Defining culture within a tourism context

Culture is about the interaction of people and results in learning and that such learning can be accumulated, assimilated and passed on. The point being that culture is observed through both social relations and material artefacts. Culture consists of behavioural patterns, knowledge and values which have been acquired and transmitted through generations, 'an organised body of conventional understandings manifest in art and artefact, which, persisting through tradition, characterises a human group' (Redfield, cited in Ogburn and Nimkoff 1964: 29).

Descriptions of a specific culture are necessarily static snapshots at a given moment. Culture should be seen as dynamic – a society that does not take on board new ideas, or adapt to changing global conditions is in danger of cultural retrocession. To underline the complexity of thinking about these matters, the following caveat should be acknowledged: 'To speak unproblematically of "traditional" culture is not permissible. All cultures continually change. What is traditional in a culture is largely a matter of internal polemic as groups within a society struggle for hegemony' (Greenwood 1989: 183). Greenwood was right in his concerns about cultural expropriation. His warning also serves as a reminder of the hypocrisy and paradoxes that surround attempts at cultural 'preservation'.

Traditional definitions of culture that have been subjected to a sustained attack over the past decade or so: an attack that has meant that the early cultural anthropologists have been dealt a blow from postmodernism wherein culture-is-nothing. While it is not my intention to discuss postmodernism at length, it is worth undertaking a brief examination of the changing perceptions of culture as they occur in late capitalist society, so an excursion into the phenomenon of postmodernism is inevitable.

Postmodernism

As indicated above, I wish to make only a few observations about postmodernism in order to identify (but not resolve) some ways in which this interpretation of culture affects the analysis of tourism within its global context. The dramatic social changes that have been undertaken in the name of 'rationalisation', 'efficiency', 'privatisation' and 'down-sizing' have meant the fragmentation of 'culture as a whole way of life' (During 1993: 4). In other words, from the locally organised and easily understood, to 'culture as organised from afar – both by the state through its education system, and by . . . the "culture industry"' (During 1993: 4). It is not difficult to ascribe tourism to this 'culture organised from afar' notion.

This fragmentation of society has been manifested in many ways, but perhaps the rise of postmodernism is its most vocal and pervasive demonstration. I again turn to During (1993), who claims that there are three grounds for suggesting that we live in a postmodern era. He notes that modernity can no longer be legitimised by the enlightenment ideas of progress and rationality 'because they take no account of cultural differences'. He continues with an assertion that 'there is no confidence that "high" or *avant-garde* art and culture has more value than "low" or popular culture', finally stating that

> it is no longer possible securely to separate the 'real' from the 'copy', or the 'natural' from the 'artificial', in a historical situation where technologies . . . which produce and disseminate information and images . . . have so much control and reach.
>
> (During 1993: 170)

This fragmentation, which may also be described as the 'tabloid society' in which emotion and feeling take precedence over content and substance, has led to a confusion about life for those in the industrialised countries of the North. MacCannell poses a conundrum that reflects postmodern times: 'In the context of a cannibalistic desire to possess *everything*, it is not especially bizarre for capitalists to want what only socialists can have, universal brotherhood in a life free of material want and social contradiction' (MacCannell 1992: 100).

Postmodernism is described by MacCannell as 'the realization of the capitalist fantasy of the socialist goal of a classless society . . . an imagined bourgeois revolution' (MacCannell 1992: 100) wherein 'classlessness' is achieved by excluding the working classes by the creation of communities where there is no place for them, in a sense itself a reflection of Thompson's notion that: 'Fragmentation of the old working-class proletarian culture meant that a politics based on a strong working-class identity was less and less significant [with] people decreasingly identifying themselves as workers' (During 1993: 4).

In his discussion on consumerism, Lee (1993) is also critical of the manifestation of postmodernism and describes postmoderns as: 'Being inadequately adapted to deal with true legitimate cultural forms, however, this . . . semi-intellectualised mode of production is more often than not turned upon those forms still in the process of a cultural consecration: "jazz, cinema . . . avant-garde underground"' (Bordieu 1984: 360). Such symbolic forms of course have the substantial advantage over fully legitimised forms of being generally far less demanding of their audience's interpretative capacities (Lee, 1993: 170).

This is indicative of the 'pure form over function' debate that rises from time to time in the discourse of art and design and feeds into the next section which discusses the relationship between consumerism and tourism.

Consumerism and tourism

For vast parts of the world, including many that are emerging tourist destinations, that is to say only just engaging in the global tourism nexus, these two aspects of postmodern life (i.e. consumerism and tourism) have no relevance for day-to-day existence which may well be dominated by survival and coping. Even so, given that most tourists come from the North, both consumerism and commoditisation impact upon the lives of destination residents.

Consumerism impacts upon destinations through tourists bringing with them the urban (and urbane?) attitudes of the consumer society that they live in, such as expectations of service levels, and that 'things have their price'. The postmodernism controversy lies at the intersection of contemporary cultural change and the political economy of commodity exchange (Shields 1991: 2). Shields, in his introduction to 'lifestyle shopping: the subject of consumption' discusses malls as being a postmodern icon and consumerism as postmodern activity. I am struck by the possibility of linking the physical and metaphysical attributes of malls to certain Third World holiday destinations and even the new generation of cruise ships . . . all developed with consumerism (rather than the consumer) at centre. The notion of 'anchor stores' becomes 'chain hotel'; other 'functional poles' could be golf course and water sports base. The architecture of modern resorts (though not necessarily high-rise) reflects the eclectic 'flashiness' of postmodern mall architecture. The feeling aimed for by developers and engendered by holiday makers at integrated resort development is reminiscent of Shields' 'conception of community' included in which (in the case of the mall) would be both the *social* function of shopping (as opposed to purchasing) and in the case of the resort or *destination*, the social function of holidaying (meaning the external interaction with other holiday makers as opposed to the internal *re* creational aspects of vacationing. There is also the notion that just as shopping malls are 'managed' and the social swirl within them manipulated (but only to some extent) by design techniques, so are integrated resorts or planned destinations. Management is in the frame, but cannot totally manipulate the crowd dynamics in the way that they were managed (in the Fordist sense) in Britain's postwar Butlin's holiday camps.

In the above, it can be seen that places of consumption are far more than accidental mixes of geographic locations and attractions (indeed this is Shields' basic thesis). They represent both postmodern social dynamics and symbolic edifices wherein everything (including recreation) is for sale – a commodity.

Discussion

While descriptions and references to the brief encounters between unequals such as in O'Rourke's (1987) 'Cannibal Tours', may be poignant reminders of the postmodern condition, and such meetings may carry severe consequences for the receiving population, the bulk of the world's tourism takes place between relative equals, i.e. as domestic or intra-regional travel, between European countries with comparative economic status or say, between the US and Europe. This means that the largest cultural problem to do with tourism in terms of scale, is that of congestion and overcrowding. In many cases this is an accepted and expected part of the holiday experience. Visits to Coney Island, Disneyland or Paris in August are not motivated by a desire 'to be alone'. Indeed, crowds and queues at the attractions, museums and fast food outlets might even be seen as part of the experience. At first glance this could place in jeopardy the general application of Greenwood's assertion that:

> The anthropological view of culture is far different from the economists' and the planners' views of culture as a 'come-on', a 'natural resource', or as a 'service'. The anthropological perspective enables us to understand why the commoditisation of local culture in the tourism industry is so fundamentally destructive and why the sale of 'culture by the pound', as it were, needs to be examined by everyone involved in tourism.
>
> (Greenwood 1989: 174)

But while both the 'host' and 'guest' will develop coping behaviours, the large, diverse populations found in the developed world means that they are much better able to understand change than a society that has had limited contact with outsiders. The cultural impact on most of the cities and places mentioned above is diluted because they are already economically, socially and culturally diverse, and well connected to the global level by well-established electronic and social networks. There is then, a fundamental difference between contacts that occur between cultures of comparable strength and generally equal wealth and those between industrial or post-industrial countries and emerging economies. The issues of power and control frame the differences. While changes are made in all tourist receiving areas, it would be reasonable to assume that much of the pressure for change at tourist spots within industrialised countries is politically intra-cultural, i.e. initiated by entrepreneurs or locally elected politicians in response to community pressure; planned and implemented by professional officers (architects, town planners, etc.) and received by the tourists as better hospitality, signage or parking facilities, etc.

The powers that shape the receiving country's economic structures and also shape its tourism industry and define and demarcate the pattern of arrivals are not only extra-cultural but located (or headquartered) in another country. In the case of less developed countries, the drive will often be from central government responding to two types of pressure.

- The financial pressure of conflicting uses for budgets. For example, the various options facing a government might be: investment in tourism in order to generate more foreign exchange and jobs; investment in education to develop a more useful and responsive workforce;

or investment in primary health so as to ensure a productive and healthy population.

• Pressures generated through the business needs of transnational tour companies (who in return are assembling a product to sell in the holiday marketplace) or their surrogates at a local level.

As indicated above however, the milieu surrounding tourism, globalisation and culture is complicated. The late Nuñez (1989: 271) noted that 'Tourists and more often their hosts are always "on stage" when they meet in face-to-face encounters.' He explains that tourists will have prepared themselves for their role by reading the literature and buying the appropriate costume. The host will 'rehearse a friendly smile' and 'assess the mood of the audience'. MacCannell too, in his analysis of tourism and 'modern' society (1989: ix) refers to 'The current structural development of society [as being] marked by the appearance everywhere of touristic space. This space can be called a "stage set", a "tourist setting", or simply a "set"' (MacCannell, 1976: 100). Biddlecomb (1981: 23) quotes a polemic from Jean-Luc Maurer 'Everything about the behaviour of the Western tourist in the Third World is entirely artificial . . . [he is] no more than an object whose functions are manipulated and controlled'. This is not a tenable position. It is both deterministic and applies a nonsensical generalisation that has no basis in fact, and is alarmingly pessimistic (even more so than Wheeller (1993) or Turner and Ash (1975)). If we accept the argument of Maurer and his like, then there is no hope: there is nothing we can do about utilising the power of tourism for the common well-being.

Doxey's (1976) work is specific to the effect on the relationship between visitors and residents at a given locale over time and as tourism increases. Smith (1989) explores this theme in the broader context of the relationship between tourism and culture indicating that different types of tourist will have different impacts upon the local culture. While the sexual metaphors of triangle (love triangle? Pubis?) illustrated in the triangular shape of the model, and penetration (a word used in the explanatory text by Smith 1989: 14) are probably unintentional, they are useful in reiterating sexual encounter as a strong motivational factor in tourism (either with other tourists as evidenced by the use of words and images in brochures aimed at the singles market, or with 'exotic natives' as implied by the prurient sub-text in the advertising of many international airlines (e.g. Singapore Girl)). The full and tragic effect of this is discussed by Minerbi (1992), with Lea (1988) providing a useful introduction. As Smith (1989) explains, the response to culture differs with the different type of tourist encountered. The proto- or incipient tourist, being few in number, and in a locale unused to tourism (virgin territory?), will require very little from the 'host' population. Consequently, it is claimed that such tourists will have little impact on culture. In a simple sense, this is clearly true. However, account must be taken of the effect these trailblazing tourists will have over time. They will probably recount tales of their experiences to friends and other potential travellers, and the appetite for tourism in the receiving area will have been whetted. The phrase 'see the place before the tourists get there' is a familiar homily to readers of journalistic travel pages. The danger in this is that it allows the explorer to deny responsibility or any role in any development that may follow.

As implied throughout this chapter, observers have shown concern about the way in which tourism and cultures interact; the ways in which cultural shows are put on for tourists. However, others would argue, as does Stanton (1989: 247–62) that in a way, this is a good thing, inasmuch that while tourists gaze upon staged authenticity their attention is diverted from the real everyday life of the local people. Similarly, while the processes of tourism can be seen as framed by cultural imperialism (in the way that it manipulates local traditions and cultures into convenient time slots and places) it could be argued that it is still a sharing process. So what if a tourist sees, even pays to see a half-hour excerpt from a ceremony that might take hours or even days to perform in traditional circumstances? This very act of the tourist involvement can help preserve the traditional culture, keeping the private culture from the pressure of tourists. In this sense, in somewhat contradictory mood, Turner and Ash (1975) hold out some hope for tourists pointing out that one motivation is the search to understand origins and developments of cultures. This, they claim, can generate new feelings in the tourist, who might reflect on their own culture in response to the new experiences. These ideas of Turner and Ash point up the major failing of their book, and one which has become more apparent over time: elitism, that (to return to the discussion of culture) some forms of culture are 'better' than others, something the postmodernists would reject wholeheartedly.

The corollary of explorer-type tourists having the least impact on culture and lifestyle in the receiving area (though I would argue that the intensity of the brief encounter is such that the impacts can be disproportionately high) is that mass tourists have the most impact on culture. Linked to this is the different psychological framework associated with those that prefer to travel in clearly defined groups to well-known and well-developed destinations that characterise the mass tourism product.

Cultural change comes about by things that are both internal and external to that culture. In other words, culture would change anyway. In this, Wood (1993) is right in asserting that there is no such thing as a 'pristine culture' waiting to be smashed by tourism in the way described in the following terms by Turner and Ash:

> The act of discovery has led directly to the destruction of something fragile and irreplaceable. Peripheral cultures only retain their air of antiquity and their ethnic individuality by virtue of their isolation from the dominant and expanding cultures of the metropoles. This is

especially true of the remote island cultures of Bali and the South Pacific. Once distance has been contracted by the building of an airport capable of taking jets, the antique, the ethnic and the pristine in their cultural and environmental aspects are immediately threatened . . . the tourist's superior economic wealth rapidly erodes the sensuous and aesthetic wealth of cultures that have developed in isolation from the Western world.

(Turner and Ash 1975: 130)

This position may have been supportable twenty years ago as social scientists and commentators first grappled with tourism and began to develop a body of knowledge about tourism and its impacts. Such elitism (familiar in a way to Boorstin's) is no longer tenable: the issues are (as Wood reminds us) far more complex. There are a number of ideas that arise from the above quote, prime among them is the use of the words 'antiquity', 'ethnic', 'pristine' and 'sensuality' in describing culture. These words could each be ascribed alternative interpretations:

- 'antiquity' can be to do with museums and that which has gone before but does not hold practical significance;
- 'ethnic' in their context seems to sum up visions of ethnic-chic directed at a normative and empowered audience of middle-class Westerners who are not 'ethnic';
- 'pristine' is a word that draws on Wood's analogy of the billiard ball smashing cultures which are uncluttered, or unpolluted, by the market-driven excesses of the West, but can also be interpreted as a reference to the fragmented post-industrial society of the West; and
- 'sensuality' serves as a reminder that Rousseau-esque images and the 'myth of primitivism' (Hiller 1991; Torgovnick 1990) still inform perceptions of 'Other' (Said 1978) by Western commentators on cultural trends.

Even though the postmodernists have criticised the traditional 'culture is everything' definition, a holistic approach to tourism planning, which is the basic premise of a *Development First* (Burns 1999b) approach, must take the traditionalists' position because it takes social structures and the way in which aspects of culture (including material culture in the form of souvenirs) are bought into the tourism system as being of fundamental importance.

Conclusions

Analysing globalisation as an economic and social phenomenon is complex. If, in the name of equitable distribution of the benefits of tourism (Strange's 'cui bono?' above), it is considered desirable to work at a local level (where culture is located), and given that local action is coming under increasing pressure from global pressures, then an understanding of the relationship between the local and the global is essential if an effective understanding of how global tourism might impact upon local cultures is to be developed.

From the foregoing, it can be seen that the discussion about culture, globalisation and tourism is important because of what tourism development eventually produces: a confused situation in which, as Selwyn (1996a) puts it, there is a need to: 'Distinguish between the myths and fantasies of tourists (authentic in some senses as these may be), on the one hand, and politico-economic and socio-cultural processes, on the other, there may in the end, as Baudrillard (1988) has warned, be no way out of the eventual wholesale Disneyfication of one part of the world built on the wasteland of the other' (Selwyn 1996a: 30).

There is a parallel to be drawn here with the iconoclastic stance of both Selwyn and Baudrillard and the politico-socio-economic notion of the 'Brazilianisation' of society whereby the top few become fabulously wealthy while the bottom few exist in abject poverty. A key element of the discussion has been that while tourism planners acknowledge cultural aspects, especially where culture clashes could damage relations between 'hosts' and 'guests,' or where culture can be seen as a potential component for product development, areas such as social institutions and the way in which society at the destination is structured are often ignored. In terms of the tourism industry's nexus with culture, coupled to the forces of globalisation may lead to an approach to development that sees tourism as a set of service provisions rather than part of a living culture. This will damage social structures as they are forced along a development path that has been determined by external forces rather than discovered by internal social progress. If capitalist exchange alone defines the tourism system at a given locale, then the purpose of culture for all the players, willing and unwilling, becomes subservient to the needs of the tourism industry; culture then becomes just another part of tourist consumption. This does not mean that an attack on consumerism is necessary. Tourism does not necessarily 'destroy' culture or even bits of culture; consumerism as a 'thing' is not 'bad' but, in planning for tourism (perhaps 'negotiating for tourism is a more apt phrase) we need to understand which bits of culture are for sale (or negotiation) and which bits are 'off limits' so to speak.

It may be inferred from the discussion above that at a global level there is something of a case for claiming that tourism is perhaps contributing towards what might be termed 'cultural homogeneity'. Tourism is assisted not only by global brand images such as Pepsi-Cola or Nike sport-wear, but also by global satellite communication such as CNN, MTV and of course the WWW. Greenwood (1989: 184) describes 'cultural dilution of all that is local and idiosyncratic'. Tourism has had a role to play in both assisting this cultural dilution or, more accurately, convergence and, paradoxically, preventing it through revitalising interest in traditions and ethnicity. One view of globalisation is that it is about the changing nature of communications (with the rise of technology) and markets (with the rise of post-*perestroika* US hegemony and consumerism). In more cynical mode, it has been described as: 'An ideological figleaf for a process in which governments have willingly abandoned many of the tools of governance and passed power from the democratically-legitimated sphere of politics to the unaccountable realm of the market' (Goldblatt 1997: 23).

The changes that arise out of both views are political and, as has been suggested in this chapter, cultural at both a local and global level. The dynamics of cultural trends operating on a global scale have both eroded and destabilised older established forms of national culture and identity which have been replaced by a 'global-local dimension [wherein] everyday, situated cultures are routinely saturated with references to the global' (O'Sullivan *et al.* 1994: 130).

This chapter has suggested that the links between tourism and globalisation can be described as cultural ones on at least two levels. On the one hand there are the obvious cultural changes and connections that are a well-acknowledged result of international travel and the growth in influence and power of consumer driven demand. On the other, there is also a growing awareness that perhaps the culture of governments is becoming 'the same' through the process of capitalism (since the fall of the Eastern bloc) and the process of globalisation which, as inferred through this chapter, is 'out of their control' (Robertson 1992: 5).

On the part of emerging economies, it might be concluded that the part of international tourism which takes place within the globalisation milieu does not come about through any sort of natural economic processes within that country. Most poor or emerging countries have little or no domestic tourism upon which to build their international tourism. Demand for any significant form of tourism (other than individual travellers 'passing through') is not even likely to come from potential tourists. Rather, the process by which such countries become a part of international tourism is likely to be through the specific actions of government, international tour and transport operators and foreign investors.

The cultural impact at a local level of tourism development framed by the global condition may be represented by the polarities of a social-interaction continuum. At one extreme tourism-induced social change can lead to development, representing increased socio-economic benefits. At the other, change can lead to dependency and reinforcement of social discrepancies based on a service economy structured to suit the needs of local elites and monopolistic transnational tourism corporations. The familiar cycle of corporations searching out 'unspoiled' destinations, encouraging development, moving on when saturation (or novelty) point has been exceeded does need reiteration here. The destination is left with an economy reliant on tourism, a workforce trained within a service culture and infrastructure oriented to supporting foreign 'play' – the options remaining open to such destinations are limited and their bargaining power lowered. They are left, so to speak, with the fleeting memory of a brief encounter.

Final observations

In addition to the conclusions about tourism's brief encounters drawn above, a range of practical policy issues need further investigation if the forces of globalisation and tourism are to be harnessed to provide culturally beneficial relationships. The observations are placed here in bullet form:

- local explanations of tourism are more useful than official definitions in identifying the local consequences of tourism;
- the poor rarely have ready access to the means of articulating problems – they have low bargaining power and this should be borne in mind during assessment of local acceptance of tourism;
- appropriation of cultural phenomena may marginalise systems of meaning, disempower sections of the community, and polarise cultural differences;
- change may result as much from perceptions of tourist potential by the powerful, as by the presence of tourists themselves;
- change in one cultural element can result in widespread distortions because of complex interrelationships within cultures;
- local control of cultural reproductions and lower scale and speed of development contribute to cultural continuity as defined by host communities;
- commoditisation of cultural elements creates new systems of values and power those without access become excluded or marginalised;
- the intimate relationship between tourist product and host can result in hosts themselves being commoditised especially in ethnic tourism;
- trickle-down effects from macro-tourism planning are insufficient to lift structural causes of chronic underdevelopment and poverty;
- cultures are not passive recipients of impacts as they have their own dynamics of change, but the conditions under which they interface with tourism can seriously constrain their options for response and adjustment; and
- tourism development plans and facilities which incorporate cultures and cultural products should include participation and control by ethnic groups. They should also seek to develop economic options as sustainable alternatives to dependence on tourism

References

Balaam, D. and Veseth, M. (1996) *Introduction to the International Political Economy*, New Jersey: Prentice-Hall.

Biddlecomb, C. (1981) *Pacific Tourism: Contrasts in Values and Expectations*, Suva: Pacific Conference of Churches.

Boniface, P. and Fowler, P. (1993) *Heritage and Tourism in the 'Global' Village*, London: Routledge.

Brewster, C. and Tyson, S. (eds) (1991) *International Comparisons in Human Resource Management*, London: Pitman.

Burns, P. (1999a) *An Introduction to Tourism and Anthropology* London, Routledge.

Burns, P. (1999b) 'Paradoxes in planning: tourism elitism or brutalism?' *Annals of Tourism Research* 26(2): 329–49.

Burns, P., Holden, A. (1995*) Tourism: a New Perspective*, Englewood Cliffs: Prentice-Hall.

Doxey G. (1976) 'When enough's enough: the natives are restless in Old Niagara', in A. Mathieson and G. Wall (1982) *Tourism Economic, Physical and Social Impacts*, Longman: Harlow.

Dunning, J. (1993) *The Globalization of Business: the Challenge of the 90s*, London: Routledge.

During, S. (ed.) (1993) *The Cultural Studies Reader*, London: Routledge.

Goldblatt, D. (1997) 'Spinning out of control', *Times Higher Educational Supplement*, 16 September.

Greenwood, D. (1989) 'Culture by the pound: an anthropological perspective on tourism as cultural commoditisation', in V. Smith *Hosts and Guests: the Anthropology of Tourism (2nd Edition)*, Philadelphia: University of Pennsylvania Press.

Hiller, S. (ed.) (1991) *The Myth of Primitivism: Perspectives on Art*, London: Routledge.

Lavigne, M. (1995) *The Economics of Transition: From Socialist Economy to Market Economy*, Basingstoke: Macmillan.

Lea, J. (1988) *Tourism Development in the Third World*, London: Routledge.

Lee, M. (1993) *Consumer Culture Reborn: the Politics of Consumption*, London: Routledge.

Levitt, T. (1983) 'The globalization of markets', *Harvard Business Review* 6(3): 92–102.

MacCannell, D. (1976) *The Tourist: a New Theory of the Leisure Class*, New York: Shocken Books.

MacCannell, D. (1989) *The Tourist*, (Second edition) New York: Random House.

MacCannell, D. (1992) *Empty Meeting Ground: the Tourist Papers*, London: Routledge.

McLuhan, M. (1960) *Explorations in Communication*, Boston: Beacon Press.

Minerbi, L. (1992) *Impacts of Tourism Development in Pacific Islands*, San Francisco: Greenpeace Pacific Campaign.

Nuñez, T. (1989) 'Touristic studies in anthropological perspective', in V. Smith, *Hosts and Guests: the Anthropology of Tourism (2nd Edition)*, Philadelphia: University of Pennsylvania Press.

O'Rourke, D. (1987) *Cannibal Tours* [Film] Canberra: O'Rourke and Associates.

O'Sullivan, T. *et al.* (1994) *Key Concepts in Communication and Cultural Studies*, London: Routledge.

Ogburn, W. and Nimkoff, M. (1964) *A Handbook of Sociology*, London: Routledge and Kegan Paul.

Ritchie, J. and Zins, M. (1978) 'Culture as a determinant of the attractiveness of a tourist region', *Annals of Tourism Research* 5: 252–67.

Ritzer, G. (1993) *The McDonaldization of Society*, Newbury Park: Pine Forge Press.

Robertson, R. (1992) *Globalization: Social Theory and Global Culture*, London: Sage.

Said, E. (1978) *Orientalism: Western Concepts of the Orient*, London: Routledge and Kegan Paul.

Selwyn, T. (1996a) (ed.) *The Tourist Image: Myths and Myth Making in Tourism*, London: Wiley.

Selwyn, T. (1996b) 'Tourism, culture and cultural conflict', in C. Fsadni and T. Selwyn (eds.) (1996) *Sustainable Tourism in Mediterranean Islands and Small Cities*, Valetta: Med-Campus.

Shields, R. (1991) *Places on the Margin: Alternative Geographies of Modernity*, London: Routledge.

Smith, V. (ed.) (1989) *Hosts and Guests: the Anthropology of Tourism (2nd Edition)*, Philadelphia: University of Pennsylvania Press.

Stanton, M. (1989) 'The Polynesian Cultural Center: a multi-ethnic model of seven Pacific cultures', in V. Smith (ed.) *Hosts and Guests: the Anthropology of Tourism (2nd Edition)*, Philadelphia: University of Pennsylvania Press.

Strange, S. (1988) *States and Markets: an Introduction to International Political Economy*, New York: Basil Blackwell.

Torgovnick, M. (1990) *Gone Primitive: Savage Intellects, Modern Lives*, Chicago: University of Chicago Press.

Turner, L. and Ash, J. (1975) *The Golden Hordes: International Tourism and the Pleasure Periphery*, London: Constable.

Wheeller, B. (1993) 'Willing victims of the ego trap', *Tourism in Focus*, 9.

Wood, R. (1993) 'Tourism, culture and the sociology of development', in Hitchcock, M. *et al.* (eds) *Tourism in South East Asia*, London: Routledge.

14 Issues of sustainable development in a developing country context

William C. Gartner

Introduction

Globalisation, measured in terms of international business activity, reached levels in the late 1980s never recorded before (Helleiner 1990) and it accelerated at an increasing rate throughout the 1990s. Tourism is a significant part of this activity as multinational businesses become more common around the world. However, as in tourism movements, much of the globalisation activity is found in the developed world. Much of the developing world, especially outside capitals or business centres, does not yet have the multinational presence one encounters throughout the developed world. This is true for much of Africa. Therefore what one encounters in Africa is for the most part locally owned and managed. While this is a position many tourism scholars view as ideal, the causes for this situation reveal some deeper development problems. These problems, some of which are presented below, lead to a situation where local people are unable to manage growth. In the long run what appeared to be a promising tourism development start may lead to an unsustainable and unsatisfactory situation.

This chapter examines one of the African countries, Ghana, and analyses its tourism development experience of recent years with respect to issues of globalisation and sustainability. In particular an argument will be offered that barriers to tourist entry, both artificial and policy imposed, allow for local development to proceed. However, is this local development, which many tourism scholars would applaud as desirable, actually sustainable? And if so is it worth it? A second and parallel line of discourse offered in this chapter relates to sustainable tourism development within the broader context of developing country issues.

Tourist arrivals in Africa have not shown any appreciable percentage gains from 1989 to 1996 although in sheer numbers the increase has been approximately 50 per cent. In 1989, 13.8 million tourists, equal to 3.22 per cent of the world total, visited the African continent. In 1996 total tourists were at 21.5 million which represented 3.62 per cent of the world total (WTO 1998). Most African travellers visit North Africa, in particular the countries of Morocco and Egypt, South Africa, or East Africa Sub-Saharan West Africa accounted for only 12 per cent of total visitors to Africa.

One could argue that there are many reasons for this relatively poor performance for the entire continent when compared to world totals. Obviously a generally negative image of the continent has had an impact. Wars fought for ethnic cleansing – or because of ages-old tribal rivalries – are common events reported in world media. Attacks against tourists, especially in Egypt and Uganda, have also received world media attention resulting in predictable declines, in the short term, in visitation. There is no doubt the autonomous image formation agents (Gartner 1993) in play here have had a serious impact on propensity to visit parts, if not all, of the African continent. However, to conclude that low levels of visitation is an image problem is much too simplistic. Other barriers (e.g. poor infrastructure, lack of receptive services, etc.) are in place making parts of Africa difficult to visit.

Sustainable tourism systems

There has been no shortage of discussion surrounding the issue of sustainable tourism systems (Gartner 1996a: 510–11; Bramwell 1991; Quesada-Mateo and Solis-Rivera 1990; United Nations 1987). What one learns from much of the discussion is that sustainable is a 'good thing' and we should work to achieve it. *Sustainable* is a value positive word. That is, it embodies all the good things one would like to achieve from development and rejects any of the potentially negative outcomes. How could anyone argue against it? On the other hand how can it be accomplished?

There has been no shortage of definitions offered to define what is meant by *sustainable*. One of the most frequently cited definitions is from *Our Common Future*, the report of the World Commission on Environment and Development (1987). It states sustainable development is 'development that meets the need of the present without compromising the ability of future generations to meet their own needs'. This is a *value positive concept* but one that this author believes to be extremely hard to achieve, if not impossible, when placed in a real world context.

One reason sustainable development is hard to define, let alone implement, is that the concept will be interpreted differently by its targeted recipients. In other words, which organisation will benefit from a 'sustainable' decision? One argument is that ideally all groups, individuals, or organisations benefit. However, when organisational goals are examined it becomes clear that this may not be possible. The conundrum that we find ourselves in then is trying to achieve a sustainable state that cannot clearly be defined for all those who are supposed to benefit from it. To understand this better a case study from Ghana will be used to explore various issues of the sustainable development argument.

Background to Ghana

The Republic of Ghana is located approximately 1,000 kilometres north of the Equator. It is a tropical country with temperatures ranging from 21–32°C and rainfall, especially in its coastal zone, averaging over 2,030 mm. Tropical forests at one time encompassed all of the southern two-thirds of the country but few extensive tracts remain today. Ghana has substantial gold reserves and is a major exporter of cocoa.

Development has been hampered by many factors both external and internal. As the first African country to achieve independence (in 1957 from the British) it has not prospered as many predicted. Internally, a burgeoning population estimated at over 20 million has retarded development. This is up from approximately 10 million in 1976 and represents a rate of growth exceeding 2.5 per cent annually. Ghana's post-independence political history has been marked by numerous attempted and successful coups with the most recent occurring in 1981. These coups have been blamed, in part, for the country's lack of a significant tourism industry (Teye 1988).

Other reasons given for Ghana's overall poor growth performance include positions advocated by the dependency school and those subscribing to the neoclassical school (Anunobi 1994). In brief the dependency theorists claim that many of the old colonies are in reality new economic colonies of the west and suffer from poor trade terms both exporting and importing. The neoclassical school cites poor infrastructure, lack of an educated and skilled workforce, weak management and low or non-existent levels of savings and capital formation for the lack of development.

There is no need to argue which school is right. One could easily find examples to support both. What is clear however is that all primary product (i.e. natural-resource based) producing countries, which includes almost all of the tropical countries, have suffered dearly from poor trade terms since, at least, the 1960s. At the end of the Second World War primary products were needed to rebuild countries and economies. Primary-product producers did very well during this period but since most of them were still under colonial rule much of the resource wealth went to the colonial powers. Shortly after independence for most African countries, which occurred in the late 1950s and accelerated throughout the 1960s, primary-product prices began to fall in relation to the import of high technology manufactured goods. It is estimated that this exchange where primary products and low-tech manufactures were exported and high-tech manufactures imported, amounted to a tax of 20–25 per cent on the export earnings of developing countries (Singer 1998a). Ghana may have suffered even more as its economy was in worse shape than the average developing country during the late 1970s and through the 1980s. Political unrest and drastic economic austerity measures plunged the country into a major depression for most of the period.

In the late 1990s, tourist arrivals to Ghana were estimated at 250,000–300,000. However, these estimates are highly suspect as the data analysis system in Ghana, operated by the Immigration and Naturalisation Service and not the Ministry of Tourism, is poorly managed. The forms one fills out upon arrival are stored in various offices and rarely if ever examined (personal experience). Regardless of how poor the statistics may be, it is generally acknowledged that most visitors to Ghana can be classified as part of the VFR (visiting friends and relatives) market or business travellers. Much of the business travel is related to donor agency projects. The lack of any large or well-organised ground transportation services catering to a pleasure tourism market is additional evidence that the majority of visitors are not there for pleasure travel purposes. What little accurate data exists indicates that almost two-thirds (64.3 per cent) of the visitors to the country are travelling alone, the majority are repeat visitors (54.5 per cent), and very few use local tour operators (2–5 per cent) (Ministry of Tourism 1996). These data support the claim that most travel is for business or VFR purposes and should not be classified as purely pleasure travel.

Trade in tourism

Most trade agencies and organisations do not deal with tourism. Part of the reason for this neglect is that tourism does not operate like other exporting businesses. Much of the debate concerning trade revolves around barriers to entry. Countries are constantly negotiating agreements that allow certain products, sometimes limited by quotas, into their country provided the exporting country reciprocates in some way. The closest thing to a trade agreement between countries for tourism is bilateral airline agreements. Usually one or more of a country's flag-carrying airlines will be granted permission to enter another country at a selected port of entry provided a reciprocal arrangement is in place. Apart from that the only other restrictions one encounters may be a ban on travel to another country imposed primarily for political purposes. The ban, now somewhat relaxed, on Americans travelling to Cuba is an example of this form of restrictive trade in tourism.

Because many developing countries have never considered the implications of their trade sanctions on tourism it is not uncommon to find countries imposing restrictive terms of trade on themselves! Edgell (1988) identifies many barriers limiting visitation including restrictive visa regulations, entry and exit taxes especially those not included in the price of transportation, passport approval delays and difficulties, and other nuisances such as government registration forms required when travelling from place to place in a country. Many of these barriers are operational for Ghana. Entry visas must be obtained prior to travel to the country and are not inexpensive compared to many other countries. Delays in processing are common, sometimes requiring fees to special services companies that are able to facilitate the process. Access to the country is primarily through international carriers in European markets. Two African-based airlines serve Ghana directly from the United States but due to poor performance they are not considered a reliable option. However, they are relatively low cost and do attract the low-end pleasure market and group tour market. Those travelling to Ghana by airlines based in Europe face high-ticket costs brought on, in part, by overflight fees paid to countries the airline flies over on its way to Ghana. Upon arrival in Ghana numerous barriers exist making movement around the country problematic. The lack of inbound operators has been mentioned but more importantly the transportation infrastructure is so poor and in many places so overcrowded, that travel is slow and relatively dangerous.

In spite of these obstacles, tourism development has taken place especially in the Central Region of Ghana. This region contains a number of significant attractions including large castles dating to the 1400s when, it is believed, the Portuguese were the first Europeans to establish trade with West Africa. The castles of Elmina (St George's Castle) and Cape Coast were important trade centres for all goods including, and most importantly, people. Many slaves to the New World were imprisoned in the dungeons of these castles as they awaited their forced journey to the Americas. These castles, as well as a small, but bio-diverse important forest reserve (approximately 1,300 square kilometres), became the focus of a tourism development initiative funded initially by the United Nations Development Programme (UNDP) with much more support provided beginning in 1991 by the United States Agency for International Development (USAID) (Gartner 1996b).

Hotel characteristics

Tourism development has occurred in the Central Region in recent history. In the 1970s a number of large hotels were in operation but by the end of the decade most were beginning to seriously degrade. A key ingredient of the tourism development formula during this period was government investment. In a major way the government of Ghana was an investor or owner of each of the large properties. The size and number of these government hotels was out of scale with respect to market share. During times of political and economic unrest demand continued to decrease such that by the time the USAID project referred to above was launched only one very poorly designed government hotel was still in operation.

There were a number of privately owned, much smaller properties, operating in the Central Region, They served primarily the domestic market and some adventure travellers. An inventory by the tourism managers of the USAID-sponsored project team identified approximately ten rooms that could be considered of international standard in 1990. Needless to say no multinational hotel companies were present in any way in the project area in the early 1990s.

In the late 1990s, the number of hotel rooms of international standard exceeded 300. How it reached this level reveals some very interesting sustainable development issues. To understand how tourism development occurs in parts of Sub-Saharan Africa, especially in Ghana, one must first understand the social system in place. Insiders' views of the social structure are revealed in books by Sarpong (1974), Hafner (1996) and to a lesser extent by Klitgaard (1990). Some of the key elements of the social system that relate to this chapter include land-ownership traditions as land is a requirement for development. There are basically three types of land ownership in Ghana with the most common outside of metropolitan areas being village-based chieftaincy controlled (Gartner 1996a: 165–6). When tourism attraction development began to take place in 1991 interest in acquiring land for hotel development occurred. Those that knew how the system operated had a decided advantage as they could obtain leases for excellent locations, However, the impetus for hotel development did not occur as a result of project activity so much as it did when the government enacted a law (Ghana Investment Promotion Center Act, 1994) which exempted hotels from paying import duties for equipment or supplies used in their primary business. Computers, air conditioners, even some vehicles were covered under this law. As a result of this action the number of new properties increased by almost 67 per cent from 9 to 15. What this Act effectively did was adopt a strategy which development theorists advocated in the 1960s and 1970s and which is being revisited. What happened was a form of import substitution that allows local businesses some measure of protection from expatriate companies while it encourages internal growth. To be effective, import substitution development is applied to selective businesses that need time to grow and in doing so require protection from larger more sophisticated multinational companies. This was the case with the development of Korea (Singer 1998b). It is not suggested here that import substitution was the reason for the Hotel Investment Act of 1994. Indeed direct evidence contradicts this as there were no restrictions placed on multinational corporations developing hotels in Ghana. Just the opposite occurred as multinational hotel company development was encouraged and did occur in the capital city, Accra. In addition, even though a few of the tourist entry barriers were reduced or eliminated some of the more onerous remained. This contradicts import substitution strategy where you not only protect infant industries but work to increase sales for them at the same time. In addition, the Act encouraged the importation of high-tech manufactured goods. However, in the absence of any locally produced comparable products this was the only option to encourage hotel development. Nonetheless for a variety of reasons, including poor communications and transportation infrastructures, the Hotel Investment Act worked as an import-substitution strategy of sorts allowing locally run hotels to develop in the Central Region.

The workforce for the hotels in the Central Region is comprised of local villagers most with only a limited amount of formal education and none with specialised degrees related to hospitality and tourism. Only one hospitality training institution exists in the country and only within the last few years has a university degree in tourism been offered. Many employees come from the owner's nuclear or extended family. This form of patronage satisfies a cultural requirement (Gartner 1999) but often prevents a business from being operated in an economically productive manner.

The privately operated hotels in the Central Region are relatively small in size (10–20 rooms). Some have substantial debt. Most are operating independently and have not seen the need to develop packages with tour operators or with any of the attractions. The attractions in the area are significant including the only canopy walkway on the continent (Kakun National Park), castles mentioned earlier with recent high-quality museums added, and a complex, colourful and welcoming culture. In the mid to late 1990s significant (at least for the Central Region) numbers of tourists visited the area to see the sites. Since the area is located approximately two hours from Accra length of stay is short as most visitors come for a day visit or stay only one night.

How sustainable are the properties described above? To answer this question one must first understand how the government of Ghana views sustainable tourism development. The rhetoric is clear – official statements made at international and local conferences emphasise sustainable development. Development that encourages low-cost charters and low-budget mass tourism are considered unacceptable. Of course it is easy to take this stand as a mass tourism market does not currently exist for Ghana. Government actions, on the other hand, lead one to believe the only way to achieve the highly optimistic projections of one million visitors by 2010 is to actually encourage forms of mass tourism. In fact the only remaining government hotel in the Central Region was completely demolished and replaced by a large, attractively presented, but entirely dysfunctional property built by proceeds derived from workers' contributions to the government's social security pension fund. The property is finally operating but only after extensive delays and huge budget overruns. By many estimates the property cost over US$20 million to build and is operating only at one-third capacity due to the structural barriers noted above limiting pleasure travel to and within the country.

At the same time that the new property was being constructed a loan programme to help existing businesses was enacted. The loan programme allowed existing hotels to secure relatively low cost, government backed loans for improvement projects. Each hotel participating in the loan programme was required to submit a business plan. Assistance in developing these plans was provided by the Ghana Tourist Board that also acted as the government agency responsible for the loan guarantee. Gartner (1999) details how this money may have been viewed by the loan recipients as part of the informal economy. A predictable outcome is a high rate of loan non-payment. Foreclosure, not a common occurrence, is a threat to some of the properties. To be sure some of the properties have benefited from the loans in the way they were intended. New additions are being constructed and new rooms and amenities (e.g. pools) are being added. Other loan recipients show no signs of improving their hotel and one wonders how the loan was used.

Sustainable?

Barriers limiting international visitation may not always be a bad thing, at least in the early development stages, especially if the reason is part of an import-substitution strategy. In the case of Ghana, especially the Central Region, structural and institutionally imposed development barriers allowed a locally developed, owned, and managed accommodations sector to emerge. Let no one be mistaken – this was not the intent of the government's imposition of structural barriers to visitation one encounters in Ghana but rather a serendipitous outcome. The creation of an attraction package of international importance occurring at the same time provided demand thereby increasing needed cash flow for local development. In other countries, such as Thailand, tourism development receives much support from government including promotion and advertising, and removal of institutional and structural barriers. This has the expected effect of increasing visitation but also encourages a multinational presence in the market as demand increases.

A major difference between development in Thailand and Ghana may be one of how each government viewed its role in local development. In Thailand's case it began to eliminate many of the barriers to development by initiating world image development campaigns, reducing barriers that it could control, and encouraging mass tourism. When the government of Ghana began to enact legislation and development action to increase local hosting capacity many access barriers still remained in place. What the government essentially did was support development of an area without a strategy to increase visitation leading to a situation where the capacity it helped to create was not filled. External marketing is largely non-existent. Some of the hotels in the Central Region are using the World Wide Web to market their properties but they are not assisted by any national effort in this regard. Infrastructure remains a major problem. Sporadic electrical outages and water delivery breakdowns continue to plague local businesses. Inbound services have shown little expansion in size or service quality. Deterioration of the attraction packages developed during the early 1990s donor agency project is reducing the Central Region's *pull*.

In spite of the problems some of the properties will survive, at least as long as their owners. A few weathered the turbulent times of the early 1980s when the country's economic system imploded. These properties are the ones with low debt service and high amenity value due to location or services provided.

A greater threat or opportunity, depending on an individual's perspective, will occur when barriers are reduced and tourism flows are increased. If Ghana can effectively deal with the numerous barriers that exist to increasing tourist flows there may come a time when corporate properties managed by multinational businesses arrive in the Central Region. It is a foregone conclusion that these properties will develop only when a predictable client base is established. There have been rumours for years that one multinational is interested in building a beach resort on one of the few remaining stretches of coast that is safe for water-based activities. However, this event is always predicated on having a sister property in the capital city that will serve as a feeder for the beach resort. If this occurs one can be assured that the distribution channel between the corporate hotels will be exclusive serving only the corporate hotels and not the existing properties. In the worst case scenario the beach resort will be so well designed and operated that it will siphon-off business that is now going to the local properties. Selling-out may be a viable option if barriers to travel to Ghana are reduced and tourist flows dramatically increase. What then of sustainability? Research conducted by Bowditch (1996) reveals that the hotels in the Central Region are opportunistic. Profit may not be the overriding concern as some of them are constrained by a variety of factors including the socio-cultural system in which they do business. However, an external sale of the property which removes them from some of the social obligations they face may appeal to many of them. In this case growth provides an opportunity to make a profit by selling.

What does the example of Ghana's Central Region tell us about sustainable tourism systems and globalisation?

- First and foremost, it reveals that tourism is indeed a system comprised of physical, social and economic elements. From an economic perspective tourism is trade and too often the trade barriers a country encounters are self-imposed. From a social perspective how businesses operate in their particular socio-cultural environment may not always be in the most productive or economically efficient way;
- Second, because of the barriers that exist it is often possible to develop locally owned tourism businesses first. Maintaining these locally owned businesses becomes problematic when government actions lead to further development that is not tied to any long-range, all-encompassing plan;
- Third, as the sustainable development argument rages over what is and what is not sustainable it becomes evermore clear that practice and ideology are not in sync. Government actions often belie their rhetoric.
- Finally, and probably one of the most important considerations, is what can be considered a sustainable tourism system in a country that

has very high levels of unemployment, poverty, illiteracy and population growth? Most scholars would argue that a tourism system that has high levels of leakage due to foreign ownership is as bad as one that has high levels of local ownership but low levels of development and ancillary economic development. Is there a middle ground?

Conclusions

This chapter's main thesis is that sustainable tourism development is not something that occurs simply because there is local ownership of resources. As is the case in Ghana's Central Region, government decisions played a major role in development; but it is argued in this chapter that those actions were without any clear idea of how to achieve a certain level of development.

Tourism is trade. Although this may seem obvious, government actions clearly show that official agencies are not really sure how to treat this type of trade. Is an import substitution strategy needed? Should export growth take precedence over import substitution? What is the role of trained human resources for tourism development? Where do you find these human resources? What trade barriers are self-imposed?

Ghana is no exception to other developing countries when it comes to its lack of understanding regarding tourism development. Yet many of these countries are actively seeking to develop a tourism industry. While terms of trade in primary commodities continue to show disappointing results for resource-dependent countries, tourism continues to gain importance, on the world economic scene. How then can tourism be developed in a sustainable way?

First and foremost, tourism must be recognised as a special type of trade and dealt with by a country's trade agency. Too often tourism development is relegated to a poorly funded Ministry of Tourism that sees its role as promotion and sometimes product development. Tourism ministries, with few exceptions, do not involve themselves with trade strategies and this is to the disadvantage of the country. Coordination between a country's trade and tourism agencies would help make for a more sustainable path to development.

Second, a country's private sector can be supported through special legislation such as discussed above but from a long-run perspective it is the human resource element that will be more important to development. In the absence of specialised human resource training an industry cannot sustain itself.

Third, *sustainable*, must be clearly defined with respect to measurable parameters. What do we actually mean when we say something is *sustainable*? Does it mean that local providers will manage 60 per cent of the hotel capacity in an area? Does it mean that management level employment in hotels, tour operations, etc. should be 99 per cent local? Does it mean that 5 per cent growth each year is achieved? To address these questions we need to return to an argument offered at the beginning of this chapter which is that sustainable tourism development is a value positive concept. To make it work, sustainable tourism development must move beyond something that is a value positive concept into the realm of an operational strategy. Only then can we get past the rhetoric and develop workable sustainable development systems. While it might be very nice to say that sustainable means we use resources today without reducing their utility value for future generations this is pretty much an unworkable concept for many of the world's developing nations. I will use an economic argument to support this statement.

The rate of self-sustaining economic growth over time can be calculated, according to the Harrod-Domar model, as the share of investment in output, which in equilibrium is equal to the share of saving, divided by the capital output ratio (Singer 1999a). The quotient is then reduced by the rate of population growth. For example, if the investment share is 9 per cent and the capital output ratio is three the rate of nominal growth is 3 per cent. If the population is 2 per cent then the effective rate of income growth is 1 per cent. Capital output ratios can be increased through efficiencies gained via technological progress and investments in human capital. Positive rates of growth are easily achieved in developed countries due to productivity increases brought on by technology and advanced educational systems plus relatively high rates of investment. However, in the developing world where savings and hence investment may be almost non-existent, capital output ratios are low due to insufficient investments in human capital over time. If high rates of population growth are common it would not be unusual to encounter negative rates of income growth. If self-sustaining economic growth is not achievable how is it possible to achieve something as ambiguous and as poorly defined as *sustainable tourism development*? In this negative economic growth situation, common to many parts of the world, resources will not be even considered for their option value but will be used as quickly as possible to achieve survival levels of development. One only needs to look at the condition of the world's tropical forests to see the reality of this statement.

Finally, what does all this mean with respect to globalisation? Due to the lack of significant human resource development over time many developing countries find themselves in a situation where the skills which they possess (e.g. agriculture) are not transferable to something as service intensive as tourism. If a country seeks tourism development and is able to overcome its self-imposed entry barriers and weak infrastructure it may find that it does not have the specialised workforce to build or manage a quality accommodation sector. What then? Given many of the other problems in Africa such as continued poor returns for primary commodities, stagnant real capital flows, foreign exchange scarcities, and an underdeveloped skill base (Helleiner 1990) one wonders how any type of significant tourism sector can be developed without major investment from multi-nationals. However, since the current situation supports local development, due primarily from the lack of a large enough market to attract multi-national interest, there remains a glimmer of hope that local providers can gain the skills necessary to remain competitive and prosper if the broader economic situation changes.

References

Anunobi, F. (1994) *International Dimensions of African Political Economy: Trends Challenges and Realities*, Lanham, Maryland: University Press of America.

Bowditch, N. (1996) *Micro-Economic Assessment of the Tourism Related Businesses in the Central Region*. Prepared Under Contract to the Tourism Center University of Minnesota.

Bramwell, W. (1991) 'Sustainability and rural tourism policy in Britain', *Tourism Recreation Research* 16(2): 49–51.

Edgell, D. (1988) 'Barriers to international travel', *Tourism Management* 9(1): 63–6.

Gartner, W. (1993) 'The image formation process', *Journal of Travel and Tourism Marketing* 2(3): 191–212.

Gartner, W. (1996a) *Tourism Development: Principles Processes Policies*, New York: Van Nostrand Reinhold.

Gartner, W. (1996b) 'An integrated tourism development project: the Central Region of Ghana', *The First International Conference on Urban and Regional Tourism: Balancing the Economy and the Ecology*, Potchefstzoom Northwest Province South Africa. Published in Proceedings.

Gartner, W. (1999) 'Small scale enterprises in the tourism industry in Ghana's Central Region', in *International Academy For The Study Of Tourism*, London: Routledge.

Ghana Investment Promotion Center Act (1994) Act 478 Ghana Investment Promotion Center Accra: Ghana.

Hafner, D. (1996) *I Was Never Here and This Never Happened*, Berkeley, CA: Ten Speed Press.

Helleiner, G. K. (1990) *The New Global Economy and the Developing Countries*, Brookfield, Vermont: Edward Elgar.

Klitgaard, R. (1990) *Tropical Gangsters*, Basic Books HarperCollins.

Ministry of Tourism Ghana (1996) *Diary Survey of International Visitors to Ghana*. Prepared for the Midwestern Universities Consortium for International Activities (MUCIA).

Quesada-Mateo, C. and Solis-Rivera, V. (1990) *Sintesis Ecodes by Carlos A. Quesada*, Memoria (Congreso De Conservaciori Para El Desarrollo Sostenible Ist San Jose Costa Rica).

Sarpong, J. (1994) *Ghana in Retrospect: Some Aspects of Ghanian Culture*, Accra: Ghana Publishing Corporation.

Singer, H. (1998a) 'How relevant is Keynesianism today for understanding problems of development' in *Growth Development and Trade: Selected Essays of Hans W. Singer*, Cheltenham, UK: Edward Elgar.

Singer, H. (1998b) 'Industrialisation: Where are we going', in *Growth Development and Trade: Selected Essays of Hans W. Singer* Cheltenham, UK: Edward Elgar.

Teye, V. (1988) 'Coups d'état and African tourism: a study of Ghana', *Annals of Tourism Research* 15: 329–56.

United Nations (1987) *Our Common Future: The World Commission on Environmental and Development*, Oxford: Oxford University Press.

WTO (1998) *Compendium of Tourism Statistics 1992–1996*, 18th edition, Madrid: Spain.

15 Conclusion

Chris Cooper and Salah Wahab

Introduction

As Hall comments in Chapter 2, globalisation is a 'concept with consequences'. Not only do such consequences apply to tourism, but also the globalisation of society is now a fundamental consideration of all industrial and emerging economies with implications at the level of individual enterprises, governments and communities. The chapters in this book have shown that there is a range of approaches to know just what it means to be a global society. Whilst each social science and business subject area adopts a different stance *viz-à-viz* globalisation, it is possible to draw together these disparate threads by considering globalisation as *boundarylessness* and the various organisational responses to it (Parker 1998). Ashkenas *et al.* (1995) see this as a new paradigm for organisational behaviour, characterised by speed, flexibility integration and innovation. In other words, the process of globalisation not only reduces borders and barriers for trade between nations, but it also renders these boundaries permeable both within and between organisations. This increased permeability of boundaries has been brought about by a series of drivers operating at all scales, not simply at the global scale, and discussed further below. Some (such as Parker 1998: 6) assert that globalisation goes beyond the idea of permeating boundaries between nations and organisations, but also crosses the 'traditional borders of time, space, scope, geography, functions, thought, cultural assumptions . . .'. This, in turn, demands both a different perspective and position to be taken on the management and operation of tourism organisations (Melin 1992). Of course, whilst tourism organisations themselves are affected by globalisation, so too they enhance and sustain the process of globalisation in terms of their own responses to the phenomenon. In many respects, as this book shows, tourism is at the forefront of the creation of a global society.

The drivers of a globalised society

As Wahab and Cooper state in Chapter 1, globalisation has been driven by different sets of converging forces operating at different times:

- First, was the internationalisation of trade from the 1950s.
- Second, the trans-nationalisation of capital flows since the 1980s.
- Finally, the globalisation of information flows (Makridakis 1989).

With each wave of forces, competition has intensified and structural adjustments have been needed. This is because drivers of globalisation are often outside the control of individual enterprises or nations, creating both opportunities and threats and demanding a response by enterprises, governments and communities. In this sense, some view the creation of a new world order resulting from the process of globalisation as a major threat to the sovereignty of nations and the ability of say, tourism destinations to determine their own destiny.

Parker (1998) identifies six inter-related drivers of globalisation:

- technology;
- economy;
- politics;
- culture;
- natural environment; and
- managing businesses globally.

Technology Viewing the evolution of businesses and economies in terms of cycles, it is clear that technological developments are a key driver of globalisation (Bradley *et al.* 1993; Dicken 1992). Although somewhat deterministic, it can be argued that, as technological revolutions have occurred, so ways of doing business and competitive landscapes have shifted (see for example Buhalis 1998; Bradley *et al.* 1993). The two key drivers of globalisation in the tourism sector are:

- *Transport technology* – reducing costs (both monetary and temporal) and acting as a 'space shrinking' technology, in turn breaking down geographical boundaries and constraints; and
- *Communication technology* – with synergy between the processing power of computers and the transmission capabilities of satellite and digital media. Here, tourism has certainly been affected by the revolution in global communications, allowing international communication, the development of global distribution systems and increasing the reach of small enterprises.

In a global society digital networks will link a multitude of actors and fuel the development of tourism managed on the basis of networks. It will also empower small and medium sized enterprises, and small and/or isolated destinations. Yet, some argue that technology in the guise of *virtual* entities will reduce the importance of physical place – always an important consideration in the 'old era' of tourism when place provided accessibility and control over the delivery of the product. Of course, technology does not take place in a vacuum, it is developed and adopted as a social process, exemplified by the fact that tourism came late to the technological revolution due to the traditional nature of management and training in the sector (Morrison 1989). As Go and Appelman note in Chapter 8, it is not the technologies themselves that drive change, but the way in which they are organised to fit successfully in the daily life of the consumer and worker.

Economic Technology has been an enabling driver of globalisation, but in turn is influenced by economic motives. Economic drivers of globalisation include the shifting patterns of production and consumption across the world, challenging traditional economic assumptions of world trade and markets. For tourism this is reflected in the rise of the new generators and destinations of international tourism, particularly the East Asia and Pacific region where the rapid rise of newly industrialising countries (NICs) is clearly evident. Examples here include China, Thailand, Taiwan, South Korea, Singapore and Taiwan. This has demanded a response by other tourism regions, such as Europe, as well as responses within the NICs, such as building new infrastructure and handling the consequences of rapid tourism growth. These changes in the economic order of the world impact upon flows of investment and labour, and encourage the rise of mergers and alliances in order to gain a geographical competitive edge. This has been made possible by technology that allows instant communication at any time of the day across the world to create a truly global society.

Politics As Wahab and Cooper state in Chapter 1, it is impossible to divorce economics from the international forces of politics. Political events have fuelled globalisation, both creating and influenced by a new world order. As the traditional core-periphery pattern no longer explains the location and success of enterprises and nations, Dicken (1992) argues for a multi-polar economy with three economic regions dominant – North America, the European Union and the NICs of South East Asia. Here, trade tensions in the global marketplace between these 'mega-markets' prompted a response by the World Trade Organisation in drawing up the General Agreement on Trade in Services (GATS). The GATS has a range of impacts upon tourism including:

- promoting free movement of labour globally;
- enabling the international development of, and access to, computer reservation systems; and
- removing barriers to overseas investment.

Elsewhere in the world the changing political map also impacts upon tourism. Examples here include the retreat of communism since 1989 and the establishment of market economies in the former Eastern Europe and Soviet Union. Not only is tourism growing rapidly both in terms of inbound and outbound flows from these countries, but also the way of doing business in tourism has changed. This often renders tourism policies outdated as Fayos-Solà and Pedro Bueno observe in Chapter 3. Hall, in Chapter 3 states that the development of regional economic and trading blocs and the growth of internationalisation are intimately related to tourism. Here the establishment of tourism trade blocs and associations is also significant – these include the North American Free Trade Agreement (NAFTA), the ASEAN trading bloc and the European Union. These developments stress the importance of multilateral agreements and reduce individual sovereign states' ability to impose unilateral actions. Flying in the face of this political development is the trend towards city-states and regionalism at a sub-national level. Each sees tourism as an important form of economic and cultural survival. Finally, political dogma is shifting as many governments work increasingly in partnership with the private sector for promotion of tourism, and pressure groups and community organisations seek, and are granted, a greater say in tourism developments. Wahab and Cooper's comments in Chapter 1 have a particular resonance here in terms of how democratic pluralism functions globally to balance the forces of the market and the state.

Culture A number of chapters in this book have noted that the consumption and adoption of global culture is significant for tourism, with criticism of many resorts as consisting of a uniform landscape of fast food restaurants, international hotels and chain stores. The homogenisation of tourism products is a problem for tourism as tourism places are increasingly commodified to reflect a global culture of consumption and it becomes difficult to differentiate them from the visitor's home surroundings (Go 1996; Keller 1996; Smeral 1998). This *Coca-Cola-isation* or *McDonald's-isation* of destinations is a consequence of the globalisation process, converging business practices and communication of both ideas and brands through media such as the Internet and television. Here we are seeing what Lee-Ross and Johns (Chapter 11) term a *cultural paradox*, as tourism products based on localness and cultural diversity are set in a sea of homogeneity. However, whilst the debate as to whether we will see a truly global culture still continues, there is no doubt that tourism is not only a conduit of cultural change but is at the vanguard of the possible creation of such a global culture.

Environment The natural environment is a global resource for tourism and non-sustainable practices in one destination impact upon others. Realisation of the finite nature of natural resources has led to growing demands for environmental accountability of all those involved in tourism – governments, companies, tourists and communities. However, as Hawkins and Holtz note in Chapter 12, traditional command and control approaches are no longer effective in environmental regulation. Instead what is needed is the concerted involvement of the public and private sectors. There is no doubt for example, that the consequences of global warming will impact upon tourism in terms of the need to shift the emphasis of product development away from beach tourism and exposure to the sun, as well emphasising the vulnerability of low-lying coastal destinations such as the Maldives. Sustainable tourism development has become the organising framework for tourism in the final decade of the twentieth century, reflecting as it does, a global concern for the exploitation of natural resources through the concept of 'eco-efficiency'. Whilst globalisation does have positive consequences for the environment, there are also a number of problematic issues (OECD 1997, 1998; Go 1996):

- The reduced effectiveness of governments acting unilaterally is felt to weaken those who regulate and prevent pollution.
- In an increasingly market-based economy, externalities will not be addressed. However, the private sector (and trans-national corporations in particular) are likely to be in the front line in terms of environmental policy and are best able to respond.
- Environmental problems do not respect traditional political boundaries, and there will be a blurring between economic and environmental boundaries.
- Rapid tourism growth in the NICs of the East Asia and the Pacific region means that governments are grappling with the impacts upon both environments and cultures.
- There is a fear that some countries will attract 'dirty' industries and in effect become pollution havens in the global marketplace.

Yet, as Hawkins and Holtz document in Chapter 12 there is an increasing number of effective initiatives here including global environmental policies, environmental tourism initiatives at all spatial scales and both environmental management and eco-labelling programmes.

Business operations Creation of a global society means that tourism businesses have the ability to operate globally and many have opted for a competitive strategy of internationalisation. Global enterprises view the world as their operating environment and establish both global strategies and global market presence. Changing business practices in response to the drivers of globalisation in turn, sustain, extend and facilitate the process of globalisation, and reshape the very boundaries that previously constrained them. In tourism, these boundary-breaking processes include:

- advertising creating global brands;
- international education and training transmitting global concepts and approaches;
- reduced transport costs;
- International communication mechanisms such as global distribution systems; and
- entertainment companies communicating global products, personalities and characters.

Indeed, as Seaton and Alford observe in Chapter 5, tourism organisations are adopting converging approaches to management, marketing, quality management (see Lee-Ross and Johns in Chapter 11) and both education and training (see Baum in Chapter 9). Parker (1998) suggests that four types of companies are crossing these boundaries:

*   trans-national Corporations (TNCs) – these are global corporations such as American Express or Disney;
*   privately-owned enterprises – firms not in public ownership who embark on an internationalisation strategy, often in the form of an alliance or merger;
*   extended family businesses – firms owned by expatriates or non-resident nationals, often highly visible millionaires in the society; Parker quotes that there are a significant number of billionaires in South East Asia who are ethnic Chinese living outside China; and
*   global gangs – the Mafia, and drug organisations thrive in a bound-aryless world as documented by Santana in Chapter 10.

The consequences of a globalised society for tourism

Given these forces driving us towards a global society, Kanter (1995) argues that the criteria of success for tourism organisations in a globalised society are changing and they will be judged according to:

*   concepts – leading-edge ideas, designs or product formulations (eco-tourism products for example are often highly creative);
*   competencies – the ability to deliver products and to transform ideas into services (here innovative tourism marketing and training of staff are important in tourism); and
*   connections – alliances and communication to lever core capabilities, create value for customers and remove boundaries – such as global airline alliances.

These are intangible assets that need to be constantly reviewed through research and collaboration. Destinations in particular can utilise these assets through:

*   training models adopting international standards and delivered with local application);
*   developing and implementing quality standards not only in accom-modation but also across attractions and other components of the destination amalgam; and
*   maintaining word class linkages such as through festivals and events.

Tourism and the service sector

Many of the forces and consequences of globalisation will benefit tourism and the service sector. Technology, information and the reduction of boundaries create new forms of service company, not only the large TNCs such as Disney, but also the small niche specialist that can take advantage of the Internet, international communications, and market positioning and targeting. Here it is clear that tourism companies will need new organisational structures, management and know-how as Wahab and Cooper state in Chapter 1. However, much of the writing on globalisation is focused not on services, but on manufacturing, and the concepts are derived from traditional economic theory (Dicken 1992). Yet the service sector's response to the challenges of globalisation is quite different. Dicken (1992) argues that services internationalise through overseas market presence whilst also demanding the right conditions within which to deliver the service – in terms of labour, technology and government regulation. Campbell and Verbeke (1994) argue that there needs to be a clear recognition of the distinctive characteristics of services as they drive the strategy adopted for internationalisation. For example:

- Service enterprises can use economies of scale in the marketing area, and in particular train personnel to market the product whilst it is being experienced.
- It is more difficult to separate the enterprises' delivery of the service from marketing as the service is 'produced where it is consumed'. This enhances the role of responsiveness at the national level, whilst ensuring strict quality control through head office marketing and quality management procedures.
- Equally, the intangibility of the service product underscores the importance of a firm's reputation and thus creates a real pressure to choose credible and legitimate partners in any alliance. Also, the enhanced scope for the delivery of all elements of the service product through alliances is heightened. Here networking flexibility is a typical service sector strategy, allowing the firm to develop networks of relationships at different levels.

Campbell and Verbeke (1994) suggest that an international strategy for service sector firms may be done in two stages:

- First, to develop a strategic capability to allow national responsiveness or centralised innovation.
- Second, to develop an administrative structure to allow networking flexibility.

Both Poon (1993) and Smeral (1998) also identify the stages that tourism is experiencing as a consequence of globalisation. They contrast the 'old' tourism with the new 'post-Fordist' tourism:

- The 'Fordist' characteristics of tourism are rigid, mass, packaged products purchased by price-sensitive consumers, as Buhalis describes in Chapter 4.
- In contrast, the post-Fordist 'new' tourism will be characterised by a more flexible delivery system of loosely-linked worldwide-acting enterprises and purchased by the 'new' tourist who is experienced and discerning, seeking value-for-money, customised quality-controlled products, but not necessarily low prices. Here, delivery of the product is 'just in time', flexibly-produced, customer-designed and uses technology to extend the 'value chain'. By this stage, the labour force is functionally flexible and technically skilled.

This post-Fordist 'new' tourism will be characterised by a number of key features in direct response to the forces of globalisation. These include: alliances and partnerships; internationalisation strategy; and trans-national corporations.

Alliances and partnerships

One of the consequences of tourism in a boundaryless world is the opportunity to work with other organisations to pool resources, overcome limitations of resources and thus to gain competitive advantage in a fast-changing environment (Dyer and Singh 1988). In Chapter 8, Go and Appelman show how this opportunity applies to SMEs as well as larger corporations. This does not necessarily mean that tourism organisations will dispense with the advantages of bounding their operations as such boundaries provide an opportunity to focus (Ashkenas *et al.* 1995). In fact, the process of forming alliances and cooperation provides a relational view of competitive advantage, to complement the traditionally held models of resource-based or industry-structure view (Dyer and Singh 1988). Dyer and Singh provide a useful checklist of determinants of inter-organisational competitiveness:

- Relation-specific assets – matching specific assets between organisations to gain competitive advantage.
- Knowledge-sharing routines and inter-organisation communication such as those promoted by international agencies (e.g. The Pacific Asia Tourism Association).
- Complementary resources and capabilities – a critical area in tourism where the product is an amalgam of resources created by differing organisations (including the public sector). The integration strategies of tourism companies are clearly evident here in the structuring of global conglomerates of tourism enterprises such as airlines (Wheatcroft 1992). Equally we see a synergy between tourism and the knowledge-based industries utilising know-how, entertainment and information.

- Effective governance – a final critical area in terms of the management of the relationship and the value chain, and one where tourism is lagging behind other sectors.

As Go and Appelman state in Chapter 8, the concept of the value chain is pivotal. Traditionally the value chain has been seen as one in which each organisation maximises its own success across the value chain. In this traditional model information, strategies and resources are not shared and the chain is inefficient (Ashkenas *et al.* 1995). However, in a globalised society the value chain represents the process by which organisations and enterprises are linked together to create products and services that have more value *combined* than *separate* (Ashkenas *et al.* 1995). In other words, organisations look outside their own boundaries and aim to strengthen the whole web, not simply their own part of it.

In tourism this concept is central to the creation of products by intermediaries. Here, globalisation is providing a new way of thinking in the tourism distribution channel as individual tourism businesses, intermediaries and tourists cease to be adversarial in the channel and move to a more cooperative model where, for example, intermediaries will invest in destinations (Crotts and Wilson 1995). This is a logical strategy given that intermediaries ultimately depend upon destinations for their product creation – in other words the recognised strategic focus should be in the value-creating system where all those involved work to co-produce value (Normann and Ramirez 1993). To quote Dyer and Singh (1988: 675): '... a pair or network of firms can develop relationships that result in sustained competitive advantage'.

This has implications for the traditional roles and relationships within the tourism sector as barriers to cooperation are removed. In tourism, these barriers include:

- legal and regulatory tradition;
- competitive confusion, particularly at the destination level;
- lack of trust;
- difficulty in letting go of control;
- slowness in learning new management skills; and
- complexity of the tourism system and product (Ashkenas *et al.* 1995: 204).

Overcoming these barriers through the formation of alliances and joint ventures demands that organisations go thorough a particular process of strategy development, partner search, negotiation and alliance operation (Pekar and Allio 1994; Crotts and Wilson 1995). As experience of such arrangements increases in tourism it is likely that non-equity alliances will also be formed, including the sharing of staffing, research and other services.

Internationalisation strategy

Globalisation is both a consequence of, and an influence upon the internationalisation of strategy, as observed by Vanhove in Chapter 6 (Melin 1992). Two well-known models of internationalisation are Vernon's (1966) product life cycle (PLC) model and the internationalisation process model (Johanson and Vahlne 1977):

- The PLC model suggests that each stage of a product's life has implications for internationalisation. For example in the launch and early growth stages, the product will be targeted at the domestic market; during later stages of growth the product may actively seek international markets, whilst in maturity, when issues of production costs become important, the manufacture of the product may be relocated to a region of say, cheaper labour. Vernon's model attempts to integrate both a national-level perspective in terms of developed and developing countries, with the firm level of aggregation. The model can be extended to consider economic stages and cycles of industrialisation and, as Dicken (1992) argues, the most recent 'neo-Fordist' phase will be characterised by flexibility.
- The internationalisation process model suggests that the process of globalisation is a sequential one from initial exporting activity to the formation of overseas production units. Here the firm progresses through a number of logical steps as it approaches each overseas market, initially targeting markets with the lowest perceived uncertainty and risk. In other words, 'psychological closeness' is an important consideration in terms of internationalisation behaviour, and one clearly exemplified in the operation of tourism markets. Weiermair and Peters (1998) deduce from this that market know-how is a critical variable determining an organisation's ability to internationalise. This is supported by Johanson and Vahlne's (1977) assertion that market commitment and market knowledge are self-reinforcing determinants of internationalisation strategy.

In addition, the size of the organisation, and also the stage of maturity of the organisation will influence the stages, as different organisational structures are needed to handle each of the above stages. However, both models have been criticised on a number of grounds, including their overly deterministic nature, their inability to include behavioural variables and the fact that they do not take into account the behaviour of TNCs. As a result, the models have been developed and extended to four stages of internationalisation:

- domestic marketing;
- experimental involvement in international markets;

- active involvement in international markets; and
- committed involvement in international markets.

More recently, Dunning (1988) has emerged with the 'eclectic' theory of internationalisation, suggesting that many perspectives are needed to understand the process. Dunning argues that the propensity of an organisation to engage in international production depends upon:

- ownership advantages relative to organisations of other nationalities in the same overseas market: these advantages include technology and access to supplies or markets;
- identification that it is profitable for the organisation to continue these ownership advantages with factor endowments located in overseas markets; and
- it being more beneficial for the organisation to use these advantages itself, rather than selling them (or their rights) to a foreign organisation. In other words there are advantages such as reduced costs.

International strategy and globalisation fundamentally affect marketing as noted by Vanhove in Chapter 6. In particular there is a need not only to manage the product, but also a need to manage the external environment, given the increasingly political nature of marketing, and the move towards 'societal marketing' (Kotler *et al.* 1996; Paliwoda and Thomas 1998). Kotler *et al.* (1996) therefore suggest adding a further two Ps to the traditional four. These are political power and public opinion formation. This recognises the influence of stakeholders and the need to broaden the understanding of how markets work.

Paliwoda and Thomas (1998) argue that organisations look to internationalising their marketing for the following reasons:

- The maturity stage of the product life cycle in a domestic market is an opportunity to penetrate lucrative overseas markets with, to them, a new product.
- Competition may be less fierce overseas than at home.
- The organisation has excess capacity.
- Geographical diversification may be a better strategy than product diversification.

Dicken (1992) argues that the clearest view of internationalisation strategy is seen in the TNCs where strategies have three key components:

- cost leadership – lowest cost producer;
- Differentiation – differentiation from competitors; and
- focus – applying these two components to a market.

Trans-national corporations (TNCs)

A globalised economy is characterised by a worldwide network of large enterprises sharing common features and advantages (OECD 1995; Dicken 1992). These are known as TNCs whose common features include:

*   large-scale direct foreign investment;
*   overseas production, joint ventures or holdings in overseas companies;
*   significant global transactions with control of economic activities in more than one country;
*   willingness to cooperate in information research production and distribution;
*   ability to take advantage of geographical differences in factor endowments; and
*   geographical flexibility to shift resources and operations.

Trans-national corporations therefore meet two criteria:

*   they have foreign direct investments in more than two countries; and
*   their corporate planning employs a global perspective impartially allocating resources worldwide (Paliwoda and Thomas 1998).

Increasingly, the trans-national corporation (TNC) can be viewed as an integrated organisational network, embedded within an external environment of suppliers, regulators and customers (Ghoshal and Bartlett 1990). In other words, TNCs embody the spirit of globalisation in that they are actively dismantling boundaries and do not depend upon their domestic markets. However, the TNC trades extensively with itself, its affiliates and its subsidiaries, effectively creating its own internal markets and thus with no loyalty to particular nation-states or tourism destinations – indeed the more international an organisation becomes the greater its internalised trade (Paliwoda and Thomas 1998).

But, as Wahab and Cooper state in Chapter 1, such TNCs trade beyond national boundaries and the reach of the regulation of national governments, striving to serve only the interests of their dominant shareholders. The advantages and disadvantages of TNCs can be summarised in Table 15.1.

In summary, the drivers of globalisation have created TNCs in the tourism sector and in turn have challenged traditional ways of thinking about trade and enterprises. The economic geography of TNCs is complex and results from of the interaction (often adversarial) between the competitive strategies of TNCs and national government policies. Indeed, it can be argued that TNCs and nation-states are the most powerful forces in shaping the globalised society and the role of tourism within it.

Table 15.1 The advantages and disadvantages of TNCs

Disadvantages of TNCs	Advantages of TNCs
Knowledge and skill transfer may be inappropriate and undermines competitive advantage of home country	Knowledge and skill transferred and industrialisation promoted
Local jobs destroyed and inappropriate jobs supported	Jobs created in the TNC and stimulated in other economic sectors
Local competition eliminated, particularly SMEs	Competition stimulated to improve standards – exposure to international hotel chains requires locally owned hotels to raise standards
Destroys local culture and imports management approaches – the Disney Corporation was criticised for imposing its culture on the French for example	Effective management and 'modern' attitudes promoted
Leakage of financial benefits to head office	Foreign exchange earned or saved by host nation
Demand distorted and social inequalities promoted	Demand stimulated locally in terms of domestic tourism
Interferes with host country politics creating a neo-colonial relationship – for example in terms of tour operators and destinations	

Sources: Livingstone (1989); Kinsey (1988) and Smeral (1998).

Conclusion

It is clear from the authors in this book that there is no real agreement as to the extent of globalisation in the tourism sector. It is not enough simply to point to tourism as an international activity; what is needed is an analysis of the extent of the changing processes, structures and behaviour resulting from the drivers of globalisation. As Fayos-Solà and Pedro Bueno state clearly in Chapter 3, on the demand side the impact of globalisation on tourism is clear; however, on the supply side the verdict is much less clear-cut. There is little real agreement about what a globalised society will actually mean for those involved in the tourism sector. Some argue that globalisation will bring significant economic benefits, whilst others see it as a harbinger of worsening social equity where the developing world becomes increasingly disadvantaged (OECD, 1997). There are also concerns that globalisation weakens environmental policy and the ability of regulators to intervene, whilst eroding the power of nation-states and destinations. Two of the major challenges of a globalised society will therefore be:

- the ability to balance a global vision in tourism (including international standards) with local demands and needs (Campbell and Verbeke 1994); and
- the need to meet the material needs of a global community without increasing inequalities and without destroying the environment (Dicken 1992).

There is no doubt that a global society will be dominated by tourism enterprises and destinations that can both meet international standards and access worldwide networks. Here, two institutions will be leading players – TNCs and nation-states – both operating in a complex and volatile technological environment and both acting as political and economic institutions.

This changing competitive landscape of tourism is driving enterprises, destinations, governments and communities to rethink their strategies and organisational structures to allow them to operate successfully in a boundaryless world. Indeed, Kanter (1995) states that the future divisions in society will not be between rich and poor, but between *cosmopolitans* with global connections and *locals* who are tied to one location. In a global society, those tourism destinations and enterprises that isolate themselves from world trends will be at a severe disadvantage. On the other hand, those who will gain from a globalised society will be tourism enterprises, individuals, and governments who embrace the new way of operating and recognise the forces of change. Whatever the shape and vision of the future globalised society, there is no doubt that tourism will be at the forefront, both as a driver of change and as a respondent. Tourism must embrace globalisation, yet treat it with a localised focus if it is to succeed (Go 1996).

References

Ashkenas, R., Ulrich, D., Jick, T. and Kerr, S. (1995) *The Boundaryless Organisation*, San Francisco: Jossey-Bass.

Bradley, S. P. Hausmann, J. A. and Nolan, R. L. (1993) *Globalisation, Technology and Competition*, Boston: Harvard Business School Press.

Buhalis, D. (1998) 'Information technology', in C. Cooper *et al. Tourism Principles and Practice*, Harlow: Addison Wesley Longman.

Campbell, A. J. and Verbeke, A. (1994) 'The globalisation of service sector multinationals', *Long Range Planning* 27(2): 95–102.

Crotts, J. C. and Wilson, D. T. (1995) 'An integrated model of buyer-seller relationships in the international travel trade', *Progress in Tourism and Hospitality Research* 1(2): 125–40.

Dicken, P. (1992) *Global Shift* (second edition) London: Paul Chapman.

Dunning, J. H. (1988) 'The eclectic paradigm of international production: a restatement and possible extensions', *Journal of International Business Studies* 19(1): 1–32.

Dyer, J. H. and Singh, H. (1988) 'The relational view: co-operative strategy and sources of inter-organisational competitive advantage', *Academy of Management Review* 23(4): 660–79.

Ghoshal, S. and Bartlett, C. A. (1990) 'The multinational corporation as an inter-organisational network', *Academy of Management Review* 15(4): 603–25.

Go, F. (1996) 'A conceptual framework for managing global tourism and hospitality marketing', *Tourism Recreation Research* 21(2): 37–43.

Johanson, J. and Vahlne, J. E. (1977) 'The internationalisation process of the firm: a model of knowledge development on increasing foreign commitments', *Journal of International Business Studies* 2: 23–32.

Kanter, R. M. (1995) 'Thinking locally in the global economy', *Harvard Business Review* September/October: 151–60.

Keller, P. (1996) 'Globalisation and tourism', *Tourist Review* 4: 6–7.

Kinsey, J. (1988) *Marketing in Developing Countries*, London: Macmillan.

Korten, D. C. (1995) *When Corporations Ruled the World*, London: Earthscan.

Kotler, P. Bowen, J. and Makens, J. (1996) *Marketing for Hospitality and Tourism* New Jersey: Prentice-Hall.

Livingstone, J. M. (1989) *The Internationalisation of Business*, London: Macmillan.

Makridakis, S. (1989) 'Management in the 21st Century', *Long Range Planning* 22(2): 37–53.

Melin, L. (1992) 'Internationalisation as a strategy process', *Strategic Management Journal*, 13(2): 99–118.

Morrison, A. (1989) *Hospitality and Travel Marketing*, New York: Delmar.

Normann, R. and Ramirez, R. (1993) 'From value chain to value constellation: designing interactive strategy', *Harvard Business Review*, July, August: 65–6.

Organisation for Economic Co-operation and Development (1995) *Local Economies and Globalisation*, Paris: OECD.

Organisation for Economic Co-operation and Development (1997) *Economic Globalisation and the Environment*, Paris: OECD.

Organisation for Economic Co-operation and Development (1998) *Globalisation and the Environment*, Paris: OECD.

Paliwoda, S. J. and Thomas, M. J. (1998) *International Marketing* (third edition) Oxford: Butterworth-Heinemann.

Parker, B. (1998) *Globalisation and Business Practice. Managing Across Boundaries*, London: Sage.

Pekar, P. and Allio, R. (1994) 'Making alliances work – guidelines for success', *Long Range Planning* 27(4): 54–65.

Poon, A. (1993) *Tourism, Technology and Competitive Strategies*, Wallingford: CAB.

Smeral, E. (1998) 'The impact of globalisation on small and medium enterprises: new challenges for tourism policies in European countries', *Tourism Management* 19(4): 371–80.

Vernon, R. (1966) 'The product cycle hypothesis in a new international environment', *Quarterly Journal of Economics* (May): 190–207.

Weiermiar, K. and Peters, P. (1998) 'The internationalization behaviour of small- and medium-sized service enterprises', *Asia Pacific Journal of Tourism Research* 2(2): 1–14.

Wheatcroft, S. (1992) 'Airlines: reaping the rewards of globalisation', *Transport* 13(6): 1–3.

Index